An Introduction to Stochastic Modeling

AN INTRODUCTION TO STOCHASTIC MODELING

Howard M. Taylor
Cornell University

Samuel Karlin
Stanford University

Academic Press, Inc.

(Harcourt Brace Jovanovich, Publishers)
Orlando   San Diego   San Francisco   New York   London
Toronto   Montreal   Sydney   Tokyo   São Paulo

Academic Press, Inc.
Orlando, Florida 32887

United Kingdom Edition Published by
Academic Press, Inc. (London) Ltd.
24/28 Oval Road, London NW1 7DX

ISBN: 0-12-684880-7

Library of Congress Catalog Card Number: 84-70475

Printed in the United States of America

Contents

★Starred sections contain material at a more difficult level than the remainder of the book. These sections may be omitted without a loss in continuity.

Preface

Stochastic processes are ways of quantifying the dynamic relationships of sequences of random events. Stochastic models play an important role in elucidating many areas of the natural and engineering sciences. They can be used to analyze the variability inherent in biological and medical processes, to deal with uncertainties affecting managerial decisions and with the complexities of psychological and social interactions, and to provide new perspectives, methodology, models, and intuition to aid in other mathematical and statistical studies.

This book is intended as a beginning text in stochastic processes for students familiar with elementary probability calculus. Its aim is to bridge the gap between basic probability knowhow and an intermediate level course in stochastic processes, for example, *A First Course in Stochastic Processes* by the present authors.

The objectives of this book are three: (1) to introduce students to the standard concepts and methods of stochastic modeling; (2) to illustrate the rich diversity of applications of stochastic processes in the sciences; and (3) to provide exercises in the application of simple stochastic analysis to appropriate problems.

The chapters are organized around several prototype classes of stochastic processes featuring Markov chains in discrete and continuous time, Poisson processes and renewal theory, the evolution of branching events, and queueing models. We have borrowed freely from the literature without explicit citations. After the concluding Chapter 9, however, we provide a list of books that incorporate more advanced discussions of several of the models set forth in this text.

To the Instructor

If possible, we recommend having students skim the first two chapters, referring as necessary to the probability review material, and starting the course with Chapter 3, on Markov chains. A one quarter course adapted to the junior–senior level could consist of a cursory (one-week) review of Chapters 1 and 2, followed in order by Chapters 3 through 6. For interested students, Chapters 7, 8, and 9 discuss other currently active areas of stochastic modeling.

Acknowledgments

Many people helped to bring this text into being. We gratefully acknowledge the help of Anna Karlin, Shelley Stevens, Karen Larsen, and Laurieann Shoemaker. Chapter 9 was enriched by a series of lectures on queueing networks given by Ralph Disney at The Johns Hopkins University in 1982. Alan Karr, Ivan Johnstone, Luke Tierney, and others besides ourselves have taught from early drafts of the text, and we have profited from their criticisms. Finally, we are grateful for improvements suggested by the several generations of students who have worked with mimeographed versions of the manuscript.

Chapter 1 ❚ Introduction

1.1 Stochastic Modeling

A quantitative description of a natural phenomenon is called a mathematical model of that phenomenon. Examples abound, from the simple equation $S = \frac{1}{2}gt^2$ describing the distance S traveled in time t by a falling object starting at rest to a complex computer program that simulates a biological population or a large industrial system.

In the final analysis, a model is judged using a single, quite pragmatic, factor, the model's *usefulness*. Some models are useful as detailed quantitative prescriptions of behavior, as for example, an inventory model that is used to determine the optimal number of units to stock. Another model in a different context may provide only general qualitative information about the relationships among and relative importance of several factors influencing an event. Such a model is useful in an equally important but quite different way. Examples of diverse types of stochastic models are spread throughout this book.

Such often mentioned attributes as realism, elegance, validity, and reproducibility are important in evaluating a model only insofar as they bear on that model's ultimate usefulness. For instance, it is both unrealistic and quite inelegant to view the sprawling city of Los Angeles as a geometrical point, a mathematical object of no size or dimension. Yet it is quite useful to do exactly that when using spherical geometry to derive a minimum distance great circle air route from New York City, another "point."

There is no such thing as the best model for a given phenomenon. The pragmatic criterion of usefulness often allows the existence of two or more models for the same event, but serving distinct purposes. Consider light. The wave form model, in which light is viewed as a continuous flow, is entirely adequate for designing eyeglass and telescope lenses. In contrast, for understanding the impact of light on the retina of the eye, the photon model, which views light as tiny discrete bundles of energy, is preferred. Neither model supersedes the other; both are relevant and useful.

The word "stochastic" derives from the Greek ($\sigma\tau o\chi\acute\alpha\zeta\epsilon\sigma\theta\alpha\iota$ to aim, to guess) and means "random" or "chance." The antonym is "sure," "deterministic," or "certain." A deterministic model predicts a single outcome from a given set of circumstances. A stochastic model predicts a set of possible outcomes weighed by their likelihoods or probabilities. A coin flipped into the air will surely return to earth somewhere. Whether it lands heads or tails is random. For a "fair" coin we consider these alternatives equally likely and assign to each the probability $\frac{1}{2}$.

However, phenomena are not in and of themselves inherently stochastic or deterministic. Rather, to model a phenomenon as stochastic or deterministic is the choice of the observer. The choice depends on the observer's purpose; the criterion for judging the choice is usefulness. Most often the proper choice is quite clear, but controversial situations do arise. If the coin once fallen is quickly covered by a book so that the outcome "heads" or "tails" remains unknown, two participants may still usefully employ probability concepts to evaluate what is a fair bet between them; that is, they may usefully view the coin as random, even though most people would consider the outcome now to be fixed or deterministic. As a less mundane example of the converse situation, changes in the level of a large population are often usefully modeled deterministically, in spite of the general agreement among observers that many chance events contribute to their fluctuations.

Scientific modeling has three components: (i) a natural phenomenon under study, (ii) a logical system for deducing implications about the phenomenon, and (iii) a connection linking the elements of the natural system under study to the logical system used to model it. If we think of these three components in terms of the great circle air route problem, the natural system is the earth with airports at Los Angeles and New York; the logical system is the mathematical subject of spherical geometry; and the two are connected by viewing the airports in the physical system as points in the logical system.

The modern approach to stochastic modeling is in a similar spirit. Nature does not dictate a unique definition of "probability," in the same way that there is no nature-imposed definition of "point" in geometry. "Probability" and "point" are terms in pure mathematics, defined only through the properties invested in them by their respective sets of axioms. (See Section 1.2.8 for a review of axiomatic probability theory.) There are, however, three general principles that are often useful in relating or connecting the abstract elements of mathematical probability theory to a real or natural phenomenon that is to be modeled. These are (i) the principle of equally likely outcomes, (ii) the principle of long run relative frequency, and (iii) the principle of odds making or subjective probabilities. Historically, these three concepts arose out of largely unsuccessful attempts to define probability in terms of physical experiences. Today, they are relevant as guidelines for the assignment of probability values in a model, and for the interpretation of the conclusions of a model in terms of the phenomenon under study.

We illustrate the distinctions between these principles with a long experiment. We will pretend that we are part of a group of people who decide to toss a coin and observe the event that the coin will fall heads up. This event is denoted by H, and the event of tails, by T.

Initially, everyone in the group agrees that $\Pr\{H\} = \frac{1}{2}$. When asked why, people give two reasons: Upon checking the coin construction, they believe that the two possible outcomes, heads and tails, are equally likely; and extrapolating from past experience, they also believe that, if the coin is tossed many times, the fraction of times that heads is observed will be close to one half.

The equally likely interpretation of probability surfaced in the works of Laplace in 1812 where the attempt was made to define the probability of an event A as the ratio of the total number of ways that A could occur to the total number of possible outcomes of the experiment. The equally likely approach is often used today to assign probabilities that reflect some notion of a total lack of knowledge about the outcome of a chance phenomenon. The principle requires judicious application if it is to be useful, however. In our coin tossing experiment, for instance, merely introducing the *possibility* that the coin could land on its edge (E) instantly results in $\Pr\{H\} = \Pr\{T\} = \Pr\{E\} = \frac{1}{3}$.

The next principle, the long run relative frequency interpretation of probability, is a basic building block in modern stochastic modeling, made precise and justified within the axiomatic structure by the Law of Large Numbers. This law asserts that the relative fraction of times in which an event occurs in a sequence of independent similar experiments approaches, in the limit, the probability of the occurrence of the event on any single trial.

The principle is not relevant in all situations, however. When the surgeon tells a patient that he has an 80-20 chance of survival, the surgeon means, most likely, that 80 percent of similar patients facing similar surgery will survive it. The patient at hand is not concerned with the long run, but in vivid contrast, is vitally concerned only in the outcome of his, the next, trial.

Returning to the group experiment, we will suppose next that the coin is flipped into the air and, upon landing, is quickly covered so that no one can see the outcome. What is $\Pr\{H\}$ now? Several in the group argue that the outcome of the coin is no longer random, that $\Pr\{H\}$ is either 0 or 1, and that although we don't know which it is, probability theory does not apply.

Others articulate a different view, that the distinction between "random" and "lack of knowledge" is fuzzy, at best, and that a person with a sufficiently large computer and sufficient information about such factors as the energy, velocity, and direction used in tossing the coin, could have predicted the outcome, heads or tails, with certainty before the toss. Therefore, even before the coin was flipped, the problem was a lack of knowledge and not some inherent randomness in the experiment.

In a related approach, several people in the group are willing to bet with each other, at even odds, on the outcome of the toss. That is, they are willing to *use* the calculus of probability to determine what is a fair bet, without considering whether the event under study is random or not. The usefulness criterion for judging a model has appeared.

While the rest of the mob were debating "random" versus "lack of knowledge," one member, Karen, looked at the coin. Her probability for heads is now different from that of everyone else. Keeping the coin covered, she announces the outcome "Tails," whereupon everyone mentally assigns the value $\Pr\{H\} = 0$. But then her companion, Mary, speaks up and says that Karen has a history of prevarication.

The last scenario explains why there are horse races; different people assign different probabilities to the same event. For this reason, probabilities used in odds making are often called *subjective* probabilities. Then, odds making forms the third principle for assigning probability values in models and for interpreting them in the real world.

The modern approach to stochastic modeling is to divorce the definition of probability from any particular type of application. Probability theory is an axiomatic structure (see Section 1.2.8), a part of pure mathematics. Its use in modeling stochastic phenomena is part of the broader realm of science and parallels the use of other branches of mathematics in modeling deterministic phenomena.

To be useful, a stochastic model must reflect all those aspects of the phenomenon under study that are relevant to the question at hand. In addition, the model must be amenable to calculation and must allow the deduction of important predictions or implications about the phenomenon.

1.1.1 Stochastic Processes

A *stochastic process* is a family of random variables X_t, where t is a parameter running over a suitable index set T. (Where convenient, we will write $X(t)$ instead of X_t.) In a common situation, the index t corresponds to discrete units of time, and the index set is $T = \{0, 1, 2, \ldots\}$. In this case, X_t might represent the outcomes at successive tosses of a coin, repeated responses of a subject in a learning experiment, or successive observations of some characteristic of a certain population. Stochastic processes for which $T = [0, \infty)$ are particularly important in applications. Here t often represents time, but different situations also frequently arise. For example, *t* may represent distance from an arbitrary origin, and X_t may count the number of defects in the interval $(0, t]$ along a thread, or the number of cars in the interval $(0, t]$ along a highway.

Stochastic processes are distinguished by their *state space*, or the range of possible values for the random variables X_t, by their index set *T*, and by the dependence relations among the random variables X_t. The most widely used classes of stochastic processes are systematically and thoroughly presented for study in the following chapters, along with the mathematical techniques of calculation and analysis that are most useful with these processes. The use of these processes as models is taught by example. Sample applications from many and diverse areas of interest are an integral part of the exposition.

1.2 Probability Review★

This section summarizes the necessary background material and establishes the book's terminology and notation. It also illustrates the level of the exposition in the following chapters. Readers who find the major part of this section's material to be familiar and easily understood should have no difficulty with what follows. Others might wish to review their probability background before continuing.

In this section statements frequently are made without proof. The reader desiring justification should consult any elementary probability text as the need arises.

1.2.1 Events and Probabilities

The reader is assumed to be familiar with the intuitive concept of an *event*. (Events are defined rigorously in Section 1.2.8, which reviews the axiomatic structure of probability theory.)

Let *A* and *B* be events. The event that at least one of *A* or *B* occurs is called the *union* of *A* and *B* and is written $A \cup B$; the event that both occur is called the *intersection* of *A* and *B* and is written $A \cap B$, or simply *AB*. This notation extends to finite and countable sequences of events. Given events $A_1, A_2, \ldots$, the event that at least one occurs is written $A_1 \cup A_2 \cup \ldots = \bigcup_{i=1}^{\infty} A_i$; the event that all occur is written $A_1 \cap A_2 \cap \ldots = \bigcap_{i=1}^{\infty} A_i$.

The probability of an event *A* is written $\Pr\{A\}$. The *certain* event, denoted by Ω, always occurs, and $\Pr\{\Omega\} = 1$. The *impossible* event, denoted

---

★Many readers will prefer to omit this review and move directly to Chapter 3, on Markov chains. They can then refer to the background material that is summarized in the remainder of this chapter and in Chapter 2 only as needed.

by $\emptyset$, never occurs, and $\Pr\{\emptyset\} = 0$. It is always the case that $0 \le \Pr\{A\} \le 1$ for any event A.

Events A, B are said to be *disjoint* if $A \cap B = \emptyset$; that is, if A and B cannot both occur. For disjoint events A, B we have the *addition law* $\Pr\{A \cup B\} = \Pr\{A\} + \Pr\{B\}$. A stronger form of the addition law is as follows: Let A_1, A_2, . . . be events with A_i and A_j disjoint whenever $i \ne j$. Then $\Pr\{\bigcup_{i=1}^{\infty} A_i\} = \Sigma_{i=1}^{\infty}\Pr\{A_i\}$. The addition law leads directly to the *Law of Total Probability:* Let A_1, A_2, . . . be disjoint events for which $\Omega = A_1 \cup A_2 \cup$ . . . . Equivalently, exactly one of the events A_1, A_2, . . . will occur. The law of total probability asserts that $\Pr\{B\} = \Sigma_{i=1}^{\infty}\Pr\{B \cap A_i\}$ for any event B. The law enables the calculation of the probability of an event B from the sometimes more easily determined probabilities $\Pr\{B \cap A_i\}$, where $i = 1$, 2, ... Judicious choice of the events A_i is prerequisite to the profitable application of the law.

Events A and B are said to be *independent* if $\Pr\{A \cap B\} = \Pr\{A\} \times \Pr\{B\}$. Events A_1, A_2, . . . are *independent* if

$$\Pr\{A_{i_1} \cap A_{i_2} \cap \cdot \cdot \cdot \cap A_{i_n}\} = \Pr\{A_{i_1}\}\Pr\{A_{i_2}\} \cdot \cdot \cdot \Pr\{A_{i_n}\}$$

for every finite set of distinct indices $i_1, i_2, . . . , i_n$.

1.2.2 Random Variables

An old-fashioned but very useful and highly intuitive definition describes a *random variable* as a variable that takes on its values by chance. In Section 1.2.8, we sketch the modern axiomatic structure for probability theory and random variables. The older definition just given serves quite adequately, however, in virtually all instances of stochastic modeling. Indeed, this older definition was the only approach available for well over a century of meaningful progress in probability theory and stochastic processes.

Most of the time we adhere to the convention of using capital letters such as X, Y, Z to denote random variables, and lower case letters such as x, y, z for real numbers. The expression $\{X \le x\}$ is the event that the random variable X assumes a value that is less than or equal to the real number x. This event may or may not occur, depending on the outcome of the experiment or phenomenon that determines the value for the random variable X. The probability that the event occurs is written $\Pr\{X \le x\}$. Allowing x to vary, this probability defines a function

$$F(x) = \Pr\{X \le x\}, \qquad -\infty < x < +\infty,$$

called the *distribution function* of the random variable X. Where several random variables appear in the same context, we may choose to distinguish their distribution functions with subscripts, writing, for example, $F_X(\xi) = \Pr\{X \le \xi\}$ and $F_Y(\xi) = \Pr\{Y \le \xi\}$, defining the distribution functions of the random variables X and Y, respectively, as functions of the real variable ξ.

The distribution function contains all the information available about a random variable before its value is determined by experiment. We have, for instance, $\Pr\{X > a\} = 1 - F(a)$, $\Pr\{a < X \le b\} = F(b) - F(a)$, and $\Pr\{X = x\} = F(x) - \lim_{\epsilon \downarrow 0} F(x - \epsilon) = F(x) - F(x-)$.

A random variable X is called *discrete* if there is a finite or denumerable set of distinct values $x_1, x_2, \ldots$ such that $a_i = \Pr\{X = x_i\} > 0$ for $i = 1, 2, \ldots$ and $\Sigma_i\, a_i = 1$. The function

$$p(x_i) = p_X(x_i) = a_i \qquad \text{for} \quad i = 1, 2, \ldots \tag{1.1}$$

is called the *probability mass function* for the random variable X and is related to the distribution function via

$$p(x_i) = F(x_i) - F(x_i-) \quad \text{and} \quad F(x) = \sum_{x_i \le x} p(x_i).$$

The distribution function for a discrete random variable is a step function that increases only in jumps, the size of the jump at x_i being $p(x_i)$.

If $\Pr\{X = x\} = 0$ for every value of x, then the random variable X is called *continuous* and its distribution function $F(x)$ is a continuous function of x. If there is a nonnegative function $f(x) = f_X(x)$ defined for $-\infty < x < \infty$ such that

$$\Pr\{a < X \le b\} = \int_a^b f(x)dx \qquad \text{for} \quad -\infty < a < b < \infty, \tag{1.2}$$

then $f(x)$ is called the *probability density function* for the random variable X. If X has a probability density function $f(x)$, then X is continuous and

$$F(x) = \int_{-\infty}^{x} f(\xi)d\xi, \qquad -\infty < x < \infty.$$

If $F(x)$ is differentiable in x, then X has a probability density function given by

$$f(x) = \frac{d}{dx} F(x) = F'(x), \qquad -\infty < x < \infty. \tag{1.3}$$

In differential form, (1.3) leads to the informal statement

$$\Pr\{x < X \le x + dx\} = F(x + dx) - F(x) = dF(x) = f(x)dx. \tag{1.4}$$

We consider (1.4) to be a shorthand version of the more precise statement

$$\Pr\{x < X \le x + \Delta x\} = f(x)\Delta x + o(\Delta x), \qquad \Delta x \downarrow 0, \tag{1.5}$$

where $o(\Delta x)$ is a generic remainder term of order less than Δx as $\Delta x \downarrow 0$. That is, $o(\Delta x)$ represents any term for which $\lim_{\Delta x \downarrow 0} o(\Delta x)/\Delta x = 0$. By the fundamental theorem of calculus, Equation (1.5) is valid whenever the probability density function is continuous at x.

While examples are known of continuous random variables that do not possess probability density functions, they do not arise in stochastic models of common natural phenomena.

1.2.3 Moments and Expected Values

If X is a discrete random variable, then its mth *moment* is given by

$$E[X^m] = \sum_i x_i^m \Pr\{X = x_i\}, \tag{1.6}$$

[where the x_i are specified in (1.1)] provided that the infinite sum converges absolutely. Where the infinite sum diverges, the moment is said not to exist. If X is a continuous random variable with probability density function $f(x)$, then its mth moment is given by

$$E[X^m] = \int_{-\infty}^{+\infty} x^m f(x)dx, \tag{1.7}$$

provided this integral converges absolutely.

The *first moment*, corresponding to $m = 1$, is commonly called the *mean* or *expected value* of X and written m_X or μ_X. The mth *central moment* of X is defined as the mth moment of the random variable $X - \mu_X$, provided μ_X exists. The first central moment is zero. The second central moment is called the *variance* of X and written σ_X^2 or $\text{Var}[X]$. We have the equivalent formulas $\text{Var}[X] = E[(X - \mu)^2] = E[X^2] - \mu^2$.

The *median* of a random variable X is any value v with the property that

$$\Pr\{X \geq v\} \geq \tfrac{1}{2} \quad \text{and} \quad \Pr\{X \leq v\} \geq \tfrac{1}{2}.$$

If X is a random variable and g is a function, then $Y = g(X)$ is also a random variable. If X is a discrete random variable with possible values x_1, x_2, . . . , then the expectation of $g(X)$ is given by

$$E[g(X)] = \sum_{i=1}^{\infty} g(x_i)\Pr\{X = x_i\} \tag{1.8}$$

provided the sum converges absolutely. If X is continuous and has the probability density function f_X, then the expected value of $g(X)$ is evaluated from

$$E[g(X)] = \int g(x)f_X(x)dx. \tag{1.9}$$

The general formula, covering both the discrete and continuous cases, is

$$E[g(X)] = \int g(x)dF_X(x) \tag{1.10}$$

where F_X is the distribution function of the random variable X. Technically speaking, the integral in (1.10) is a Lebesgue-Stieltjes integral. We do not require knowledge of such integrals in this text, but interpret (1.10) to signify (1.8) when X is a discrete random variable, and to represent (1.9) when X possesses a probability density f_X.

Let $F_Y(y) = \Pr\{Y \leq y\}$ denote the distribution function for $Y = g(X)$. When X is a discrete random variable, then

$$E[Y] = \sum_j y_j \Pr\{Y = y_j\}$$

$$= \sum_i g(x_i)\Pr\{X = x_i\}$$

if $y_i = g(x_i)$, and provided the second sum converges absolutely. In general

$$E[Y] = \int y \, dF_Y(y)$$

$$= \int g(x) dF_X(x). \tag{1.11}$$

If X is a discrete random variable, then so is $Y = g(X)$. It may be, however, that X is a continuous random variable while Y is discrete (the reader should provide an example). Even so, one may compute $E[Y]$ from either form in (1.11) with the same result.

1.2.4 Joint Distribution Functions

Given a pair (X, Y) of random variables, their *joint distribution function* is the function F_{XY} of two real variables given by

$$F_{XY}(x, y) = F(x, y) = \Pr\{X \le x \text{ and } Y \le y\}.$$

Usually the subscripts X, Y will be omitted, unless ambiguity is possible. A joint distribution function F_{XY} is said to possess a (joint) probability density if there exists a function f_{XY} of two real variables for which

$$F_{XY}(x, y) = \int_{-\infty}^{x} \int_{-\infty}^{y} f_{XY}(\xi, \eta) d\eta d\xi \qquad \text{for all} \quad x, y.$$

The function $F_X(x) = \lim_{y \to \infty} F(x, y)$ is a distribution function, called the *marginal distribution function* of X. Similarly, $F_Y(y) = \lim_{x \to \infty} F(x, y)$ is the marginal distribution function of Y. If the distribution function F possesses the joint density function f, then the marginal density functions for X and Y are given, respectively, by

$$f_X(x) = \int_{-\infty}^{+\infty} f(x, y) dy \quad \text{and} \quad f_Y(y) = \int_{-\infty}^{+\infty} f(x, y) dx.$$

If X and Y are jointly distributed, then $E[X + Y] = E[X] + E[Y]$, provided only that all these moments exist.

Independence

If it happens that $F(x, y) = F_X(x) \times F_Y(y)$ for every choice of x, y, then the random variables X and Y are said to be *independent*. If X and Y are independent and possess a joint density function $f(x, y)$, then necessarily $f(x, y) = f_X(x) f_Y(y)$ for all x, y.

Given jointly distributed random variables X and Y having means μ_X and μ_Y and finite variances, the *covariance* of X and Y, written σ_{XY} or Cov$[X, Y]$ is the product moment $\sigma_{XY} = E[(X - \mu_X)(Y - \mu_Y)] = E[XY] - \mu_X\mu_Y$, and X and Y are said to be *uncorrelated* if their covariance is zero, that is, $\sigma_{XY} = 0$. Independent random variables having finite variances are uncorrelated, but the converse is not true; there are uncorrelated random variables that are not independent.

Dividing the covariance σ_{XY} by the standard deviations σ_X and σ_Y defines the *correlation coefficient* $\rho = \sigma_{XY}/\sigma_X\sigma_Y$ for which $-1 \le \rho \le +1$.

The joint distribution function of any finite collection $X_1, \ldots, X_n$ of random variables is defined as the function

$$F(x_1, \ldots, x_n) = F_{X_1, \ldots, X_n}(x_1, \ldots, x_n)$$
$$= \Pr\{X_1 \le x_1, \ldots, X_n \le x_n\}.$$

If $F(x_1, \ldots, x_n) = F_{X_1}(x_1) \ldots F_{X_n}(x_n)$ for all values of $x_1, \ldots, x_n$, then the random variables $X_1, \ldots, X_n$ are said to be independent.

A joint distribution function $F(x_1, \ldots, x_n)$ is said to have a probability density function $f(\xi_1, \ldots, \xi_n)$ if

$$F(x_1, \ldots, x_n) = \int_{-\infty}^{x_1} \ldots \int_{-\infty}^{x_n} f(\xi_1, \ldots, \xi_n) d\xi_n \ldots d\xi_1,$$

for all values of $x_1, \ldots, x_n$.

Expectation

For jointly distributed random variables $X_1, \ldots, X_n$ and arbitrary functions $h_1, \ldots, h_m$ of n variables each, then

$$E[\sum_{j=1}^{m} h_j(X_1, \ldots, X_n)] = \sum_{j=1}^{m} E[h_j(X_1, \ldots, X_n)]$$

provided only that all these moments exist.

1.2.5 Sums and Convolutions

If X and Y are independent random variables having distribution functions F_X and F_Y, respectively, then the distribution function of their sum $Z = X + Y$ is the *convolution* of F_X and F_Y:

$$F_Z(z) = \int_{-\infty}^{+\infty} F_X(z - \xi) dF_Y(\xi) = \int_{-\infty}^{+\infty} F_Y(z - \eta) dF_X(\eta). \qquad (1.12)$$

If we specialize to the situation where X and Y have the probability densities f_X and f_Y, respectively, then the density function f_Z of the sum $Z = X + Y$ is the convolution of the densities f_X and f_Y:

$$f_Z(z) = \int_{-\infty}^{\infty} f_X(z - \eta) f_Y(\eta) d\eta = \int_{-\infty}^{+\infty} f_Y(z - \xi) f_X(\xi) d\xi. \qquad (1.13)$$

Where X and Y are nonnegative random variables, the range of integration is correspondingly reduced to

$$f_Z(z) = \int_{0}^{z} f_X(z - \eta) f_Y(\eta) d\eta = \int_{0}^{z} f_Y(z - \xi) f_X(\xi) d\xi \qquad \text{for} \quad z \ge 0. \qquad (1.14)$$

If X and Y are independent and have respective variances σ_X^2 and σ_Y^2, then the variance of the sum $Z = X + Y$ is the sum of the variances: $\sigma_Z^2 = \sigma_X^2 + \sigma_Y^2$. More generally, if $X_1, \ldots, X_n$ are independent random variables having variances $\sigma_1^2, \ldots, \sigma_n^2$, respectively, then the variance of the sum $Z = X_1 + \ldots + X_n$ is $\sigma_Z^2 = \sigma_1^2 + \ldots + \sigma_n^2$.

1.2.6 Change of Variable

Suppose that X is a random variable with probability density function f_X and that g is a strictly increasing differentiable function. Then $Y = g(X)$ defines a random variable and the event $\{Y \leq y\}$ is the same as the event $\{X \leq g^{-1}(y)\}$ where g^{-1} is the inverse function to g; i.e., $y = g(x)$ if and only if $x = g^{-1}(y)$. Thus we obtain the correspondence $F_Y(y) = \Pr\{Y \leq y\} = \Pr\{X \leq g^{-1}(y)\} = F_X(g^{-1}(y))$ between the distribution function of Y and that of X. Recall the differential calculus formula

$$\frac{dg^{-1}}{dy} = \frac{1}{g'(x)} = \frac{1}{dg/dx} \qquad \text{where} \quad y = g(x)$$

and use this in the chain rule of differentiation to obtain

$$f_Y(y) = \frac{dF_Y(y)}{dy} = \frac{dF_X(g^{-1}(y))}{dy} = f_X(x)\frac{1}{g'(x)} \qquad \text{where} \quad y = g(x).$$

The formula

$$f_Y(y) = \frac{1}{g'(x)} f_X(x) \qquad \text{where} \quad y = g(x), \tag{1.15}$$

expresses the density function for Y in terms of the density for X when g is strictly increasing and differentiable.

1.2.7 Conditional Probability

For any events A and B, the *conditional probability* of A given B is written $\Pr\{A|B\}$ and defined by

$$\Pr\{A|B\} = \frac{\Pr\{A \cap B\}}{\Pr\{B\}} \qquad \text{if} \quad \Pr\{B\} > 0, \tag{1.16}$$

and is left undefined if $\Pr\{B\} = 0$. [When $\Pr\{B\} = 0$, the right side of (1.16) is the indeterminate quantity $\frac{0}{0}$.]

In stochastic modeling, conditional probabilities are rarely procured via (1.16), but instead are dictated as primary data by the circumstances of the application, and then (1.16) is applied in its equivalent multiplicative form

$$\Pr\{A \cap B\} = \Pr\{A|B\}\Pr\{B\} \tag{1.17}$$

to compute other probabilities. (An example follows shortly.) Central in this role is the *Law of Total Probability*, which results from substituting

$\Pr\{A \cap B_i\} = \Pr\{A|B_i\}\Pr\{B_i\}$ into $\Pr\{A\} = \sum_{i=1}^{\infty}\Pr\{A \cap B_i\}$, where $\Omega = B_1 \cup B_2 \cup \ldots$ and $B_i \cap B_j = \emptyset$ if $i \neq j$ (cf. page 6), to yield

$$\Pr\{A\} = \sum_{i=1}^{\infty}\Pr\{A|B_i\}\Pr\{B_i\}. \tag{1.18}$$

Example   Gold and silver coins are allocated among three urns labeled I, II, III according to the following table.

Urn	Number of Gold Coins	Number of Silver Coins
I	4	8
II	3	9
III	6	6

An urn is selected at random, all urns being equally likely, and then a coin is selected at random from that urn. Using the notation I, II, III for the events of selecting urns I, II, and III, respectively, and G for the event of selecting a gold coin, then the problem description provides the following probabilities and conditional probabilities as data:

$$\Pr\{I\} = \tfrac{1}{3} \qquad \Pr\{G|I\} = \tfrac{4}{12}$$
$$\Pr\{II\} = \tfrac{1}{3} \qquad \Pr\{G|II\} = \tfrac{3}{12}$$
$$\Pr\{III\} = \tfrac{1}{3} \qquad \Pr\{G|III\} = \tfrac{6}{12}$$

and we *calculate* the probability of selecting a gold coin according to (1.18) *viz.*

$$\Pr\{G\} = \Pr\{G|I\}\Pr\{I\} + \Pr\{G|II\}\Pr\{II\} + \Pr\{G|III\}\Pr\{III\}$$
$$= \tfrac{4}{12}(\tfrac{1}{3}) + \tfrac{3}{12}(\tfrac{1}{3}) + \tfrac{6}{12}(\tfrac{1}{3}) = \tfrac{13}{36}.$$

As seen here, more often than not conditional probabilities are given as data and are not the end result of calculation.

Discussion of conditional distributions and conditional expectation merits an entire chapter (Chapter 2).

1.2.8 Review of Axiomatic Probability Theory⋆

For the most part this book studies random variables only through their distributions. In this spirit, we defined a random variable as a variable that takes on its values by chance. For some purposes, however, a little more precision and structure is needed.

⋆The material included in this review of axiomatic probability theory is not used in the remainder of the book. It is included in this review chapter only for the sake of completeness.

Recall that the basic elements of probability theory are

(1) the *sample space*, a set Ω whose elements ω correspond to the possible outcomes of an experiment;
(2) the *family of events*, a collection $\mathcal{F}$ of subsets A of Ω: we say that the event A *occurs* if the outcome ω of the experiment is an element of A; and
(3) the *probability measure*, a function P defined on $\mathcal{F}$ and satisfying

(a) $0 = P[\emptyset] \le P[A] \le P[\Omega] = 1 \qquad$ for $\quad A \in \mathcal{F}$
$(\emptyset = $ the empty set$)$

and

(b) $P[\overset{\infty}{\underset{n=1}{\cup}} A_n] = \sum_{n=1}^{\infty} P[A_n],$ \hfill (1.19)

if the events $A_1, A_2 \ldots$ are disjoint, i.e., if $A_i \cap A_j = \emptyset$ when $i \neq j$.

The triple $(\Omega, \mathcal{F}, P)$ is called a *probability space*.

Example When there are only a denumerable number of possible outcomes, say $\Omega = \{\omega_1, \omega_2, \ldots\}$, we may take $\mathcal{F}$ to be the collection of all subsets of Ω. If $p_1, p_2, \ldots$ are nonnegative numbers with $\sum_n p_n = 1$, the assignment

$$P[A] = \sum_{\omega_i \in A} p_i$$

determines a probability measure defined on $\mathcal{F}$.

It is not always desirable, consistent, or feasible to take the family of events as the collection of *all* subsets of Ω. Indeed, when Ω is nondenumerably infinite, it may not be possible to define a probability measure on the collection of all subsets maintaining the properties of (1.19). In whatever way we prescribe $\mathcal{F}$ such that (1.19) holds, the family of events $\mathcal{F}$ should satisfy

(a) $\emptyset$ is in $\mathcal{F}$ and Ω is in $\mathcal{F}$;
(b) A^c is in $\mathcal{F}$ whenever A is in $\mathcal{F}$, where $A^c = \{\omega \in \Omega; \omega \notin A\}$ is the complement of A; and \hfill (1.20)
(c) $\cup_{n=1}^{\infty} A_n$ is in $\mathcal{F}$ whenever A_n is in $\mathcal{F}$ for $n = 1, 2, \ldots$.

A collection $\mathcal{F}$ of subsets of a set Ω satisfying (1.20) is called a σ-*algebra*. If $\mathcal{F}$ is a σ-algebra, then

$$\overset{\infty}{\underset{n=1}{\cap}} A_n = \left(\overset{\infty}{\underset{n=1}{\cup}} A_n^c \right)^c$$

is in $\mathcal{F}$ whenever A_n is in $\mathcal{F}$ for $n = 1, 2, \ldots$. Manifestly, as a consequence we find that finite unions and finite intersections of members of $\mathcal{F}$ are maintained in $\mathcal{F}$.

In this framework, a real random variable X is a real-valued function defined on Ω fulfilling certain "measurability" conditions given here. The distribution function of the random variable X is formally given by

$$\Pr\{a < X \le b\} = P[\{\omega: a < X(\omega) \le b\}]. \qquad (1.21)$$

In words, the probability that the random variable X takes a value in $(a, b]$ is calculated as the probability of the set of outcomes ω for which $a < X(\omega) \le b$. If relation (1.21) is to have meaning, X cannot be an arbitrary function on Ω, but must satisfy the condition that

$\{\omega: a < X(\omega) \le b\}$ is in $\mathcal{F}$ for all real $a < b$,

since $\mathcal{F}$ embodies the only sets A for which $P[A]$ is defined. In fact, by exploiting the properties (1.20) of the σ-algebra $\mathcal{F}$, we find that it is enough to require

$\{\omega: X(\omega) \le x\}$ is in $\mathcal{F}$ for all real x.

Let $\mathcal{A}$ be any σ-algebra of subsets of Ω. We say that X is *measurable with respect to* $\mathcal{A}$, or more briefly *$\mathcal{A}$-measurable*, if

$\{\omega: X(\omega) \le x\}$ is in $\mathcal{A}$ for all real x.

Thus, every real random variable is by definition $\mathcal{F}$-measurable. There may, in general, be smaller σ-algebras with respect to which X is also measurable.

The σ-algebra *generated* by a random variable X is defined to be the smallest σ-algebra with respect to which X is measurable. It is denoted by $\mathcal{F}(X)$ and consists exactly of those sets A that are in every σ-algebra $\mathcal{A}$ for which X is $\mathcal{A}$-measurable. For example, if X has only denumerably many possible values $x_1, x_2, \ldots$, the sets

$$A_i = \{\omega: X(\omega) = x_i\}, \qquad i = 1, 2, \ldots$$

form a countable *partition* of Ω, i.e.,

$$\Omega = \bigcup_{i=1}^{\infty} A_i,$$

and

$$A_i \cap A_j = \emptyset \qquad \text{if} \quad i \ne j,$$

and then $\mathcal{F}(X)$ includes precisely $\emptyset$, Ω, and every set that is the union of some of the A_i's.

Example For the reader completely unfamiliar with this framework, the following simple example will help illustrate the concepts. The experiment consists in tossing a nickel and a dime and observing "heads" or "tails." We take Ω to be

$$\Omega = \{(H, H), (H, T), (T, H), (T, T)\},$$

where, for example, (H, T) stands for the outcome "nickel = heads, and dime = tails." We will take the collection of all subsets of Ω as the family of events. Assuming each outcome in Ω to be equally likely, we arrive at the probability measure:

$A \in \mathcal{F}$	$P[A]$	$A \in \mathcal{F}$	$P[A]$
$\emptyset$	0	Ω	1
$\{(H, H)\}$	$\frac{1}{4}$	$\{(H, T), (T, H), (T, T)\}$	$\frac{3}{4}$
$\{(H, T)\}$	$\frac{1}{4}$	$\{(H, H), (T, H), (T, T)\}$	$\frac{3}{4}$
$\{(T, H)\}$	$\frac{1}{4}$	$\{(H, H), (H, T), (T, T)\}$	$\frac{3}{4}$
$\{(T, T)\}$	$\frac{1}{4}$	$\{(H, H), (H, T), (T, H)\}$	$\frac{3}{4}$
$\{(H, H), (H, T)\}$	$\frac{1}{2}$	$\{(T, H), (T, T)\}$	$\frac{1}{2}$
$\{(H, H), (T, H)\}$	$\frac{1}{2}$	$\{(H, T), (T, T)\}$	$\frac{1}{2}$
$\{(H, H), (T, T)\}$	$\frac{1}{2}$	$\{(H, T), (T, H)\}$	$\frac{1}{2}$

The event "the nickel is heads" is $\{(H, H), (H, T)\}$ and has, according to the table, probability $\frac{1}{2}$, as it should.

Let X_n be 1 if the nickel is heads, and 0 otherwise, let X_d be the corresponding random variable for the dime, and let $Z = X_n + X_d$ be the total number of heads. As functions on Ω, we have

$\omega \in \Omega$	$X_n(\omega)$	$X_d(\omega)$	$Z(\omega)$
(H, H)	1	1	2
(H, T)	1	0	1
(T, H)	0	1	1
(T, T)	0	0	0

Finally, the σ-algebras generated by X_n and Z are

$$\mathcal{F}(X_n) = \emptyset, \Omega, \{(H, H), (H, T)\}, \{(T, H), (T, T)\},$$

and

$$\mathcal{F}(Z) = \emptyset, \Omega, \{(H, H)\}, \{(H, T), (T, H)\}, \{(T, T)\},$$
$$\{(H, T), (T, H), (T, T)\}, \{(H, H), (T, T)\},$$
$$\{(H, H), (H, T), (T, H)\}.$$

$\mathcal{F}(X_n)$ contains four sets and $\mathcal{F}(Z)$ contains eight. Is X_n measurable with respect to $\mathcal{F}(Z)$, or vice versa?

Every pair X, Y of random variables determines a σ-algebra called the σ-algebra generated by X, Y. It is the smallest σ-algebra with respect to which both X and Y are measurable. This σ-algebra comprises exactly those sets A that are in every σ-algebra $\mathcal{A}$ for which X and Y are both $\mathcal{A}$-

measurable. If both X and Y assume only denumerably many possible values, say $x_1, x_2, \ldots$ and $y_1, y_2, \ldots$, respectively, then the sets

$$A_{ij} = \{\omega: X(\omega) = x_i, \, Y(\omega) = y_j\}, \qquad i, j = 1, 2, \ldots$$

present a countable partition of Ω and $\mathscr{F}(X, Y)$ consists precisely of $\emptyset$, Ω, and every set that is the union of some of the A_{ij}'s. Observe that X is measurable with respect to $\mathscr{F}(X, Y)$, and thus $\mathscr{F}(X) \subset \mathscr{F}(X, Y)$.

More generally, let $\{X(t); \, t \in T\}$ be any family of random variables. Then the σ-algebra generated by $\{X(t); \, t \in T\}$ is the smallest σ-algebra with respect to which every random variable $X(t)$, $t \in T$, is measurable. It is denoted by $\mathscr{F}\{X(t); \, t \in T\}$.

A special role is played by a distinguished σ-algebra of sets of real numbers. The σ-algebra of *Borel sets* is the σ-algebra generated by the identity function $f(x) = x$, for $x \in (-\infty, \infty)$. Alternatively, the σ-algebra of Borel sets is the smallest σ-algebra containing every interval of the form $(a, b]$, $-\infty \le a \le b < +\infty$. A real-valued function of a real variable is said to be *Borel measurable* if it is measurable with respect to the σ-algebra of Borel sets.

Problems I.2

1. Let A and B be arbitrary, not necessarily disjoint, events. Use the law of total probability to verify the formula

$$\Pr\{A\} = \Pr\{AB\} + \Pr\{AB^c\}$$

where B^c is the complementary event to B. (That is, B^c occurs if and only if B does not occur.)

2. Let A and B be arbitrary, not necessarily disjoint, events. Establish the general addition law

$$\Pr\{A \cup B\} = \Pr\{A\} + \Pr\{B\} - \Pr\{AB\}.$$

Hint: Apply the result of Problem 1 to evaluate $\Pr\{AB^c\} = \Pr\{A\} - \Pr\{AB\}$. Then apply the addition law to the disjoint events AB and AB^c, noting that $A = (AB) \cup (AB^c)$.

3. Let A, B, and C be arbitrary events. Establish the addition law

$$\Pr\{A \cup B \cup C\} = \Pr\{A\} + \Pr\{B\} + \Pr\{C\} - \Pr\{AB\} - \Pr\{AC\}$$
$$- \Pr\{BC\} + \Pr\{ABC\}.$$

Hint: Let $D = A \cup B$. Then apply the result of Problem 2 to the events D and C, and then to $D = A \cup B$.

4. (a) Plot the distribution function

$$F(x) = \begin{cases} 0 & \text{for} \quad x \le 0, \\ x^3 & \text{for} \quad 0 < x < 1, \\ 1 & \text{for} \quad x \ge 1. \end{cases}$$

(b) Determine the corresponding density function $f(x)$ in the three regions (i) $x \le 0$, (ii) $0 < x < 1$ and (iii) $1 \le x$.
(c) What is the mean of the distribution?
(d) If X is a random variable following the distribution specified in (a), evaluate $\Pr\{\frac{1}{4} \le X \le \frac{3}{4}\}$.

5. Let Z be a discrete random variable having possible values 0, 1, 2, and 3 and probability mass function

$$\begin{array}{ll} p(0) = \frac{1}{4} & p(2) = \frac{1}{8} \\ p(1) = \frac{1}{2} & p(3) = \frac{1}{8} \end{array}$$

(a) Plot the corresponding distribution function.
(b) Determine the mean $E[Z]$.
(c) Evaluate the variance $\text{Var}[Z]$.

6. Let X and Y be independent random variables having distribution functions F_X and F_Y, respectively.
(a) Define $Z = \max\{X, Y\}$ to be the larger of the two. Show that $F_Z(z) = F_X(z)F_Y(z)$ for all z.
(b) Define $W = \min\{X, Y\}$ to be the smaller of the two. Show that $F_W(w) = 1 - [1 - F_X(w)][1 - F_Y(w)]$ for all w.

7. Suppose X is a random variable having the probability density function

$$f(x) = \begin{cases} Rx^{R-1} & \text{for} \quad 0 \le x \le 1, \\ 0 & \text{elsewhere,} \end{cases}$$

where $R > 0$ is a fixed parameter.
(a) Determine the distribution function $F_X(x)$.
(b) Determine the mean $E[X]$.
(c) Determine the variance $\text{Var}[X]$.

8. A random variable V has the distribution function

$$F(v) = \begin{cases} 0 & \text{for} \quad v < 0, \\ 1 - (1 - v)^A & \text{for} \quad 0 \le v \le 1, \\ 1 & \text{for} \quad v > 1, \end{cases}$$

where $A > 0$ is a parameter. Determine the density function, mean, and variance.

9. Determine the distribution function, mean, and variance corresponding to the triangular density

$$f(x) = \begin{cases} x & \text{for } 0 \le x \le 1, \\ 2 - x & \text{for } 1 \le x \le 2, \\ 0 & \text{elsewhere.} \end{cases}$$

10. Let $1\{A\}$ be the indicator random variable associated with an event A, defined to be one if A occurs, and zero, otherwise. Define A^c, the complement of event A, to be the event that occurs when A does not occur. Show
    (a) $1\{A^c\} = 1 - 1\{A\}$
    (b) $1\{A \cap B\} = 1\{A\}1\{B\} = \min\{1\{A\}, 1\{B\}\}$
    (c) $1\{A \cup B\} = \max\{1\{A\}, 1\{B\}\}$

11. Thirteen cards numbered $1, \ldots, 13$ are shuffled and dealt one at a time. Say a *match* occurs on deal k if the kth card revealed is card number k. Let N be the total number of matches that occur in the thirteen cards. Determine $E[N]$.
    Hint: Write $N = 1\{A_1\} + \ldots + 1\{A_{13}\}$ where A_k is the event that a match occurs on deal k.

12. Let N cards carry the distinct numbers $x_1, \ldots, x_n$. If two cards are drawn at random without replacement, show that the correlation coefficient ρ between the numbers appearing on the two cards is $-1/(N-1)$.

13. A population having N distinct elements is sampled with replacement. Because of repetitions, a random sample of size r may contain fewer than r distinct elements. Let S_r be the sample size necessary to get r distinct elements. Show that

$$E[S_r] = N\left(\frac{1}{N} + \frac{1}{N-1} + \cdots + \frac{1}{N-r+1}\right).$$

14. A fair coin is tossed until the first time that the same side appears twice in succession. Let N be the number of tosses required.
    (a) Determine the probability mass function for N.
    (b) Let A be the event that N is even and B be the event that $N \le 6$. Evaluate $\Pr\{A\}$, $\Pr\{B\}$, and $\Pr\{AB\}$.

15. Two players, A and B, take turns on a gambling machine until one of them scores a success, the first to do so being the winner. Their probabilities for success on a single play are p for A and q for B, and successive plays are independent.
    (a) Determine the probability that A wins the contest given that A plays first.
    (b) Determine the mean number of plays required, given that A wins.

16. A pair of dice are tossed. If the two outcomes are equal, the dice are tossed again, and the process repeated. If the dice are unequal, their sum is recorded. Determine the probability mass function for the sum.

17. Let U and W be jointly distributed random variables. Show that U and W are independent if

$$\Pr\{U > u \text{ and } W > w\} = \Pr\{U > u\}\Pr\{W > w\} \qquad \text{for all} \quad u, w.$$

18. Suppose X is a random variable with finite mean μ and variance σ^2, and $Y = a + bX$ for certain constants a, $b \neq 0$. Determine the mean and variance for Y.

19. Determine the mean and variance for the probability mass function

$$p(k) = \frac{2(n - k)}{n(n - 1)} \qquad \text{for} \quad k = 1, 2, \ldots, n.$$

20. Random variables X and Y are independent and have the probability mass functions:

$$p_X(0) = \tfrac{1}{2} \qquad p_Y(1) = \tfrac{1}{3}$$
$$p_X(3) = \tfrac{1}{2} \qquad p_Y(2) = \tfrac{1}{3}$$
$$p_Y(3) = \tfrac{1}{2}$$

Determine the probability mass function of the sum $Z = X + Y$.

21. Random variables U and V are independent and have the probability mass functions:

$$p_U(0) = \tfrac{1}{3} \qquad p_V(1) = \tfrac{1}{2}$$
$$p_U(2) = \tfrac{1}{3} \qquad p_V(2) = \tfrac{1}{2}$$
$$p_U(4) = \tfrac{1}{3}$$

Determine the probability mass function of the sum $W = U + V$.

22. Let U, V, and W be independent random variables with equal variances σ^2. Define $X = U + W$ and $Y = V - W$. Find the covariance between X and Y.

23. Let X and Y be independent random variables each with the uniform probability density function

$$f(x) = \begin{cases} 1 & \text{for } 0 < x < 1, \\ 0 & \text{elsewhere.} \end{cases}$$

Find the joint probability density function of U and V, where $U = \max\{X, Y\}$ and $V = \min\{X, Y\}$.

1.3 The Major Discrete Distributions

The most important discrete probability distributions and their relevant properties are summarized in this section. The exposition is brief since most readers will be familiar with this material from an earlier course in probability.

1.3.1 Bernoulli Distribution

A random variable X following the Bernoulli distribution with parameter p has only two possible values, 0 and 1, and the probability mass function is $p(1) = p$ and $p(0) = 1 - p$, where $0 < p < 1$. The mean and variance are $E[X] = p$ and $\text{Var}[X] = p(1 - p)$, respectively.

Bernoulli random variables occur frequently as indicators of events. The *indicator* of an event A is the random variable

$$\mathbf{1}(A) = \mathbf{1}_A = \begin{cases} 1 & \text{if } A \text{ occurs,} \\ 0 & \text{if } A \text{ does not occur.} \end{cases} \quad (1.22)$$

Then $\mathbf{1}_A$ is a Bernoulli random variable with parameter $p = E[\mathbf{1}_A] = \Pr\{A\}$.

The simple expedient of using indicators often reduces formidable calculations into trivial ones. For example, let $\alpha_1, \alpha_2, \ldots, \alpha_n$ be arbitrary real numbers and $A_1, A_2, \ldots, A_n$ be events, and consider the problem of showing that

$$\sum_{i=1}^{n}\sum_{j=1}^{n} \alpha_i\alpha_j\Pr\{A_i \cap A_j\} \geq 0. \quad (1.23)$$

Attacked directly, the problem is difficult. But bringing in the indicators $\mathbf{1}(A_i)$ and observing that

$$0 \leq \{\sum_{i=1}^{n}\alpha_i\mathbf{1}(A_i)\}^2 = \{\sum_{i=1}^{n}\alpha_i\mathbf{1}(A_i)\}\{\sum_{j=1}^{n}\alpha_j\mathbf{1}(A_j)\}$$

$$= \sum_{i=1}^{n}\sum_{j=1}^{n}\alpha_i\alpha_j\mathbf{1}(A_i)\mathbf{1}(A_j) = \sum_{i=1}^{n}\sum_{j=1}^{n}\alpha_i\alpha_j\mathbf{1}(A_i \cap A_j)$$

gives, after taking expectations,

$$0 \leq E\left[\{\sum_{i=1}^{n}\alpha_i\mathbf{1}(A_i)\}^2\right] = \sum_{i=1}^{n}\sum_{j=1}^{n}\alpha_i\alpha_j E[\mathbf{1}(A_i \cap A_j)]$$

$$= \sum_{i=1}^{n}\sum_{j=1}^{n}\alpha_i\alpha_j\Pr\{A_i \cap A_j\},$$

and the demonstration of (1.23) is complete.

1.3.2 Binomial Distribution

Consider independent events $A_1, A_2, \ldots, A_n$, all having the same probability $p = \Pr\{A_i\}$ of occurrence. Let Y count the total number of events among $A_1, \ldots, A_n$ that occur. Then Y has a binomial distribution with parameters n and p. The probability mass function is

$$p_Y(k) = \Pr\{Y = k\}$$

$$= \frac{n!}{k!(n-k)!} p^k(1-p)^{n-k} \quad \text{for} \quad k = 0, 1, \ldots, n. \tag{1.24}$$

Writing Y as a sum of indicators in the form $Y = \mathbf{1}(A_1) + \ldots + \mathbf{1}(A_n)$ makes it easy to determine the moments

$$E[Y] = E[\mathbf{1}(A_1)] + \cdots + E[\mathbf{1}(A_n)] = np,$$

and using independence, we can also determine that

$$\text{Var}[Y] = \text{Var}[\mathbf{1}(A_1)] + \cdots + \text{Var}[\mathbf{1}(A_n)] = np(1-p).$$

Briefly, we think of a binomial random variable as counting the number of "successes" in n independent trials where there is a constant probability p of success on any single trial.

1.3.3 Geometric and Negative Binomial Distributions

Let $A_1, A_2, \ldots$ be independent events having a common probability $p = \Pr\{A_i\}$ of occurrence. Say that trial k is a success (S) or failure (F) according as A_k occurs or not, and let Z count the number of *failures* prior to the first success. To be precise, $Z = k$ if and only if $\mathbf{1}(A_1) = 0, \ldots, \mathbf{1}(A_k) = 0$ and $\mathbf{1}(A_{k+1}) = 1$. Then Z has a geometric distribution with parameter p. The probability mass function is

$$p_Z(k) = p(1-p)^k \quad \text{for} \quad k = 0, 1, \ldots \tag{1.25}$$

and the first two moments are

$$E[Z] = \frac{1-p}{p}; \quad \text{Var}[Z] = \frac{1-p}{p^2}.$$

Sometimes the term "geometric distribution" is used in referring to the probability mass function

$$p_{Z'}(k) = p(1-p)^{k-1} \quad \text{for} \quad k = 1, 2, \ldots \tag{1.26}$$

This is merely the distribution of the random variable $Z' = 1 + Z$, the number of *trials* until the first success. Hence $E[Z'] = 1 + E[Z] = 1/p$, and $\text{Var}[Z'] = \text{Var}[Z] = (1-p)/p^2$.

Now fix an integer $r \geq 1$ and let W_r count the number of failures observed before the rth success in $A_1, A_2, \ldots$. Then W_r has a *negative binomial* distribution with parameters r and p. The event $W_r = k$ calls for (A) exactly $r-1$ successes in the first $k + r - 1$ trials, followed by, (B) a success on trail $k + r$. The probability for (A) is obtained from a binomial distribution and the probability for (B) is simply p, which leads to the following probability mass function for W_r:

$$p(k) = \Pr\{W_r = k\} = \frac{(k + r - 1)!}{(r - 1)!k!} \, p^r(1 - p)^k, \qquad k = 0, 1, \ldots \quad (1.27)$$

Another way of writing W_r is as the sum $W_r = Z_1 + \ldots + Z_r$ where $Z_1, \ldots, Z_r$ are independent random variables, each having the geometric distribution of (1.25). This formulation readily yields the moments

$$E[W_r] = \frac{r(1 - p)}{p}; \qquad \text{Var}[W_r] = \frac{r(1 - p)}{p^2}. \qquad (1.28)$$

1.3.4 The Poisson Distribution

If distributions were graded on a scale of one to ten, the Poisson clearly merits a ten. It plays a role in the class of discrete distributions that parallels in some sense that of the normal distribution in the continuous class. The Poisson distribution occurs often in natural phenomena, for powerful and convincing reasons (The Law of Rare Events, see p. 23). At the same time the Poisson distribution has many elegant and surprising mathematical properties that make analysis a pleasure.

The Poisson distribution with parameter $\lambda > 0$ has the probability mass function

$$p(k) = \frac{\lambda^k e^{-\lambda}}{k!} \qquad \text{for} \quad k = 0, 1, \ldots \qquad (1.29)$$

Using the series expansion

$$e^\lambda = 1 + \lambda + \frac{\lambda^2}{2!} + \frac{\lambda^3}{3!} + \cdots \qquad (1.30)$$

we see that $\Sigma_{k \geq 0} \, p(k) = 1$. The same series helps calculate the mean via

$$\sum_{k=0}^{\infty} kp(k) = \sum_{k=1}^{\infty} k \, \frac{\lambda^k e^{-\lambda}}{k!} = \lambda e^{-\lambda} \sum_{k=1}^{\infty} \frac{\lambda^{k-1}}{(k - 1)!} = \lambda.$$

The same trick works on the variance, beginning with

$$\sum_{k=0}^{\infty} k(k - 1)p(k) = \sum_{k=2}^{\infty} k(k - 1) \, \frac{\lambda^k e^{-\lambda}}{k!} = \lambda^2 e^{-\lambda} \sum_{k=2}^{\infty} \frac{\lambda^{k-2}}{(k - 2)!} = \lambda^2.$$

Written in terms of a random variable X having the Poisson distribution with parameter λ, we have just calculated $E[X] = \lambda$ and $E[X(X - 1)] = \lambda^2$ whence $E[X^2] = E[X(X - 1)] + E[X] = \lambda^2 + \lambda$ and $\text{Var}[X] = E[X^2] - \{E[X]\}^2 = \lambda$. That is, the mean and variance are both the same and equal to the parameter λ of the Poisson distribution.

The simplest form of the Law of Rare Events asserts that the binomial distribution with parameters n and p converges to the Poisson with parameter λ if $n \to \infty$ and $p \to 0$ in such a way that $\lambda = np$ remains constant. In words, given an indefinitely large number of independent trials, where success on each trial occurs with the same arbitrarily small probability, then the total number of successes will follow, approximately, a Poisson distribution.

The proof is a relatively simple manipulation of limits. We begin by writing the binomial distribution in the form

$$\Pr\{X = k\} = \frac{n!}{k!(n-k)!}\, p^k (1-p)^{n-k}$$

$$= n(n-1) \cdots (n-k+1) \frac{p^k(1-p)^n}{k!(1-p)^k}$$

and then substitute $p = \lambda/n$ to get

$$\Pr\{X = k\} = n(n-1) \cdots (n-k+1) \frac{\left(\dfrac{\lambda}{n}\right)^k \left(1 - \dfrac{\lambda}{n}\right)^n}{k!\left(1 - \dfrac{\lambda}{n}\right)^k}$$

$$= 1\left(1 - \frac{1}{n}\right) \cdots \left(1 - \frac{k-1}{n}\right) \frac{\lambda^k \left(1 - \dfrac{\lambda}{n}\right)^n}{k!\left(1 - \dfrac{\lambda}{n}\right)^k}.$$

Now let $n \to \infty$ and observe that

$$1\left(1 - \frac{1}{n}\right) \cdots \left(1 - \frac{k-1}{n}\right) \to 1 \qquad \text{as } n \to \infty;$$

$$\left(1 - \frac{\lambda}{n}\right)^n \to e^{-\lambda} \qquad \text{as } n \to \infty;$$

and

$$\left(1 - \frac{\lambda}{n}\right)^k \to 1 \qquad \text{as } n \to \infty$$

to obtain the Poisson distribution

$$\Pr\{X = k\} = \frac{\lambda^k e^{-\lambda}}{k!} \qquad \text{for } k = 0, 1, \ldots$$

in the limit. Extended forms of the Law of Rare Events are presented in Chapter 5.

Example *You Be the Judge*   In a purse snatching incident, a woman described her assailant as being seven feet tall and wearing an orange hat, red shirt, green trousers, and yellow shoes. A short while later and a few blocks away a person fitting that description was seen and charged with the crime.

In court, the prosecution argued that the characteristics of the assailant were so rare as to make the evidence overwhelming that the defendant was the criminal.

The defense argued that the description of the assailant was rare, and that therefore the number of people fitting the description should follow a Poisson distribution. Since one person fitting the description was found, the best estimate for the parameter is $\mu = 1$. Finally they argued that the relevant computation is the conditional probability that there is at least one other person at large fitting the description given that one was observed. The defense calculated

$$\Pr\{X \geq 2 | X \geq 1\} = \frac{1 - \Pr\{X = 0\} - \Pr\{X = 1\}}{1 - \Pr\{X = 0\}}$$
$$= \frac{1 - e^{-1} - e^{-1}}{1 - e^{-1}} = .4180,$$

and since this figure is rather large, they argued that the circumstantial evidence arising out of the unusual description was too weak to satisfy the "beyond a reasonable doubt" criterion for guilt in criminal cases.

1.3.5 The Multinomial Distribution

This is a joint distribution of r variables in which only nonnegative integer values $0, \ldots, n$ are possible. The joint probability mass function is

$$\Pr\{X_1 = k_1, \ldots, X_r = k_r\}$$

$$= \begin{cases} \dfrac{n!}{k_1! \ldots k_r!} p_1^{k_1} \cdots p_r^{k_r} & \text{if } k_1 + \cdots + k_r = n, \\ \\ 0 & \text{otherwise,} \end{cases} \qquad (1.31)$$

where $p_i > 0$ for $i = 1, \ldots, r$ and $p_1 + \ldots + p_r = 1$.

Some moments are $E[X_i] = np_i$, $\mathrm{Var}[X_i] = np_i(1 - p_i)$, and $\mathrm{Cov}[X_i X_j] = -np_i p_j$.

The multinomial distribution generalizes the binomial. Consider an experiment having a total of r possible outcomes, and let the corresponding probabilities be $p_1, \ldots, p_r$, respectively. Now perform n independent replications of the experiment and let X_i record the total number of times that the ith type outcome is observed in the n trials. Then $X_1, \ldots, X_r$ has the multinomial distribution given in (1.31).

Problems 1.3

1. The discrete uniform distribution on $\{1, \ldots, n\}$ corresponds to the probability mass function

$$p(k) = \begin{cases} \dfrac{1}{n} & \text{for} \quad k = 1, \ldots, n \\ 0 & \text{elsewhere.} \end{cases}$$

(a) Determine the mean and variance.
(b) Suppose X and Y are independent random variables, each having the discrete uniform distribution on $\{0, \ldots, n\}$. Determine the probability mass function for the sum $Z = X + Y$.
(c) Under the assumptions of (b), determine the probability mass function for the minimum $U = \min\{X, Y\}$.

2. Suppose that X has a discrete uniform distribution on the integers 0, 1, ..., 9, and Y is independent and has the probability distribution $\Pr\{Y = k\} = a_k$ for $k = 0, 1, \ldots$ What is the distribution of $Z = X + Y \pmod{10}$, their sum modulo 10?

3. The *mode* of a probability mass function $p(k)$ is any value $k^\star$ for which $p(k^\star) \geq p(k)$ for all k. Determine the mode(s) for
(a) The Poisson distribution with parameter $\lambda > 0$.
(b) The binomial distribution with parameters n and p.

4. Let X be a Poisson random variable with parameter λ. Determine the probability that X is odd.

5. Let U be a Poisson random variable with mean μ. Determine the expected value of the random variable $V = 1/(1 + U)$.

6. Let $Y = N - X$ where X has a binomial distribution with parameters N and p. Evaluate the product moment $E[XY]$ and the covariance $\text{Cov}[X, Y]$.

7. Suppose (X_1, X_2, X_3) has a multinomial distribution with parameters M and $\pi_i > 0$ for $i = 1, 2, 3$, with $\pi_1 + \pi_2 + \pi_3 = 1$.
(a) Determine the marginal distribution for X_1.
(b) Find the distribution for $N = X_1 + X_2$.
(c) What is the conditional probability $\Pr\{X_1 = k | N = n\}$ for $0 \leq k \leq n$?

8. Let X and Y be independent Poisson distributed random variables having means μ and v, respectively. Evaluate the convolution of their mass functions to determine the probability distribution of their sum $Z = X + Y$.

9. Let X and Y be independent binomial random variables having parameters (N, p) and (M, p), respectively. Let $Z = X + Y$.
   (a) Argue that Z has a binomial distribution with parameters $(N + M, p)$ by writing X and Y as appropriate sums of Bernoulli random variables.
   (b) Validate the result in (a) by evaluating the necessary convolution.

10. Suppose that X and Y are independent random variables with the geometric distribution

$$p(k) = (1 - \pi)\pi^k \qquad \text{for} \quad k = 0, 1, \ldots$$

Perform the appropriate convolution to identify the distribution of $Z = X + Y$ as a negative binomial.

11. Determine numerical values to three decimal places for $\Pr\{X = k\}$, $k = 0, 1, 2$ when
    (a) X has a binomial distribution with parameters $n = 10$ and $p = 0.1$.
    (b) X has a binomial distribution with parameters $n = 100$ and $p = 0.01$.
    (c) X has a Poisson distribution with parameter $\lambda = 1$.

12. Let X and Y be independent random variables sharing the geometric distribution whose mass function is

$$p(k) = (1 - \pi)\pi^k \qquad \text{for} \quad k = 0, 1, \ldots$$

where $0 < \pi < 1$. Let $U = \min\{X, Y\}$, $V = \max\{X, Y\}$ and $W = V - U$. Determine the joint probability mass function for U and W and show that U and W are independent.

13. Suppose that the telephone calls coming into a certain switchboard during a minute time interval follow a Poisson distribution with mean $\lambda = 4$. If the switchboard can handle at most 6 calls per minute, what is the probability that the switchboard will receive more calls than it can handle during a specified minute interval?

14. Suppose that a sample of 10 is taken from a day's output of a machine that normally produces 5 percent defective parts. If 100 percent of a day's production is inspected whenever the sample of 10 gives 2 or more defectives, then what is the probability that 100 percent of a day's production will be inspected? What assumptions did you make?

15. Suppose that a random variable Z has the geometric distribution

$$p_Z(k) = p(1 - p)^k \qquad \text{for} \quad k = 0, 1, \ldots,$$

where $p = 0.10$.
    (a) Evaluate the mean and variance of Z.
    (b) What is the probability that Z strictly exceeds 10?

16. Suppose that X is a Poisson distributed random variable with mean $\lambda = 2$. Determine $\Pr\{X \leq \lambda\}$.

1.4 Important Continuous Distributions

For future reference, this section catalogs several continuous distributions and some of their properties.

1.4.1 The Normal Distribution

The *normal distribution* with parameters μ and $\sigma^2 > 0$ is given by the familiar bell-shaped probability density function

$$\phi(x; \mu, \sigma^2) = \frac{1}{\sqrt{2\pi}\,\sigma}\, e^{-(x-\mu)^2/2\sigma^2}, \qquad -\infty < x < \infty. \qquad (1.32)$$

The density function is symmetric about the point μ and the parameter σ^2 is the variance of the distribution. The case $\mu = 0$ and $\sigma^2 = 1$ is referred to as the *standard normal distribution*. If X is normally distributed with mean μ and variance σ^2, then $Z = (X - \mu)/\sigma$ has a standard normal distribution. By this means, probability statements about arbitrary normal random variables can be reduced to equivalent statements about standard normal random variables. The standard normal density and distribution functions are given respectively by

$$\phi(\xi) = \frac{1}{\sqrt{2\pi}}\, e^{-\xi^2/2}, \qquad -\infty < \xi < \infty, \qquad (1.33)$$

and

$$\Phi(x) = \int_{-\infty}^{x} \phi(\xi)d\xi, \qquad -\infty < x < \infty. \qquad (1.34)$$

The *central limit theorem* explains in part the wide prevalence of the normal distribution in nature. A simple form of this aptly named result concerns the partial sums $S_n = \xi_1 + \ldots + \xi_n$ of independent and identically distributed summands $\xi_1, \xi_2, \ldots$ having finite means $\mu = E[\xi_k]$ and finite variances $\sigma^2 = \text{Var}[\xi_k]$. In this case, the central limit theorem asserts that

$$\lim_{n\to\infty} \Pr\left\{ \frac{S_n - n\mu}{\sigma\sqrt{n}} \leq x \right\} = \Phi(x) \qquad \text{for all} \quad x. \qquad (1.35)$$

The precise statement of the theorem's conclusion is given by Equation (1.35). Intuition is sometimes enhanced by the looser statement that, for large n, then S_n is approximately normally distributed with mean $n\mu$ and variance $n\sigma^2$.

In practical terms we expect the normal distribution to arise whenever the numerical outcome of an experiment results from numerous small additive effects, all operating independently, and where no single or small group of effects is dominant.

The Lognormal Distribution

If the natural logarithm of a nonnegative random variable V is normally distributed, then V is said to have a lognormal distribution. Conversely, if X is normally distributed with mean μ and variance σ^2, then $V = e^X$ defines a lognormally distributed random variable. The change-of-variable formula (1.15) applies to give the density function for V to be

$$f_V(v) = \frac{1}{\sqrt{2\pi}\,\sigma v} \exp\left\{ -\frac{1}{2}\left(\frac{\ln v - \mu}{\sigma}\right)^2 \right\}, \qquad v \geq 0. \qquad (1.36)$$

The mean and variance are, respectively,

$$\begin{aligned} E[V] &= \exp\{\mu + \tfrac{1}{2}\sigma^2\}, \\ \mathrm{Var}[V] &= \exp\{2(\mu + \tfrac{1}{2}\sigma^2)\}[\exp\{\sigma^2\} - 1]. \end{aligned} \qquad (1.37)$$

1.4.2 The Exponential Distribution

A nonnegative random variable T is said to have an exponential distribution with parameter $\lambda > 0$ if the probability density function is

$$f_T(t) = \begin{cases} \lambda e^{-\lambda t} & \text{for} \quad t \geq 0, \\ 0 & \text{for} \quad t < 0. \end{cases} \qquad (1.38)$$

The corresponding distribution function is

$$F_T(t) = \begin{cases} 1 - e^{-\lambda t} & \text{for} \quad t \geq 0, \\ 0 & \text{for} \quad t < 0, \end{cases} \qquad (1.39)$$

and the mean and variance are given, respectively, by

$$E[T] = \frac{1}{\lambda} \quad \text{and} \quad \mathrm{Var}[T] = \frac{1}{\lambda^2}.$$

Note that the parameter is the reciprocal of the mean and *not* the mean itself.

The exponential distribution is fundamental in the theory of continuous time Markov chains (see Chapter 5), due in major part to its *memoryless property,* as now explained. Think of T as a lifetime and, given that the unit has survived up to time t, ask for the conditional distribution of the remaining life $T - t$. Equivalently, for $x > 0$ determine the conditional probability $\Pr\{T - t > x \mid T > t\}$. Directly applying the definition of conditional probability (see Section 1.2.7), we obtain

$$\Pr\{T - t > x | T > t\} = \frac{\Pr\{T > t + x, \, T > t\}}{\Pr\{T > t\}}$$

$$= \frac{\Pr\{T > t + x\}}{\Pr\{T > t\}} \qquad \text{(because } x > 0)$$

$$= \frac{e^{-\lambda(t+x)}}{e^{-\lambda t}} \qquad \text{[from (1.39)]}$$

$$= e^{-\lambda x}. \tag{1.40}$$

There is no memory in the sense that $\Pr\{T - t > x | T > t\} = e^{-\lambda x} = \Pr\{T > x\}$, and an item that has survived for t units of time has a remaining lifetime that is the same as that for a new item.

To view the memoryless property somewhat differently, we introduce the *hazard rate* or *failure rate* $r(s)$ associated with a nonnegative random variable S having continuous density $g(s)$ and distribution function $G(s) < 1$. The failure rate is defined by

$$r(s) = \frac{g(s)}{1 - G(s)} \qquad \text{for} \quad s > 0. \tag{1.41}$$

We obtain the interpretation by calculating (see Section 1.2.2)

$$\Pr\{s < S \le s + \Delta s | s < S\} = \frac{\Pr\{s < S \le s + \Delta s\}}{\Pr\{s < S\}}$$

$$= \frac{g(s)\Delta s}{1 - G(s)} + o(\Delta s) \qquad \text{[from (1.4)]}$$

$$= r(s)\Delta s + o(\Delta s).$$

An item that has survived to time s will then fail in the interval $(s, s + \Delta s]$ with conditional probability $r(s)\Delta s + o(\Delta s)$, thus motivating the name "failure rate."

We can invert (1.41) by integrating

$$-r(s) = \frac{-g(s)}{1 - G(s)} = \frac{d[1 - G(s)]/ds}{1 - G(s)} = \frac{d\{\ln[1 - G(s)]\}}{ds}$$

to obtain

$$-\int_0^t r(s)ds = \ln[1 - G(t)]$$

or

$$G(t) = 1 - \exp\{-\int_0^t r(s)ds\}, \qquad t \ge 0$$

which gives the distribution function explicitly in terms of the hazard rate.

The exponential distribution is uniquely the continuous distribution with the *constant* failure rate $r(t) \equiv \lambda$. (See Problem 3 at the end of this section for the discrete analog.) The failure rate does not vary in time, another reflection of the memoryless property.

Section 1.5 contains several exercises concerning the exponential distribution. In addition to providing practice in relevant algebraic and calculus manipulations, these exercises are designed to enhance the reader's intuition concerning the exponential law.

1.4.3 The Uniform Distribution

A random variable U is uniformly distributed over the interval $[a, b]$, where $a < b$, if it has the probability density function

$$f_U(u) = \begin{cases} \dfrac{1}{b - a} & \text{for } a \leq u \leq b, \\ \\ 0 & \text{elsewhere.} \end{cases} \tag{1.42}$$

The uniform distribution extends the notion of "equally likely" to the continuous case. The distribution function is

$$F_U(x) = \begin{cases} 0 & \text{for } u \leq a, \\ \dfrac{x - a}{b - a} & \text{for } a < x \leq b, \\ 1 & \text{for } x > b, \end{cases} \tag{1.43}$$

and the mean and variance are, respectively,

$$E[U] = \frac{1}{2}(a + b) \quad \text{and} \quad \text{Var}[U] = \frac{(b - a)^2}{12}.$$

The uniform distribution on the unit interval $[0, 1]$, for which $a = 0$ and $b = 1$, is most prevalent.

1.4.4 The Gamma Distribution

The gamma distribution with parameters $\alpha > 0$ and $\lambda > 0$ has probability density function

$$f(x) = \frac{\lambda}{\Gamma(\alpha)} (\lambda x)^{\alpha - 1} e^{-\lambda x} \quad \text{for } x > 0. \tag{1.44}$$

Given an integer number α of independent exponentially distributed random variables $Y_1, \ldots, Y_\alpha$ having common parameter λ, then their sum $X_\alpha = Y_1 + \ldots + Y_\alpha$ has the gamma density of (1.44), from which we obtain the moments

$$E[X_\alpha] = \frac{\alpha}{\lambda} \quad \text{and} \quad \text{Var}[X_\alpha] = \frac{\alpha}{\lambda^2},$$

these moment formulas holding for noninteger α as well.

1.4.5 The Beta Distribution

The Beta density with parameters $\alpha > 0$ and $\beta > 0$ is given by

$$f(x) = \begin{cases} \dfrac{\Gamma(\alpha + \beta)}{\Gamma(\alpha)\Gamma(\beta)} \, x^{\alpha-1}(1 - x)^{\beta-1} & \text{for} \quad 0 < x < 1, \\[2ex] 0 & \text{elsewhere.} \end{cases} \tag{1.45}$$

The mean and variance are, respectively,

$$E[X] = \frac{\alpha}{\alpha + \beta} \quad \text{and} \quad \text{Var}[X] = \frac{\alpha\beta}{(\alpha + \beta)^2(\alpha + \beta + 1)}.$$

(The Gamma and Beta functions are defined and briefly discussed in Section 1.6.)

1.4.6 The Joint Normal Distribution

Let σ_X, σ_Y, μ_X, μ_Y and ρ be real constants subject to $\sigma_X > 0$, $\sigma_Y > 0$ and $-1 < \rho < 1$. For real variables x, y define

$$Q(x, y) = \frac{1}{1 - \rho^2}\left\{\left(\frac{x - \mu_X}{\sigma_X}\right)^2 \right.$$
$$\left. - 2\rho\left(\frac{x - \mu_X}{\sigma_X}\right)\left(\frac{y - \mu_Y}{\sigma_Y}\right) + \left(\frac{y - \mu_Y}{\sigma_Y}\right)^2\right\}. \tag{1.46}$$

The joint normal (or bivariate normal) distribution for random variables X, Y is defined by the density function

$$\phi_{X,Y}(x, y) = \frac{1}{2\pi\sigma_X\sigma_Y\sqrt{1 - \rho^2}} \exp\left\{-\frac{1}{2} Q(x, y)\right\},$$
$$-\infty < x, y < \infty. \tag{1.47}$$

The moments are given by

$$E[X] = \mu_X, \qquad E[Y] = \mu_Y,$$
$$\text{Var}[X] = \sigma_X^2, \qquad \text{Var}[Y] = \sigma_Y^2,$$

and

$$\text{Cov}[X, Y] = E[(X - \mu_X)(Y - \mu_Y)] = \rho\sigma_X\sigma_Y.$$

The dimensionless parameter ρ is called the *correlation coefficient*. When ρ is positive, then positive values of X are (stochastically) associated with positive values of Y. When ρ is negative, then positive values of X are associated with negative values of Y. If $\rho = 0$, then X and Y are independent random variables.

Linear Combinations of Normally Distributed Random Variables

Suppose X and Y have the bivariate normal density (1.47), and let $Z = aX + bY$ for arbitrary constants a, b. Then Z is normally distributed with mean

$$E[Z] = a\mu_X + b\mu_Y$$

and variance

$$\text{Var}[Z] = a^2\sigma_X^2 + 2ab\rho\sigma_X\sigma_Y + b^2\sigma_Y^2.$$

Problems 1.4

1. Suppose that U has a uniform distribution on the interval $[0, 1]$. Derive the density function for the random variables
   (a) $Y = -\ln(1 - U)$
   (b) $W_n = U^n$ for $n \geq 1$
   Hint: Refer to Section 1.2.6.

2. Given independent exponentially distributed random variables S and T with common parameter λ, determine the probability density function of the sum $R = S + T$ and identify its type by name.

3. Let Z be a random variable with the geometric probability mass function

   $$p(k) = (1 - \pi)\pi^k, \qquad k = 0, 1, \ldots,$$

   where $0 < \pi < 1$.
   (a) Show that Z has a constant failure rate in the sense that $\Pr[Z = k | Z \geq k\} = 1 - \pi$ for $k = 0, 1, \ldots$.
   (b) Suppose Z' is a discrete random variable whose possible values are $0, 1, \ldots$ and for which $\Pr\{Z' = k | Z \geq k\} = 1 - \pi$ for $k = 0, 1, \ldots$. Show that the probability mass function for Z' is $p(k)$.

4. Evaluate the moment $E[e^{\lambda Z}]$, where λ is an arbitrary real number and Z is a random variable following a standard normal distribution, by integrating

   $$E[e^{\lambda Z}] = \int_{-\infty}^{+\infty} e^{\lambda z} \frac{1}{\sqrt{2\pi}} e^{-z^2/2} \, dz.$$

   Hint: Complete the square $-\frac{1}{2}z^2 + \lambda z = -\frac{1}{2}[(z - \lambda)^2 - \lambda^2]$ and use the fact that

   $$\int_{-\infty}^{+\infty} \frac{1}{\sqrt{2\pi}} e^{-(z-\lambda)^2/2} \, dz = 1.$$

5. Let W be an exponentially distributed random variable with parameter θ and mean $\mu = 1/\theta$.
   (a) Determine $\Pr\{W > \mu\}$.
   (b) What is the mode of the distribution?

6. Let X and Y be independent random variables uniformly distributed over the interval $[\theta - \frac{1}{2}, \theta + \frac{1}{2}]$ for some fixed θ. Show that $W = X - Y$ has a distribution that is independent of θ with density function

$$
f_W(w) = \begin{cases} 1 + w & \text{for} \quad -1 \le w < 0, \\ 1 - w & \text{for} \quad 0 \le w \le 1, \\ 0 & \text{for} \quad |w| > 1. \end{cases}
$$

7. Suppose that the diameters of bearings are independent normally distributed random variables with mean $\mu_B = 1.005$ inch and variance $\sigma_B^2 = (0.003)^2$ inch^2. The diameters of shafts are independent normally distributed random variables having mean $\mu_S = 0.995$ inch and variance $\sigma_S^2 = (0.004)^2$ inch^2.

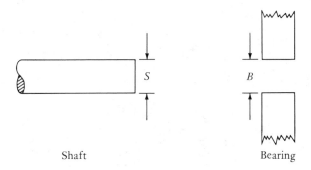

Shaft   Bearing

Let S be the diameter of a shaft taken at random and let B be the diameter of a bearing.
(a) What is the probability $\Pr\{S > B\}$ of interference?
(b) What is the probability of one or less interferences in 20 random shaft-bearing pairs?
Hint: The clearance, defined by $C = B - S$, is normally distributed (why?) and interference occurs only if $C < 0$.

8. If X follows an exponential distribution with parameter $\lambda = 2$, then what is the mean of X? Determine $\Pr\{X > 2\}$.

1.5 Some Elementary Exercises

We have collected in this section a number of exercises that go beyond what is usually covered in a first course in probability.

1.5.1 Tail Probabilities

In mathematics, what is a "trick" upon first encounter becomes a basic tool when familiarity through use is established. In dealing with nonnegative random variables, we can often simplify the analysis by the trick of approaching the problem through the upper tail probabilities of the form $\Pr\{X > x\}$. Consider the following example.

A jar has n chips numbered 1, 2, . . . , n. A person draws a chip, returns it, draws another, returns it, and so on, until a chip is drawn that has been drawn before. Let X be the number of drawings. Find the probability distribution for X.

It's easiest to compute $\Pr\{X > k\}$ first. Then, $\Pr\{X > 1\} = 1$, since at least two draws are always required. The event $\{X > 2\}$ occurs when distinct numbers appear on the first two draws, whence $\Pr\{X > 2\} = (n/n)$ $[(n - 1)/n]$. Continuing in this manner, we obtain

$$\Pr\{X > k\} = 1\left(1 - \frac{1}{n}\right)\left(1 - \frac{2}{n}\right) \cdots \left(1 - \frac{k-1}{n}\right),$$

$$\text{for} \quad k = 1, \ldots, n - 1. \quad (1.48)$$

Finally

$$\Pr\{X = k\} = \Pr\{X > k - 1\} - \Pr\{X > k\}$$

$$= \left[\left(1 - \frac{1}{n}\right) \cdots \left(1 - \frac{k-2}{n}\right)\right]$$

$$- \left[\left(1 - \frac{1}{n}\right) \cdots \left(1 - \frac{k-2}{n}\right)\left(1 - \frac{k-1}{n}\right)\right]$$

$$= \left(1 - \frac{1}{n}\right) \cdots \left(1 - \frac{k-2}{n}\right)\left[1 - \left(1 - \frac{k-1}{n}\right)\right]$$

$$= \frac{k-1}{n}\left(1 - \frac{1}{n}\right) \cdots \left(1 - \frac{k-2}{n}\right),$$

$$\text{for} \quad k = 2, \ldots, n + 1.$$

Now try deriving $\Pr\{X = k\}$ directly, for comparison with the "trick" approach.

The usefulness of the upper tail probabilities is enhanced by the formula

$$E[X] = \sum_{k=0}^{\infty} \Pr\{X > k\} = \sum_{k=1}^{\infty} \Pr\{X \geq k\}, \quad (1.49)$$

valid for nonnegative integer valued random variables X. To establish (1.49), abbreviate the notation by using $p(k) = \Pr\{X = k\}$, and rearrange the terms in $E[X] = \sum_{k \geq 0} kp(k)$ as follows:

$$E[X] = 0p(0) + 1p(1) + 2p(2) + 3p(3) + \cdots$$
$$= p(1) + p(2) + p(3) + p(4) + \cdots$$
$$+ p(2) + p(3) + p(4) + \cdots$$
$$+ p(3) + p(4) + \cdots$$
$$+ p(4) + \cdots$$
$$\vdots$$

$$= \Pr\{X \geq 1\} + \Pr\{X \geq 2\} + \Pr\{X \geq 3\} + \cdots$$

$$= \sum_{k=1}^{\infty} \Pr\{X \geq k\},$$

thus establishing (1.49).

For the chip drawing problem, the mean number of draws required is, then,

$$E[X] = \Pr\{X > 0\} + \Pr\{X > 1\} + \cdots + \Pr\{X > n\}$$

since $\Pr\{X > k\} = 0$ for $k > n$. Then substituting (1.48) into (1.49) leads directly to

$$E[X] = 2 + \left(1 - \frac{1}{n}\right) + \left(1 - \frac{1}{n}\right)\left(1 - \frac{2}{n}\right) + \cdots$$
$$+ \left(1 - \frac{1}{n}\right)\left(1 - \frac{2}{n}\right) \cdots \left(1 - \frac{n-1}{n}\right).$$

Now let X be a *nonnegative* continuous random variable with density $f(x)$ and distribution function $F(x)$. The analog to (1.49) is

$$E[X] = \int_0^{\infty} [1 - F(z)]dz, \tag{1.50}$$

obtained by interchanging an order of integration as follows:

$$E[X] = \int_0^{\infty} x f(x)dx = \int_0^{\infty}\left(\int_0^x dz\right) f(x)dx$$
$$= \int_0^{\infty}\left[\int_z^{\infty} f(x)dx\right]dz = \int_0^{\infty} [1 - F(z)]dz.$$

Interchanging the order of integration where the limits are variables often proves difficult for many students. The trick of using indicator functions to make the limits of integration constant may simplify matters. In the preceding interchange, let

$$\mathbf{1}(z < x) = \begin{cases} 1 & \text{if } 0 \leq z < x, \\ 0 & \text{otherwise,} \end{cases}$$

and then

$$\int\limits_0^\infty \left[\int\limits_0^x dz\right] f(x)dx = \int\limits_0^\infty \left[\int\limits_0^\infty \mathbf{1}(z < x) f(x)dz\right]dx$$

$$= \int\limits_0^\infty \left[\int\limits_0^\infty \mathbf{1}(z < x) f(x)dx\right]dz = \int\limits_0^\infty \left[\int\limits_z^\infty f(x)dx\right]dx.$$

As an application of (1.50), let $X_c = \min\{c, X\}$ for some positive constant c. For example, suppose X is the failure time of a certain piece of equipment. A planned replacement policy is put in use that calls for replacement of the equipment upon its failure or upon its reaching age c, whichever occurs first. Then

$$X_c = \min\{c, X\} = \begin{cases} X & \text{if } X \le c, \\ \\ c & \text{if } X > c \end{cases}$$

is the time for replacement.

Now

$$\Pr\{X_c > z\} = \begin{cases} 1 - F(z) & \text{if } 0 \le z < c, \\ \\ 0 & \text{if } c \le z, \end{cases}$$

whence we obtain

$$E[X_c] = \int\limits_0^c [1 - F(z)]dz$$

which is decidedly shorter than

$$E[X_c] = \int\limits_0^c xf(x)dx + c[1 - F(c)].$$

Observe that X_c is a random variable whose distribution is partly continuous and partly discrete, thus establishing by example that such distributions do occur in practical applications.

1.5.2 The Exponential Distribution

This exercise is designed to foster intuition about the exponential distribution, as well as to provide practice in algebraic and calculus manipulations relevant to stochastic modeling.

Let X_0 and X_1 be independent exponentially distributed random variables with respective parameters λ_0 and λ_1 so that

$$\Pr\{X_i > t\} = e^{-\lambda_i t} \qquad \text{for } t \ge 0, i = 0, 1.$$

Let

$$N = \begin{cases} 0 & \text{if } X_0 \le X_1, \\ \\ 1 & \text{if } X_1 < X_0; \end{cases}$$
$$U = \min\{X_0, X_1\} = X_N;$$
$$M = 1 - N;$$
$$V = \max\{X_0, X_1\} = X_M;$$

and

$$W = V - U = |X_0 - X_1|.$$

In this context, we derive the following:

(a)
$$\Pr\{N = 0 \text{ and } U > t\} = e^{-(\lambda_0 + \lambda_1)t} \left(\frac{\lambda_0}{\lambda_0 + \lambda_1} \right).$$

The event $\{N = 0 \text{ and } U > t\}$ is exactly the event $\{t < X_0 \le X_1\}$, whence

$$\Pr\{N = 0, U > t\} = \Pr\{t < X_0 < X_1\}$$
$$= \iint_{t < x_0 < x_1} \lambda_0 e^{-\lambda_0 x_0} \lambda_1 e^{-\lambda_1 x_1} dx_1 \, dx_0$$
$$= \int_t^\infty \left(\int_{x_0}^\infty \lambda_1 e^{-\lambda_1 x_1} \, dx_1 \right) \lambda_0 e^{-\lambda_0 x_0} \, dx_0$$
$$= \int_t^\infty e^{-\lambda_1 x_0} \lambda_0 e^{-\lambda_0 x_0} \, dx_0$$
$$= \frac{\lambda_0}{\lambda_0 + \lambda_1} \int_t^\infty (\lambda_0 + \lambda_1) e^{-(\lambda_0 + \lambda_1)x_0} \, dx_0$$
$$= \frac{\lambda_0}{\lambda_0 + \lambda_1} e^{-(\lambda_0 + \lambda_1)t}.$$

(b)
$$\Pr\{N = 0\} = \frac{\lambda_0}{\lambda_0 + \lambda_1} \quad \text{and} \quad \Pr\{N = 1\} = \frac{\lambda_1}{\lambda_0 + \lambda_1}.$$

We use the result in (a) as follows:

$$\Pr\{N = 0\} = \Pr\{N = 0, U > 0\} = \frac{\lambda_0}{\lambda_0 + \lambda_1} \qquad \text{from (a).}$$

Obviously, $\Pr\{N = 1\} = 1 - \Pr\{N = 0\} = \lambda_1/(\lambda_0 + \lambda_1)$.

(c)
$$\Pr\{U > t\} = e^{-(\lambda_0 + \lambda_1)t}, \qquad t \ge 0.$$

Upon adding the result in (a),

$$\Pr\{N = 0 \text{ and } U > t\} = e^{-(\lambda_0+\lambda_1)t}\frac{\lambda_0}{\lambda_0 + \lambda_1},$$

to the corresponding quantity associated with $N = 1$,

$$\Pr\{N = 1 \text{ and } U > t\} = e^{-(\lambda_0+\lambda_1)t}\frac{\lambda_1}{\lambda_0 + \lambda_1},$$

we obtain the desired result via

$$\Pr\{U > t\} = \Pr\{N = 0, U > t\} + \Pr\{N = 1, U > t\}$$
$$= e^{-(\lambda_0+\lambda_1)t}\left(\frac{\lambda_0}{\lambda_0 + \lambda_1} + \frac{\lambda_1}{\lambda_0 + \lambda_1}\right)$$
$$= e^{-(\lambda_0+\lambda_1)t}.$$

At this point observe that U and N are independent random variables. This follows because (a), (b), and (c) together give

$$\Pr\{N = 0 \text{ and } U > t\} = \Pr\{N = 0\} \times \Pr\{U > t\}.$$

Think about this remarkable result for a moment. Suppose X_0 and X_1 represent lifetimes and $\lambda_0 = 0.001$ while $\lambda_1 = 1$. The mean lifetimes are $E[X_0] = 1000$ and $E[X_1] = 1$. Suppose we observe that the time of the first death is rather small, say, $U = \min\{X_0, X_1\} = \frac{1}{2}$. In spite of vast disparity between the mean lifetimes, the observation that $U = \frac{1}{2}$ provides no information about which of the two units, 0 or 1, was first to die! This apparent paradox is yet another, more subtle, manifestation of the memoryless property unique to the exponential density.

We continue with the exercise.

(d) $$\Pr\{W > t | N = 0\} = e^{-\lambda_1 t}, \qquad t \geq 0.$$

The event $\{W > t \text{ and } N = 0\}$ for $t \geq 0$ corresponds exactly to the event $\{t < X_1 - X_0\}$. Thus

$$\Pr\{W > t \text{ and } N = 0\} = \Pr\{X_1 - X_0 > t\}$$

$$= \iint\limits_{x_1-x_0>t} \lambda_0 e^{-\lambda_0 x_0}\lambda_1 e^{-\lambda_1 x_1}\, dx_0 dx_1$$

$$= \int_0^\infty \left(\int_{x_0+t}^\infty \lambda_1 e^{-\lambda_1 x_1}\, dx_1\right) \lambda_0 e^{-\lambda_0 x_0}\, dx_0$$

$$= \int_0^\infty e^{-\lambda_1(x_0+t)} \lambda_0 e^{-\lambda_0 x_0}\, dx_0$$

$$= \frac{\lambda_0}{\lambda_0 + \lambda_1} e^{-\lambda_1 t}\int_0^\infty (\lambda_0 + \lambda_1)e^{-(\lambda_0+\lambda_1)x_0}\, dx_0$$

$$= \frac{\lambda_0}{\lambda_0 + \lambda_1} e^{-\lambda_1 t}$$

$$= \Pr\{N = 0\}e^{-\lambda_1 t} \qquad \text{[from (b)]}.$$

Then, using the basic definition of conditional probability (Section 1.2.7), we obtain

$$\Pr\{W > t | N = 0\} = \frac{\Pr\{W > t, N = 0\}}{\Pr\{N = 0\}} = e^{-\lambda_1 t}, \qquad t \geq 0$$

as desired.

Of course a parallel formula holds conditional on $N = 1$:

$$\Pr\{W > t | N = 1\} = e^{-\lambda_0 t}, \qquad t \geq 0,$$

and using the law of total probability we obtain the distribution of W in the form

$$\Pr\{W > t\} = \Pr\{W > t, N = 0\} + \Pr\{W > t, N = 1\}$$

$$= \frac{\lambda_0}{\lambda_0 + \lambda_1} e^{-\lambda_1 t} + \frac{\lambda_1}{\lambda_0 + \lambda_1} e^{-\lambda_0 t}, \qquad t \geq 0,$$

(e) U and $W = V - U$ are independent random variables.

To establish this final consequence of the memoryless property, it suffices to show that

$$\Pr\{U > u \text{ and } W > w\} = \Pr\{U > u\}\Pr\{W > w\} \qquad \text{for all} \quad u \geq 0, w \geq 0.$$

Determining first

$$\Pr\{N = 0, U > u, W > w\} = \Pr\{u < X_0 < X_1 - w\}$$

$$= \int\!\!\!\int_{u < x_0 < x_1 - w} \lambda_0 e^{-\lambda_0 x_0} \lambda_1 e^{-\lambda_1 x_1} \, dx_0 dx_1$$

$$= \int_u^\infty \left(\int_{x_0 + w}^\infty \lambda_1 e^{-\lambda_1 x_1} \, dx_1 \right) \lambda_0 e^{-\lambda_0 x_0} \, dx_0$$

$$= \int_u^\infty e^{-\lambda_1 (x_0 + w)} \lambda_0 e^{-\lambda_0 x_0} \, dx_0$$

$$= \left(\frac{\lambda_0}{\lambda_0 + \lambda_1} \right) e^{-\lambda_1 w} \int_u^\infty (\lambda_0 + \lambda_1) e^{-(\lambda_0 + \lambda_1) x_0} \, dx_0$$

$$= \left(\frac{\lambda_0}{\lambda_0 + \lambda_1} \right) e^{-\lambda_1 w} e^{-(\lambda_0 + \lambda_1) u},$$

and then, by symmetry,

$$\Pr\{N = 1, U > u, W > w\} = \left(\frac{\lambda_1}{\lambda_0 + \lambda_1} \right) e^{-\lambda_0 w} e^{-(\lambda_0 + \lambda_1) u},$$

and finally adding the two expressions, we obtain

$$\Pr\{U > u,\ W > w\} = \left[\left(\frac{\lambda_0}{\lambda_0 + \lambda_1}\right)e^{-\lambda_1 w} + \left(\frac{\lambda_1}{\lambda_0 + \lambda_1}\right)e^{-\lambda_0 w}\right]e^{-(\lambda_0 + \lambda_1)u}$$

$$= \Pr\{W > w\}\Pr\{U > u\}, \qquad u,\ w \geq 0.$$

The calculation is complete.

Problems 1.5

1. Let $X_0, X_1, \ldots$ be independent and identically distributed random variables having the cumulative distribution function $F(x) = \Pr\{X \leq x\}$. For a fixed number ξ, let N be the first index k for which $X_k > \xi$. That is, $N = 1$ if $X_1 > \xi$; $N = 2$ if $X_1 \leq \xi$ and $X_2 > \xi$; etc. Determine the probability mass function for N.

2. Let $X_1, X_2, \ldots, X_n$ be independent random variables, all exponentially distributed with the same parameter λ. Determine the distribution function for the minimum $Z = \min\{X_1, \ldots, X_n\}$.

3. Suppose that X is a discrete random variable having the geometric distribution whose probability mass function is

$$p(k) = p(1 - p)^k \qquad \text{for} \quad k = 0, 1, \ldots$$

   (a) Determine the upper tail probabilities $\Pr\{X > k\}$ for $k = 0, 1, \ldots$
   (b) Evaluate the mean via $E[X] = \sum_{k \geq 0} \Pr\{X > k\}$.

4. Let V be a continuous random variable taking both positive and negative values and whose mean exists. Derive the formula

$$E[V] = \int_0^\infty [1 - F_V(v)]dv - \int_{-\infty}^0 F_V(v)dv.$$

5. Show that

$$E[W^2] = \int_0^\infty 2y[1 - F_W(y)]dy$$

   for a nonnegative random variable W.

6. Determine the upper tail probabilities $\Pr\{V > t\}$ and mean $E[V]$ for a random variable V having the exponential density

$$f_V(v) = \begin{cases} 0 & \text{for } v < 0, \\ \lambda e^{-\lambda v} & \text{for } v \geq 0, \end{cases}$$

   where λ is a fixed positive parameter.

7. Let $X_1, X_2, \ldots, X_n$ be independent random variables that are exponentially distributed with respective parameters $\lambda_1, \lambda_2, \ldots, \lambda_n$.

Identify the distribution of the minimum $V = \min\{X_1, X_2, \ldots, X_n\}$.
Hint: For any real number v, the event $\{V > v\}$ is equivalent to $\{X_1 > v, X_2 > v, \ldots, X_n > v\}$.

8. Let $U_1, U_2, \ldots, U_n$ be independent uniformly distributed random variables on the unit interval $[0, 1]$. Define the minimum $V_n = \min\{U_1, U_2, \ldots, U_n\}$.
   (a) Show that $\Pr\{V_n > v\} = (1 - v)^n$ for $0 \le v \le 1$.
   (b) Let $W_n = nV_n$. Show that $\Pr\{W_n > w\} = [1 - (w/n)]^n$ for $0 \le w \le n$, and thus

$$\lim_{n \to \infty} \Pr\{W_n > w\} = e^{-w} \qquad \text{for} \quad w \ge 0.$$

1.6 Useful Functions, Integrals, and Sums

Collected here for later reference are some calculations and formulas that are especially pertinent in probability modeling.

We begin with several exponential integrals, the first and simplest being

$$\int e^{-x}dx = -e^{-x}. \tag{1.51}$$

When we use integration by parts, the second integral that we introduce reduces to the first in the manner

$$\int xe^{-x}dx = -xe^{-x} + \int e^{-x}dx = -e^{-x}(1 + x). \tag{1.52}$$

Then (1.51) and (1.52) are the special cases of $\alpha = 1$ and $\alpha = 2$, respectively, in the general formula, valid for any real number α for which the integrals are defined, given by

$$\int x^{\alpha-1}e^{-x}dx = -x^{\alpha-1}e^{-x} + (\alpha - 1)\int x^{\alpha-2}e^{-x}dx. \tag{1.53}$$

Fixing the limits of integration leads to the Gamma function, defined by

$$\Gamma(\alpha) = \int_0^\infty x^{\alpha-1}e^{-x}dx, \qquad \text{for} \quad \alpha > 0. \tag{1.54}$$

From (1.53) it follows that

$$\Gamma(\alpha) = (\alpha - 1)\Gamma(\alpha - 1) \tag{1.55}$$

and therefore for any integers k,

$$\Gamma(k) = (k - 1)(k - 2) \cdots 2 \cdot \Gamma(1). \tag{1.56}$$

An easy consequence of (1.51) is the evaluation $\Gamma(1) = 1$, which with (1.56) shows that the Gamma function at integral arguments is a generalization of the factorial function, and

$$\Gamma(k) = (k - 1)! \qquad \text{for} \quad k = 1, 2, \ldots. \tag{1.57}$$

A more difficult integration shows that

$$\Gamma(\tfrac{1}{2}) = \sqrt{\pi} \tag{1.58}$$

which with (1.56) provides

$$\Gamma(n + \tfrac{1}{2}) = \frac{1 \times 3 \times 5 \times \cdots \times (2n - 1)}{2^n}\sqrt{\pi},$$
$$\text{for} \quad n = 0, 1, \ldots . \tag{1.59}$$

Stirling's formula is the following important asymptotic evaluation of the factorial function:

$$n! = n^n e^{-n}(2\pi n)^{1/2} e^{r(n)/12n} \tag{1.60}$$

in which

$$1 - \frac{1}{12n + 1} < r(n) < 1. \tag{1.61}$$

We sometimes write this in the looser form

$$n! \sim n^n e^{-n}(2\pi n)^{1/2} \quad \text{as } n \to \infty, \tag{1.62}$$

the symbol "$\sim$" signifying that the ratio of the two sides in (1.62) approaches 1 as $n \to \infty$. For the binomial coefficient $\binom{n}{k} = n!/k!(n - k)!$ we then obtain

$$\binom{n}{k} \sim \frac{(n - k)^k}{k!} \quad \text{as } n \to \infty \tag{1.63}$$

as a consequence of (1.62) and the exponential limit

$$e^{-k} = \lim_{n\to\infty}\left(1 - \frac{k}{n}\right)^n.$$

The integral

$$B(m, n) = \int_0^1 x^{m-1}(1 - x)^{n-1}dx, \tag{1.64}$$

which converges when m and n are positive, defines the *Beta* function, related to the Gamma function by

$$B(m, n) = \frac{\Gamma(m)\Gamma(n)}{\Gamma(m + n)} \quad \text{for} \quad m > 0, \quad n > 0. \tag{1.65}$$

For nonnegative integral values m, n then

$$B(m + 1, n + 1) = \int_0^1 x^m(1 - x)^n dx = \frac{m!n!}{(m + n + 1)!}. \tag{1.66}$$

For $n = 1, 2, \ldots$, the binomial theorem provides the evaluation

$$(1 - x)^n = \sum_{k=0}^n (-1)^k \binom{n}{k}x^k, \quad \text{for} \quad -\infty < x < \infty. \tag{1.67}$$

The formula may be generalized for nonintegral n by appropriately generalizing the binomial coefficient, defining for any real number α,

$$\binom{\alpha}{k} = \frac{\alpha(\alpha - 1) \cdot \cdot \cdot (\alpha - k + 1)}{k!} \qquad \text{for} \quad k = 1, 2, \ldots$$

$$= 1 \qquad \text{for} \quad k = 0. \qquad (1.68)$$

As a special case, for any positive integer n, then

$$\binom{-n}{k} = (-1)^k \frac{n(n + 1) \cdot \cdot \cdot (n + k - 1)}{k!}$$

$$= (-1)^k \binom{n + k - 1}{k}. \qquad (1.69)$$

The general binomial theorem, valid for all real α, is

$$(1 - x)^\alpha = \sum_{k=0}^{\infty} (-1)^k \binom{\alpha}{k} x^k \qquad \text{for} \quad -1 < x < 1. \qquad (1.70)$$

When $\alpha = -n$ for a positive integer n, then we obtain a group of formulas useful in dealing with geometric series. For a positive integer n, in view of (1.69) and (1.70), we have

$$(1 - x)^{-n} = \sum_{k=0}^{\infty} \binom{n + k - 1}{k} x^k \qquad \text{for} \quad |x| < 1. \qquad (1.71)$$

The familiar formula

$$\sum_{k=0}^{\infty} x^k = 1 + x + x^2 + \ldots = \frac{1}{1 - x} \qquad \text{for} \quad |x| < 1 \qquad (1.72)$$

for the sum of a geometric series results from (1.71) with $n = 1$. The cases $n = 2$ and $n = 3$ yield the formulas

$$\sum_{k=0}^{\infty} (k + 1)x^k = 1 + 2x + 3x^2 + \ldots$$

$$= \frac{1}{(1 - x)^2} \qquad \text{for} \quad |x| < 1, \qquad (1.73)$$

$$\sum_{k=0}^{\infty} (k + 2)(k + 1)x^k = \frac{2}{(1 - x)^3} \qquad \text{for} \quad |x| < 1. \qquad (1.74)$$

Sums of Numbers

The following sums of powers of integers have simple expressions:

$$1 + 2 + \cdot \cdot \cdot + n = \frac{n(n + 1)}{2}$$

$$1 + 2^2 + \cdot \cdot \cdot + n^2 = \frac{n(n + 1)(2n + 1)}{6}$$

$$1 + 2^3 + \cdot \cdot \cdot + n^3 = \frac{n^2(n + 1)^2}{4}.$$

	Conditional Probability and
Chapter 2	Conditional Expectation

2.1 The Discrete Case

The conditional probability $\Pr\{A|B\}$ of the event A given the event B is defined by

$$\Pr\{A|B\} = \frac{\Pr\{A \text{ and } B\}}{\Pr\{B\}} \quad \text{if} \quad \Pr\{B\} > 0, \tag{2.1}$$

and is not defined, or is assigned an arbitrary value, when $\Pr\{B\} = 0$. Let X and Y be random variables that can attain only countably many different values, say $0, 1, 2, \ldots$. The *conditional probability mass function* $p_{X|Y}(x|y)$ of X given $Y = y$ is defined by

$$p_{X|Y}(x|y) = \frac{\Pr\{X = x \text{ and } Y = y\}}{\Pr\{Y = y\}} \quad \text{if} \quad \Pr\{Y = y\} > 0,$$

and is not defined, or is assigned an arbitrary value, whenever $\Pr\{Y = y\} = 0$. In terms of the joint and marginal probability mass functions $p_{XY}(x, y)$ and $p_Y(y) = \Sigma_x p_{XY}(x, y)$, respectively, the definition is

$$p_{X|Y}(x|y) = \frac{p_{XY}(x, y)}{p_Y(y)} \quad \text{if} \quad p_Y(y) > 0; \quad x, y = 0, 1, \ldots \tag{2.2}$$

Observe that $p_{X|Y}(x|y)$ is a probability mass function in x for each fixed y, i.e., $p_{X|Y}(x|y) \geq 0$ and $\Sigma_\xi p_{X|Y}(\xi|y) = 1$, for all x, y. The law of total probability takes the form

$$\Pr\{X = x\} = \sum_{y=0}^{\infty} p_{X|Y}(x|y) p_Y(y). \tag{2.3}$$

Notice in (2.3) that the points y where $p_{X|Y}(x|y)$ is not defined are exactly those values for which $p_Y(y) = 0$, and hence, do not affect the computation. The lack of a complete prescription for the conditional probability mass function, a nuisance in some instances, is always consistent with subsequent calculations.

Example Let X have a binomial distribution with parameters p and N, where N has a binomial distribution with parameters q and M. What is the marginal distribution of X?

We are given the conditional probability mass function

$$p_{X|N}(k|n) = \binom{n}{k} p^k (1 - p)^{n-k}, \qquad k = 0, 1, \ldots, n$$

and the marginal distribution

$$p_N(n) = \binom{M}{n} q^n (1 - q)^{M-n}, \qquad n = 0, 1, \ldots, M.$$

We apply the law of total probability in the form of (2.3) to obtain

$$\Pr\{X = k\} = \sum_{n=0}^{M} p_{X|N}(k|n) p_N(n)$$

$$= \sum_{n=k}^{M} \frac{n!}{k!(n-k)!} p^k (1 - p)^{n-k} \frac{M!}{n!(M-n)!} q^n (1 - q)^{M-n}$$

$$= \frac{M!}{k!} p^k (1 - q)^M \left(\frac{q}{1-q}\right)^k \sum_{n=k}^{M} \frac{1}{(n-k)!(M-n)!} (1 - p)^{n-k}$$

$$\times \left(\frac{q}{1-q}\right)^{n-k}$$

$$= \frac{M!}{k!(M-k)!} (pq)^k (1 - q)^{M-k} \left[1 + \frac{q(1-p)}{1-q}\right]^{M-k}$$

$$= \frac{M!}{k!(M-k)!} (pq)^k (1 - pq)^{M-k}, \qquad k = 0, 1, \ldots, M.$$

In words, X has a binomial distribution with parameters M and pq.

Example Suppose X has a binomial distribution with parameters p and N where N has a Poisson distribution with mean λ. What is the marginal distribution for X?

Proceeding as in the previous example but now using

$$p_N(n) = \frac{\lambda^n e^{-\lambda}}{n!}, \qquad n = 0, 1, \ldots,$$

we obtain

$$\Pr\{X = k\} = \sum_{n=0}^{\infty} p_{X|N}(k|n)p_N(n)$$

$$= \sum_{n=k}^{\infty} \frac{n!}{k!(n-k)!} p^k(1-p)^{n-k} \frac{\lambda^n e^{-\lambda}}{n!}$$

$$= \frac{\lambda^k e^{-\lambda} p^k}{k!} \sum_{n=k}^{\infty} \frac{[\lambda(1-p)]^{n-k}}{(n-k)!}$$

$$= \frac{(\lambda p)^k e^{-\lambda}}{k!} e^{\lambda(1-p)}$$

$$= \frac{(\lambda p)^k e^{-\lambda p}}{k!} \qquad \text{for} \quad k = 0, 1, \ldots$$

In words, X has a Poisson distribution with mean λp.

Example Suppose X has a negative binomial distribution with parameters p and N, where N has the geometric distribution

$$p_N(n) = (1 - \beta)\beta^{n-1} \qquad \text{for} \quad n = 1, 2, \ldots$$

What is the marginal distribution for X?

We are given the conditional probability mass function

$$p_{X|N}(k|n) = \binom{n+k-1}{k} p^n(1-p)^k, \qquad k = 0, 1, \ldots$$

Using the law of total probability, we obtain

$$\Pr\{X = k\} = \sum_{n=0}^{\infty} p_{X|N}(k|n)p_N(n)$$

$$= \sum_{n=1}^{\infty} \frac{(n+k-1)!}{k!(n-1)!} p^n(1-p)^k(1-\beta)\beta^{n-1}$$

$$= (1 - \beta)(1 - p)^k p \sum_{n=1}^{\infty} \binom{n+k-1}{k} (\beta p)^{n-1}$$

$$= (1 - \beta)(1 - p)^k p(1 - \beta p)^{-k-1}$$

$$= \left(\frac{p - \beta p}{1 - \beta p}\right)\left(\frac{1 - p}{1 - \beta p}\right)^k \qquad \text{for} \quad k = 0, 1, \ldots$$

We recognize the marginal distribution of X as being of geometric form.

Let g be a function for which the expectation of $g(X)$ is finite. We define the *conditional* expected value of $g(X)$, given $Y = y$, by the formula

$$E[g(X)|Y = y] = \sum_x g(x)p_{X|Y}(x|y) \qquad \text{if} \quad p_Y(y) > 0, \qquad (2.4)$$

and the conditional mean is not defined at values y for which $p_Y(y) = 0$. The law of total probability for conditional expectation reads

$$E[g(X)] = \sum_y E[g(X)|Y = y]p_Y(y). \qquad (2.5)$$

The conditional expected value $E[g(X)|Y = y]$ is a function of the real variable y. If we evaluate this function at the random variable Y, we obtain a random variable that we denote by $E[g(X)|Y]$. The law of total probability in (2.5) now may be written in the form

$$E[g(X)] = E\{E[g(X)|Y]\}. \tag{2.6}$$

Since the conditional expectation of $g(X)$ given $Y = y$ is the expectation with respect to the conditional probability mass function $p_{X|Y}(x|y)$, conditional expectations behave in many ways like ordinary expectations. The following list summarizes some properties of conditional expectations. In this list, with or without affixes, X and Y are jointly distributed random variables; c is a real number; g is a function for which $E[|g(X)|] < \infty$; h is a bounded function; and v is a function of two variables for which $E[|v(X, Y)|] < \infty$. The properties are

$$E[c_1 g_1(X_1) + c_2 g_2(X_2)|Y = y]$$
$$= c_1 E[g_1(X_1)|Y = y] + c_2 E[g_2(X_2)|Y = y]; \tag{2.7}$$

$$\text{if } g \geq 0, \text{ then } E[g(X)|Y = y] \geq 0; \tag{2.8}$$

$$E[v(X, Y)|Y = y] = E[v(X, y)|Y = y]; \tag{2.9}$$

$$E[g(X)|Y = y] = E[g(X)] \quad \text{if } X \text{ and } Y \text{ are independent}; \tag{2.10}$$

$$E[g(X)h(Y)|Y = y] = h(y)E[g(X)|Y = y]; \text{ and} \tag{2.11}$$

$$E[g(X)h(Y)] = \sum_y h(y)E[g(X)|Y = y]p_Y(y)$$
$$= E\{h(Y)E[g(X)|Y]\}. \tag{2.12}$$

As a consequence of (2.7), (2.11), and (2.12), with either $g \equiv 1$ or $h \equiv 1$, we obtain,

$$E[c|Y = y] = c, \tag{2.13}$$

$$E[h(Y)|Y = y] = h(y), \tag{2.14}$$

$$E[g(X)] = \sum_y E[g(X)|Y = y]p_Y(y) = E\{E[g(X)|Y]\}. \tag{2.15}$$

Problems 2.1

1. Let X be a Poisson random variable with parameter λ. Find the conditional mean of X given that X is odd.

2. Suppose U and V are independent and follow the geometric distribution

$$p(k) = \rho(1 - \rho)^k \quad \text{for} \quad k = 0, 1, \ldots .$$

Define the random variable $Z = U + V$.

(a) Determine the joint probability mass function $p_{U,Z}(u, z) = \Pr\{U = u, Z = z\}$.

(b) Determine the conditional probability mass function for U given that $Z = n$.

3. Let M have a binomial distribution with parameters N and p. Conditioned on M, the random variable X has a binomial distribution with parameters M and π.

(a) Determine the marginal distribution for X.

(b) Determine the covariance between X and $Y = M - X$.

4. A card is picked at random from N cards labeled 1, 2, . . . , N, and the number that appears is X. A second card is picked at random from cards numbered 1, 2, . . . , X and its number is Y. Determine the conditional distribution of X given $Y = y$, for $y = 1, 2, \ldots$.

5. Let X and Y denote the respective outcomes when two fair dice are thrown. Let $U = \min\{X, Y\}$, $V = \max\{X, Y\}$ and $S = U + V$, $T = V - U$.

(a) Determine the conditional probability mass function for U given $V = v$.

(b) Determine the joint mass function for S and T.

6. Suppose that X has a binomial distribution with parameters $p = \frac{1}{2}$ and N where N is also random and follows a binomial distribution with parameters $q = \frac{1}{4}$ and $M = 20$. What is the mean of X?

7. A nickel is tossed 20 times in succession. Every time that the nickel comes up heads, a dime is tossed. Let X count the number of heads appearing on tosses of the dime. Determine $\Pr\{X = 0\}$.

8. A dime is tossed repeatedly until a head appears. Let N be the trial number on which this first head occurs. Then a nickel is tossed N times. Let X count the number of times that the nickel comes up tails. Determine $\Pr\{X = 0\}$, $\Pr\{X = 1\}$, and $E[X]$.

9. The probability that an airplane accident that is due to structural failure is correctly diagnosed is .85 and the probability that an airplane accident that is not due to structural failure is incorrectly diagnosed as being due to structural failure is .35. If 30 percent of all airplane accidents are due to structural failure, then find the probability that an airplane accident is due to structural failure given that it has been diagnosed as due to structural failure.

10. Initially an urn contains one red and one green ball. A ball is drawn at random from the urn, observed, and then replaced. If this ball is red, then an additional red ball is placed in the urn. If the ball is green, then a green ball is added. A second ball is drawn. Find the conditional

probability that the first ball was red given that the second ball drawn was red.

2.2 The Dice Game Craps

An analysis of the dice game known as craps provides an educational example of the use of conditional probability in stochastic modeling. In craps, two dice are rolled and the sum of their uppermost faces is observed. If the sum has value 2, 3, or 12, the player loses immediately. If the sum is 7 or 11, the player wins. If the sum is 4, 5, 6, 8, 9, or 10, then further rolls are required to resolve the game. In the case where the sum is 4, for example, the dice are rolled repeatedly until either a sum of 4 reappears or a sum of 7 is observed. If the 4 appears first, the roller wins; if the seven appears first, he loses.

Consider repeated rolls of the pair of dice and let Z_n for $n = 0, 1, \ldots$ be the sum observed on the nth roll. Then $Z_1, Z_2, \ldots$ are independent identically distributed random variables. If the dice are fair, the probability mass function is

$$
\begin{array}{ll}
p_Z(2) = \frac{1}{36} & p_Z(8) = \frac{5}{36} \\
p_Z(3) = \frac{2}{36} & p_Z(9) = \frac{4}{36} \\
p_Z(4) = \frac{3}{36} & p_Z(10) = \frac{3}{36} \\
p_Z(5) = \frac{4}{36} & p_Z(11) = \frac{2}{36} \\
p_Z(6) = \frac{5}{36} & p_Z(12) = \frac{1}{36} \\
p_Z(7) = \frac{6}{36} &
\end{array}
\tag{2.16}
$$

Let A denote the event that the player wins the game. By the law of total probability

$$
\Pr\{A\} = \sum_{k=2}^{12} \Pr\{A|Z_0 = k\} p_Z(k).
\tag{2.17}
$$

Because $Z_0 = 2, 3,$ or 12 calls for an immediate loss, then $\Pr\{A|Z_0 = k\} = 0$ for $k = 2, 3,$ or 12. Similarly, $Z_0 = 7$ or 11 results in an immediate win and thus $\Pr\{A|Z_0 = 7\} = \Pr\{A|Z_0 = 11\} = 1$. It remains to consider the values $Z_0 = 4, 5, 6, 8, 9,$ and 10 which call for additional rolls. Since the logic remains the same in each of these cases, we will argue only the case in which $Z_0 = 4$. Abbreviate with $\alpha = \Pr\{A|Z_0 = 4\}$. Then α is the probability that in successive rolls $Z_1, Z_2, \ldots$ of a pair of dice, a sum of 4 appears before a sum of 7. Denote this event by B, and again bring in the law of total probability. Then

$$
\alpha = \Pr\{B\} = \sum_{k=2}^{12} \Pr\{B|Z_1 = k\} p_Z(k).
\tag{2.18}
$$

Now $\Pr\{B|Z_1 = 4\} = 1$ while $\Pr\{B|Z_1 = 7\} = 0$. If the first roll results in anything other than a 4 or a 7, the problem is repeated in a statistically identical setting. That is, $\Pr\{B|Z_1 = k\} = \alpha$ for $k \neq 4$ or 7. Substitution into (2.18) results in

$$\alpha = p_Z(4) \times 1 + p_Z(7) \times 0 + \sum_{k \neq 4,7} p_Z(k) \times \alpha$$

$$= p_Z(4) + [1 - p_Z(4) - p_Z(7)]\alpha$$

or

$$\alpha = \frac{p_Z(4)}{p_Z(4) + p_Z(7)}. \qquad (2.19)$$

The same result may be secured by means of a longer, more computational, method. One may partition the event B into disjoint elemental events by writing

$$B = \{Z_1 = 4\} \cup \{Z_1 \neq 4 \text{ or } 7, Z_2 = 4\}$$

$$\cup \{Z_1 \neq 4 \text{ or } 7, Z_2 \neq 4 \text{ or } 7, Z_3 = 4\} \cup \ldots$$

and then

$$\Pr\{B\} = \Pr\{Z_1 = 4\} + \Pr\{Z_1 \neq 4 \text{ or } 7, Z_2 = 4\}$$
$$+ \Pr\{Z_1 \neq 4 \text{ or } 7, Z_2 \neq 4 \text{ or } 7, Z_3 = 4\} + \cdots.$$

Now use the independence of $Z_1, Z_2, \ldots$ and sum a geometric series to secure

$$\Pr\{B\} = p_Z(4) + [1 - p_Z(4) - p_Z(7)]p_Z(4)$$

$$+ [1 - p_Z(4) - p_Z(7)]^2 p_Z(4) + \cdots$$

$$= \frac{p_Z(4)}{p_Z(4) + p_Z(7)}$$

in agreement with (2.19).

Extending the result just obtained to the other cases having more than one roll, we have

$$\Pr\{A|Z_0 = k\} = \frac{p_Z(k)}{p_Z(k) + p_Z(7)} \qquad \text{for} \quad k = 4, 5, 6, 8, 9, 10.$$

Finally, substitution into (2.17) yields the total win probability

$$\Pr\{A\} = p_Z(7) + p_Z(11) + \sum_{k=4,5,6,8,9,10} \frac{p_Z(k)^2}{p_Z(k) + p_Z(7)}. \qquad (2.20)$$

The numerical values for $p_Z(k)$ given in (2.16) together with (2.20) determine the win probability

$$\Pr\{A\} = .49292929 \ldots .$$

Having explained the computations, let us go on to a more interesting question. Suppose that the dice are not perfect cubes but are shaved so as to be slightly thinner in one dimension than in the other two. The numbers that appear on opposite faces on a single die always sum to 7. That is, 1 is opposite 6, 2 is opposite 5, and 3 is opposite 4. Suppose it is the 3-4 dimension that is smaller than the other two. See Figure 2.1. This will cause 3 and 4 to appear more frequently than the other faces 1, 2, 5, and 6. To see this, think of the extreme case in which the 3-4 dimension is very thin, leading to a 3 or 4 on almost all tosses. Letting Y denote the result of tossing a single shaved die, we postulate that the probability mass function is given by

$$p_Y(3) = p_Y(4) = \tfrac{1}{6} + 2\epsilon \equiv p_+$$
$$p_Y(1) = p_Y(2) = p_Y(5) = p_Y(6) = \tfrac{1}{6} - \epsilon \equiv p_-$$

where $\epsilon > 0$ is a small quantity depending on the amount by which the die has been biased.

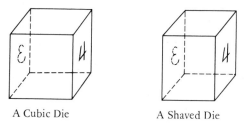

A Cubic Die   A Shaved Die

Figure 2.1 A cubic die versus a die that has been shaved down in one dimension

If both dice are shaved in the same manner, the mass function for their sum can be determined in a straightforward manner from the following joint table:

		Die #1					
Die #2		1 p_-	2 p_-	3 p_+	4 p_+	5 p_-	6 p_-
1	p_-	p_-^2	p_-^2	p_+p_-	p_+p_-	p_-^2	p_-^2
2	p_-	p_-^2	p_-^2	p_+p_-	p_+p_-	p_-^2	p_-^2
3	p_+	p_+p_-	p_+p_-	p_+^2	p_+^2	p_+p_-	p_+p_-
4	p_+	p_+p_-	p_+p_-	p_+^2	p_+^2	p_+p_-	p_+p_-
5	p_-	p_-^2	p_-^2	p_+p_-	p_+p_-	p_-^2	p_-^2
6	p_-	p_-^2	p_-^2	p_+p_-	p_+p_-	p_-^2	p_-^2

It is easily seen that the probability mass function for the sum of the dice is

$$p(2) = p_-^2 = p(12),$$
$$p(3) = 2p_-^2 = p(11),$$
$$p(4) = p_-(p_- + 2p_+) = p(10),$$
$$p(5) = 4p_+p_- = p(9),$$
$$p(6) = p_-^2 + (p_+ + p_-)^2 = p(8),$$
$$p(7) = 4p_-^2 + 2p_+^2.$$

To obtain a numerical value to compare to the win probability .492929 . . . associated with fair dice, let us arbitrarily set $\epsilon = .02$ so that $p_- = .146666$ . . . and $p_+ = .206666$ . . . . Then routine substitutions according to the table lead to

$$p(2) = p(12) = .02151111 \qquad p(5) = p(9) = .12124445$$
$$p(3) = p(11) = .04302222 \qquad p(6) = p(8) = .14635556 \quad (2.21)$$
$$p(4) = p(10) = .08213333 \qquad p(7) = .17146667$$

and the win probability becomes $\Pr\{A\} = .5029237$.

The win probability of .4929293 with fair dice is unfavorable, that is, is less than $\frac{1}{2}$. With shaved dice, the win probability is favorable, now being .5029237. What appears to be a slight change becomes, in fact, quite significant when a large number of games are played. See Section 3.5.

Problems 2.2

1. Verify the win probability of .5029237 by substituting from (2.21) into (2.20).

2. Determine the win probability when the dice are shaved on the 1–6 faces and $p_+ = .206666$ . . . and $p_- = .146666$ . . . .

3. Let $X_1, X_2, \ldots$ be independent identically distributed positive random variables whose common distribution function is F. We interpret $X_1, X_2, \ldots$ as successive bids on an asset offered for sale. Suppose that the policy is followed of accepting the first bid that exceeds some prescribed number A. Formally, the accepted bid is X_N where

$$N = \min\{k \ge 1 : X_k > A\}.$$

Set $\alpha = \Pr\{X_1 > A\}$ and $M = E[X_N]$.
(a) Argue the equation

$$M = \int_A^\infty x \, dF(x) + (1 - \alpha)M$$

by considering the possibilities: Either the first bid is accepted, or it is not.

(b) Solve for M, thereby obtaining

$$M = \alpha^{-1} \int_A^\infty x dF(x).$$

(c) When X_1 has an exponential distribution with parameter λ, use the memoryless property to deduce $M = A + \lambda^{-1}$.
(d) Verify this result by calculation in (b).

4. Consider a pair of dice that are unbalanced by the addition of weights in the following manner: Die #1 has a small piece of lead placed near the four side, causing the appearance of the outcome 3 more often than usual, while die #2 is weighted near the three side, causing the outcome 4 to appear more often than usual. We assign the probabilities

Die #1
$p(1) = p(2) = p(5) = p(6) = .166667$
$p(3) = .186666$
$p(4) = .146666$

Die #2
$p(1) = p(2) = p(5) = p(6) = .166667$
$p(4) = .186666$
$p(3) = .146666$

Determine the win probability if the game of craps is played with these loaded dice.

2.3 Random Sums

Sums of the form $X = \xi_1 + \ldots + \xi_N$, where N is random, arise frequently and in varied contexts. Our study of random sums begins with a crisp definition and a precise statement of the assumptions effective in this section, followed by some quick examples.

We postulate a sequence $\xi_1, \xi_2, \ldots$ of independent and identically distributed random variables. Let N be a discrete random variable, independent of $\xi_1, \xi_2, \ldots$ and having the probability mass function $p_N(n) = \Pr\{N = n\}$ for $n = 0, 1, \ldots$. Define the random sum X by

$$X = \begin{cases} 0 & \text{if } N = 0, \\ \xi_1 + \cdots + \xi_N & \text{if } N > 0. \end{cases} \tag{2.22}$$

We save space by abbreviating (2.22) to simply $X = \xi_1 + \ldots + \xi_N$, understanding that $X = 0$ whenever $N = 0$.

Examples

(a) *Queueing*   Let N be the number of customers arriving at a service facility in a specified period of time, and let ξ_i be the service time required by the ith customer. Then $X = \xi_1 + \ldots + \xi_N$ is the total demand for service time.

(b) *Risk Theory*   Suppose that a total of N claims arrives at an insurance company in a given week. Let ξ_i be the amount of the ith claim. Then the total liability of the insurance company is $X = \xi_1 + \ldots + \xi_N$.

(c) *Population Models*   Let N be the number of plants of a given species in a specified area, and let ξ_i be the number of seeds produced by the ith plant. Then $X = \xi_1 + \ldots + \xi_N$ gives the total number of seeds produced in the area.

(d) *Biometrics*   A wildlife sampling scheme traps a random number N of a given species. Let ξ_i be the weight of the ith specimen. Then $X = \xi_1 + \ldots + \xi_N$ is the total weight captured.

When $\xi_1, \xi_2, \ldots$ are discrete random variables, the necessary background in conditional probability is covered in Section 2.1. In order to study the random sum $X = \xi_1 + \ldots + \xi_N$ when $\xi_1, \xi_2, \ldots$ are continuous random variables, we need to extend our knowledge of conditional distributions.

2.3.1 Conditional Distributions: The Mixed Case

Let X and N be jointly distributed random variables and suppose that the possible values for N are the discrete set $n = 0, 1, 2, \ldots$. Then the elementary definition of conditional probability (2.1) applies to define the *conditional distribution function* $F_{X|N}(x|n)$ of the random variable X given that $N = n$ to be

$$F_{X|N}(x|n) = \frac{\Pr\{X \leq x \text{ and } N = n\}}{\Pr\{N = n\}} \qquad \text{if} \quad \Pr\{N = n\} > 0, \quad (2.23)$$

and the conditional distribution function is not defined at values of n for which $\Pr\{N = n\} = 0$. It is elementary to verify that $F_{X|N}(x|n)$ is a probability distribution function in x at each value of n for which it is defined.

The case in which X is a discrete random variable is covered in Section 2.1. Now let us suppose that X is continuous and that $F_{X|N}(x|n)$ is differentiable in x at each value of n for which $\Pr\{N = n\} > 0$. We define the *conditional probability density function* $f_{X|N}(x|n)$ for the random variable X given that $N = n$ by setting

$$f_{X|N}(x|n) = \frac{d}{dx} F_{X|N}(x|n) \qquad \text{if} \quad \Pr\{N = n\} > 0. \quad (2.24)$$

Again, $f_{X|N}(x|n)$ is a probability density function in x at each value of n for which it is defined. Moreover, the conditional density as defined in (2.24) has the appropriate properties, for example

$$\Pr\{a \le X < b, N = n\} = \int_a^b f_{X|N}(x|n)p_N(n)dx \qquad (2.25)$$

for $a < b$ and where $p_N(n) = \Pr\{N = n\}$. The law of total probability leads to the marginal probability density function for X via

$$f_X(x) = \sum_{n=0}^{\infty} f_{X|N}(x|n)p_N(n). \qquad (2.26)$$

Suppose that g is a function for which $E[|g(X)|] < \infty$. The conditional expectation of $g(X)$ given that $N = n$ is defined by

$$E[g(X)|N = n] = \int g(x)f_{X|N}(x|n)dx. \qquad (2.27)$$

Stipulated thus, $E[g(X)|N = n]$ satisfies the properties listed in (2.7) to (2.15) for the joint discrete case. For example, the law of total probability is

$$E[g(X)] = \sum_{n=0}^{\infty} E[g(X)|N = n]p_N(n) = E\{E[g(X)|N]\}. \qquad (2.28)$$

2.3.2 The Moments of a Random Sum

Let us assume that ξ_k and N have the finite moments:

$$\begin{aligned} E[\xi_k] &= \mu, & \mathrm{Var}[\xi_k] &= \sigma^2, \\ E[N] &= \nu, & \mathrm{Var}[N] &= \tau^2, \end{aligned} \qquad (2.29)$$

and determine the mean and variance for $X = \xi_1 + \ldots + \xi_N$ as defined in (2.22). The derivation provides practice in manipulating conditional expectations, and the results,

$$E[X] = \mu\nu, \qquad \mathrm{Var}[X] = \nu\sigma^2 + \mu^2\tau^2, \qquad (2.30)$$

are useful and important. The properties of conditional expectation listed in (2.7) to (2.15) justify the steps in the determination.

If we begin with the mean $E[X]$, then

$$\begin{aligned} E[X] &= \sum_{n=0}^{\infty} E[X|N = n]p_N(n) & &\text{[by (2.15)]} \\ &= \sum_{n=1}^{\infty} E[\xi_1 + \cdots + \xi_N|N = n]p_N(n) & &\text{(definition of } X) \\ &= \sum_{n=1}^{\infty} E[\xi_1 + \cdots + \xi_n|N = n]p_N(n) & &\text{[by (2.9)]} \\ &= \sum_{n=1}^{\infty} E[\xi_1 + \cdots + \xi_n]p_N(n) & &\text{[by (2.10)]} \\ &= \mu\sum_{n=1}^{\infty} np_N(n) = \mu\nu. \end{aligned}$$

To determine the variance, we begin with the elementary step

$$\begin{aligned} \mathrm{Var}[X] &= E[(X - \mu\nu)^2] = E[(X - N\mu + N\mu - \nu\mu)^2] \qquad (2.31) \\ &= E[(X - N\mu)^2] + E[\mu^2(N - \nu)^2] + 2E[\mu(X - N\mu)(N - \nu)]. \end{aligned}$$

Then

$$E[(X - N\mu)^2] = \sum_{n=0}^{\infty} E[(X - N\mu)^2 | N = n]p_N(n)$$

$$= \sum_{n=1}^{\infty} E[(\xi_1 + \cdots + \xi_n - n\mu)^2 | N = n]p_N(n)$$

$$= \sigma^2 \sum_{n=1}^{\infty} np_N(n) = v\sigma^2,$$

and

$$E[\mu^2(N - v)^2] = \mu^2 E[(N - v)^2] = \mu^2 \tau^2,$$

while

$$E[\mu(X - N\mu)(N - v)] = \mu \sum_{n=0}^{\infty} E[(X - n\mu)(n - v) | N = n]p_N(n)$$

$$= \mu \sum_{n=0}^{\infty} (n - v) \, E[(X - n\mu) | N = n]p_N(n)$$

$$= 0$$

(because $E[(X - n\mu) | N = n] = E[\xi_1 + \cdots + \xi_n - n\mu] = 0$).

Then (2.31) with the subsequent three calculations validates the variance of X as stated in (2.30).

Example   The number of offspring of a given species is a random variable having probability mass function $p(k)$ for $k = 0, 1, \ldots$. A population begins with a single parent who produces a random number N of progeny, each of which independently produces offspring according to $p(k)$ to form a second generation. Then the total number of descendants in the second generation may be written $X = \xi_1 + \ldots + \xi_N$, where ξ_k is the number of progeny of the kth offspring of the original parent. Let $E[N] = E[\xi_k] = \mu$ and $\text{Var}[N] = \text{Var}[\xi_k] = \sigma^2$. Then

$$E[X] = \mu^2 \quad \text{and} \quad \text{Var}[X] = \mu\sigma^2(1 + \mu).$$

2.3.3 The Distribution of a Random Sum

Suppose that the summands $\xi_1, \xi_2, \ldots$ are continuous random variables having a probability density function $f(z)$. For $n \geq 1$, the probability density function for the fixed sum $\xi_1 + \ldots + \xi_n$ is the n-fold convolution of the density $f(z)$, denoted by $f^{(n)}(z)$ and recursively defined by

$$f^{(1)}(z) = f(z)$$

and

$$f^{(n)}(z) = \int f^{(n-1)}(z - u)f(u)du \qquad \text{for} \quad n > 1. \tag{2.32}$$

(See Section 1.2.5 for a discussion of convolutions.) Because N and ξ_1,

$\xi_2, \ldots$ are independent, then $f^{(n)}(z)$ is also the conditional density function for $X = \xi_1 + \ldots + \xi_N$ given that $N = n \geq 1$. Let us suppose that $\Pr\{N = 0\} = 0$. Then, by the law of total probability as expressed in (2.26), X is continuous and has the marginal density function

$$f_X(x) = \sum_{n=1}^{\infty} f^{(n)}(x) p_N(n). \tag{2.33}$$

Remark   When $N = 0$ can occur with positive probability, then $X = \xi_1 + \ldots + \xi_N$ is a random variable having both continuous and discrete components to its distribution. Assuming that $\xi_1, \xi_2, \ldots$ are continuous with probability density function $f(z)$, then

$$\Pr\{X = 0\} = \Pr\{N = 0\} = p_N(0)$$

while for $0 < a < b$ or $a < b < 0$, then

$$\Pr\{a < X < b\} = \int_a^b \left\{ \sum_{n=1}^{\infty} f^{(n)}(z) p_N(n) \right\} dz. \tag{2.34}$$

Example   *A Geometric Sum of Exponential Random Variables*   In the following computational example, suppose

$$f(z) = \begin{cases} \lambda e^{-\lambda z} & \text{for} \quad z \geq 0, \\ 0 & \text{for} \quad z < 0, \end{cases}$$

and

$$p_N(n) = \beta(1 - \beta)^{n-1} \qquad \text{for} \quad n = 1, 2, \ldots .$$

For $n \geq 1$, the n-fold convolution of $f(z)$ is the gamma density

$$f^{(n)}(z) = \begin{cases} \dfrac{\lambda^n}{(n-1)!} z^{n-1} e^{-\lambda z} & \text{for} \quad z \geq 0, \\ 0 & \text{for} \quad z < 0. \end{cases}$$

(See Section 1.4.4 for discussion.)

The density for $X = \xi_1 + \ldots + \xi_N$ is given, according to (2.26), by

$$\begin{aligned} f_X(x) &= \sum_{n=1}^{\infty} f^{(n)}(x) p_N(n) \\ &= \sum_{n=1}^{\infty} \frac{\lambda^n}{(n-1)!} z^{n-1} e^{-\lambda z} \beta(1 - \beta)^{n-1} \\ &= \lambda \beta e^{-\lambda z} \sum_{n=1}^{\infty} \frac{[\lambda(1-\beta)z]^{n-1}}{(n-1)!} \\ &= \lambda \beta e^{-\lambda z} e^{\lambda(1-\beta)z} \\ &= \lambda \beta e^{-\lambda \beta z}, \qquad z \geq 0. \end{aligned}$$

Surprise! X has an exponential distribution with parameter $\lambda\beta$.

Example   *Stock Price Changes*   Stochastic models for price fluctuations of publicly traded assets were developed as early as 1900.

Let Z denote the difference in price of a single share of a certain stock between the close of one trading day and the close of the next. For an actively traded stock, a large number of transactions takes place in a single day, and the total daily price change is the sum of the changes over these individual transactions. If we assume that price changes over successive transactions are independent random variables having a common finite variance,[*] then the central limit theorem applies. The price change over a large number of transactions should follow a normal or Gaussian distribution.

A variety of empirical studies have supported this conclusion. For the most part, these studies involved price changes over a fixed number of transactions. Other studies found discrepancies in that both very small and very large price changes occurred more frequently in the data than normal theory suggested that they would. At the same time, intermediate size price changes were underrepresented in the data. For the most part these studies examined price changes over fixed durations containing a random number of transactions.

A natural question arises: Does the random number of transactions in a given day provide a possible explanation for the departures from normality that are observed in data of daily price changes? Let us model the daily price change in the form

$$Z = \xi_0 + \xi_1 + \cdots + \xi_N = \xi_0 + X \qquad (2.35)$$

where $\xi_0, \xi_1, \ldots$ are independent normally distributed random variables with common mean zero and variance σ^2 and N has a Poisson distribution with mean v.

We interpret N as the number of transactions during the day, ξ_i for $i \geq 1$ as the price change during the ith transaction, and ξ_0 as an initial price change arising between the close of the market on one day and the opening of the market on the next. (An obvious generalization would allow the distribution of ξ_0 to differ from that of $\xi_1, \xi_2, \ldots$.)

Conditioned on $N = n$, the random variable $Z = \xi_0 + \xi_1 + \ldots + \xi_n$ is normally distributed with mean zero and variance $(n + 1)\sigma^2$. The conditional density function is

$$\phi_n(z) = \frac{1}{\sqrt{2\pi(n + 1)}\sigma} \exp\left\{ -\frac{1}{2}\frac{z^2}{(n + 1)\sigma^2} \right\}.$$

---

[*]Rather strong economic arguments in support of these assumptions can be given. The independence follows from concepts of a "perfect market," and the common variance from notions of time stationarity.

Since the probability mass function for N is

$$p_N(n) = \frac{\lambda^n e^{-\lambda}}{n!}, \qquad n = 0, 1, \ldots,$$

using (2.33) we determine the probability density function for the daily price change to be

$$f_Z(z) = \sum_{n=0}^{\infty} \phi_n(z) \, \frac{\lambda^n e^{-\lambda}}{n!}.$$

The formula for the density $f_Z(z)$ does not simplify. Nevertheless, numerical calculations are possible. When $\lambda = 1$ and $\sigma^2 = \frac{1}{2}$, then (2.30) shows that the variance of the daily price change Z in the model (2.35) is $\mathrm{Var}[Z] = (1 + \lambda)\sigma^2 = 1$. Thus comparing the density $f_Z(z)$ when $\lambda = 1$ and $\sigma^2 = \frac{1}{2}$ to a normal density with mean zero and variance one sheds some light on the question at hand.

The calculations were carried out and are shown in Figure 2.2.

The departure from normality that is exhibited by the random sum in Figure 2.2 is consistent with the departure from normality shown by stock price changes over fixed time intervals. Of course our calculations do not *prove* that the observed departure from normality is *caused* by the random number of transactions in a fixed time interval. Rather the calculations show only that such an explanation is consistent with the data and is, therefore, a possible cause.

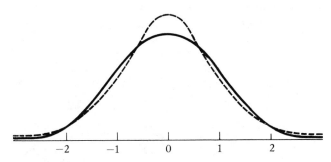

Figure 2.2   A standard normal density (solid line) as compared to a density for a random sum (dashed line). Both densities have a zero mean and unit variance.

Problems 2.3

1. The following experiment is performed: An observation is made of a Poisson random variable N with parameter λ. Then N independent Bernoulli trials are performed, each with probability p of success. Let Z be the total number of successes observed in the N trials.

(a) Formulate Z as a random sum and thereby determine its mean and variance.
(b) What is the distribution of Z?

2. For each given p, let Z have a binomial distribution with parameters p and N. Suppose that N is itself binomially distributed with parameters q and M. Formulate Z as a random sum and show that Z has a binomial distribution with parameters pq and M.

3. Suppose that ξ_1, ξ_2, . . . are independent and identically distributed with $\Pr\{\xi_k = \pm 1\} = \frac{1}{2}$. Let N be independent of ξ_1, ξ_2, . . . and follow the geometric probability mass function

$$p_N(k) = \alpha(1 - \alpha)^k \qquad \text{for} \quad k = 0, 1, . . . ,$$

where $0 < \alpha < 1$. Form the random sum $Z = \xi_1 + . . . + \xi_N$.
(a) Determine the mean and variance of Z.
(b) Evaluate the higher moments $m_3 = E[Z^3]$ and $m_4 = E[Z^4]$.
Hint: Express Z^4 in terms of the ξ_i's where $\xi_i^2 = 1$ and $E[\xi_i \xi_j] = 0$.

4. Suppose ξ_1, ξ_2, . . . are independent and identically distributed random variables having mean μ and variance σ^2. Form the random sum $S_N = \xi_1 + . . . + \xi_N$.
(a) Derive the mean and variance of S_N when N has a Poisson distribution with parameter λ.
(b) Determine the mean and variance of S_N when N has a geometric distribution with mean $\lambda = (1 - p)/p$.
(c) Compare the behaviors in (a) and (b) as $\lambda \to \infty$.

5. To form a slightly different random sum, let ξ_0, ξ_1, . . . be independent identically distributed random variables and let N be a nonnegative integer valued random variable, independent of ξ_0, ξ_1, ... The first two moments are

$$E[\xi_k] = \mu, \qquad \text{Var}[\xi_k] = \sigma^2,$$
$$E[N] = \nu, \qquad \text{Var}[N] = \tau^2.$$

Determine the mean and variance of the random sum $Z = \xi_0 + . . . + \xi_N$.

6. The number of accidents occuring in a factory in a week is a Poisson random variable with mean 2. The number of individuals injured in different accidents are independently distributed, each with mean 3 and variance 4. Determine the mean and variance of the number of individuals injured in a week.

2.4 Conditioning on a Continuous Random Variable⋆

Let X and Y be jointly distributed continuous random variables with joint probability density function $f_{X,Y}(x, y)$. We define the conditional probability density function $f_{X|Y}(x|y)$ for the random variable X given that $Y = y$ by the formula

$$f_{X|Y}(x|y) = \frac{f_{X,Y}(x, y)}{f_Y(y)} \qquad \text{if } f_Y(y) > 0, \qquad (2.36)$$

and the conditional density is not defined at values y for which $f_Y(y) = 0$. The conditional distribution function for X given $Y = y$ is defined by

$$F_{X|Y}(x|y) = \int_{-\infty}^{x} f_{X|Y}(\xi|y)d\xi \qquad \text{if } f_Y(y) > 0. \qquad (2.37)$$

Finally, given a function g for which $E[|g(X)|] < \infty$, the conditional expectation of $g(X)$ given that $Y = y$ is defined to be

$$E[g(X)|Y = y] = \int g(x)f_{X|Y}(x|y)dx \qquad \text{if } f_Y(y) > 0. \qquad (2.38)$$

The definitions given in (2.36) to (2.38) are a significant extension of our elementary notions of conditional probability because they allow us to condition on certain events having zero probability. To understand the distinction, try to apply the elementary formula

$$\Pr\{A|B\} = \frac{\Pr\{A \text{ and } B\}}{\Pr\{B\}} \qquad \text{if } \Pr\{B\} > 0 \qquad (2.39)$$

to evaluate the conditional probability $\Pr\{a < X \le b|Y = y\}$. We set $A = \{a < X \le b\}$ and $B = \{Y = y\}$. But Y is a continuous random variable and thus $\Pr\{B\} = \Pr\{Y = y\} = 0$ and (2.39) cannot be applied. Equation (2.37) saves the day, yielding

$$\Pr\{a < X \le b|Y = y\} = F_{X|Y}(b|y) - F_{X|Y}(a|y) = \int_a^b f_{X|Y}(\xi|y)d\xi, \quad (2.40)$$

provided only that the density $f_Y(y)$ is strictly positive at the point y.

To emphasize the important advance being made, we consider the following simple problem. A woman arrives at a bus stop at a time Y that is uniformly distributed between 0 (noon) and 1. Independently, the bus arrives at a time Z that is also uniformly distributed between 0 and 1. Given that the woman arrives at time $Y = .20$, what is the probability that she misses the bus?

On the one hand, the answer $\Pr\{Z < Y|Y = .20\} = .20$ is obvious. On the other hand, this elementary question cannot be answered by the

---

⋆The reader may wish to defer reading this section until encountering Chapter 7, on renewal processes, where conditioning on a continuous random variable first appears.

elementary conditional probability formula (2.39) because the event $\{Y = .20\}$ has zero probability. To apply (2.36), start with the joint density function

$$f_{Z,Y}(z, y) = \begin{cases} 1 & \text{for } 0 \le z, \quad y \le 1, \\ 0 & \text{elsewhere,} \end{cases}$$

and change variables according to $X = Y - Z$. Then

$$f_{X,Y}(x, y) = 1 \quad \text{for } 0 \le y \le 1, \quad y - 1 \le x \le y,$$

and, applying (2.36), we find that

$$f_{X|Y}(x|.20) = \frac{f_{X,Y}(x, .20)}{f_Y(.20)} = 1 \quad \text{for } -.80 \le x \le .20.$$

Finally

$$\Pr\{Z < Y | Y = .20\} = \Pr\{X > 0 | Y = .20\} = \int_0^\infty f_{X|Y}(x|.20)dx = .20.$$

The conditional density function that is prescribed by (2.36) possesses all of the properties that are called for by our intuition and basic concept of conditional probability. In particular, one can calculate the probability of joint events by the formula

$$\Pr\{a < X < b, c < Y < d\} = \int_c^d \left\{ \int_a^b f_{X|Y}(x|y)dx \right\} f_Y(y)dy \qquad (2.41)$$

which becomes the law of total probability by setting $c = -\infty$ and $d = +\infty$:

$$\Pr\{a < X < b\} = \int_{-\infty}^{+\infty} \left\{ \int_a^b f_{X|Y}(x|y)dx \right\} f_Y(y)dy. \qquad (2.42)$$

For the same reasons, the conditional expectation as defined in (2.38) satisfies the requirements listed in (2.7) to (2.11). The property (2.12), adapted to a continuous random variable Y, is written

$$E[g(X)h(Y)] = E\{h(Y)E[g(X)|Y]\}$$
$$= \int h(y)E[g(X)|Y = y]f_Y(y)dy, \qquad (2.43)$$

valid for any bounded function h, and assuming $E[|g(X)|] < \infty$. When $h(y) \equiv 1$, we recover the law of total probability in the form

$$E[g(X)] = E\{E[g(X)|Y]\} = \int E[g(X)|Y = y]f_Y(y)dy. \qquad (2.44)$$

Both the discrete and continuous cases of (2.43) and (2.44) are contained in the expressions

$$E[g(X)h(Y)] = E\{h(Y)E[g(X)|Y]\} = \int h(y)E[g(X)|Y = y]dF_Y(y), \quad (2.45)$$

and

$$E[g(X)] = E\{E[g(X)|Y]\} = \int E[g(X)|Y = y]dF_Y(y). \qquad (2.46)$$

[See the discussion following (1.9) in Chapter 1 for an explanation of the symbolism in (2.45) and (2.46).]

The following exercises provide practice in deriving conditional probability density functions and in manipulating the law of total probability.

Example Suppose X and Y are jointly distributed random variables having the density function

$$f_{XY}(x, y) = \frac{1}{y} e^{-(x/y)-y} \qquad \text{for} \quad x, y > 0.$$

We first determine the marginal density for y, obtaining

$$f_Y(y) = \int_0^\infty f_{XY}(x, y)dx$$

$$= e^{-y} \int_0^\infty y^{-1} e^{-(x/y)} dx = e^{-y} \qquad \text{for} \quad y > 0.$$

Then

$$f_{X|Y}(x|y) = \frac{f_{XY}(x, y)}{f_Y(y)} = y^{-1} e^{-(x/y)} \qquad \text{for} \quad x, y > 0.$$

That is, conditional on $Y = y$, the random variable X has an exponential distribution with parameter $1/y$. It is easily seen that $E[X|Y = y] = y$.

Example For each given p, let X have a binomial distribution with parameters p and N. Suppose that p is uniformly distributed on the interval $[0, 1]$. What is the resulting distribution of X?

We are given the marginal distribution for p and the conditional distribution for X. Applying the law of total probability and the Beta integral (Section 1.7), we obtain

$$\Pr\{X = k\} = \int_0^1 \Pr\{X = k|p = \xi\} f_p(\xi)d\xi$$

$$= \int_0^1 \frac{N!}{k!(N - k)!} \xi^k (1 - \xi)^{N-k} d\xi$$

$$= \frac{N!}{k!(N - k)!} \frac{k!(N - k)!}{(N + 1)!} = \frac{1}{N + 1}$$

$$= \frac{1}{N + 1} \qquad \text{for} \quad k = 0, 1, \ldots, N.$$

That is, X is uniformly distributed on the integers $0, 1, \ldots, N$.

When p has the beta distribution with parameters r and s, then similar calculations give

$$\Pr\{X = k\} = \frac{N!}{k!(N - k)!} \frac{\Gamma(r + s)}{\Gamma(r)\Gamma(s)} \int_0^1 \xi^{r-1}(1 - \xi)^{s-1}\xi^k(1 - \xi)^{N-k}d\xi$$

$$= \binom{N}{k} \frac{\Gamma(r + s)\Gamma(r + k)\Gamma(s + N - k)}{\Gamma(r)\Gamma(s)\Gamma(N + r + s)}$$

$$\text{for} \quad k = 0, 1, \ldots, N.$$

Example A random variable Y follows the exponential distribution with parameter θ. Given that $Y = y$, the random variable X has a Poisson distribution with mean y. Applying the law of total probability then yields

$$\Pr\{X = k\} = \int_0^\infty \frac{y^k e^{-y}}{k!} \theta e^{-\theta y}dy$$

$$= \frac{\theta}{k!}\int_0^\infty y^k e^{-(1+\theta)y}dy$$

$$= \frac{\theta}{k!(1 + \theta)^{k+1}}\int_0^\infty u^k e^{-u}du$$

$$= \frac{\theta}{(1 + \theta)^{k+1}} \quad \text{for} \quad k = 0, 1, \ldots$$

Suppose that Y has the gamma density

$$f_Y(y) = \frac{\theta}{\Gamma(\alpha)} (\theta y)^{\alpha-1}e^{-\theta y}, \quad y \geq 0.$$

Then similar calculations yield

$$\Pr\{X = k\} = \int_0^\infty \frac{y^k e^{-y}}{k!} \frac{\theta}{\Gamma(\alpha)} (\theta y)^{\alpha-1}e^{-\theta y}dy$$

$$= \frac{\theta^\alpha}{k!\Gamma(\alpha)(1 + \theta)^{k+\alpha}}\int_0^\infty u^{k+\alpha-1}e^{-u}du$$

$$= \frac{\Gamma(k + \alpha)}{k!\Gamma(\alpha)} \left(\frac{\theta}{1 + \theta}\right)^\alpha \left(\frac{1}{1 + \theta}\right)^k, \quad k = 0, 1, \ldots$$

This is the negative binomial distribution.

Problems 2.4

1. Suppose X and Y are independent random variables, each exponentially distributed with parameter λ. Determine the probability density function for $Z = X/Y$.

2. Let U be uniformly distributed over the interval $[0, L]$ where L follows the gamma density $f_L(x) = xe^{-x}$ for $x \geq 0$. What is the joint density function of U and $V = L - U$?

3. Suppose that the outcome X of a certain chance mechanism depends on a parameter p according to $\Pr\{X = 1\} = p$ and $\Pr\{X = 0\} = 1 - p$ where $0 \leq p \leq 1$. Suppose that p is chosen at random, uniformly distributed over the unit interval $[0, 1]$, and then that two independent outcomes X_1 and X_2 are observed. What is the unconditional correlation coefficient between X_1 and X_2?
   Note: Conditionally independent random variables may become dependent if they share a common parameter.

4. Let N have a Poisson distribution with parameter $\lambda > 0$. Suppose that, conditioned on $N = n$, the random variable X is binomially distributed with parameters $N = n$ and p. Set $Y = N - X$. Show that X and Y have Poisson distributions with respective parameters λp and $\lambda(1 - p)$ *and that X and Y are independent.*
   Note: Conditionally dependent random variables may become independent through randomization.

5. Let X have a Poisson distribution with parameter $\lambda > 0$. Suppose λ itself is random, following an exponential density with parameter θ.
   (a) What is the marginal distribution of X?
   (b) Determine the conditional density for λ given $X = k$.

6. Suppose X and Y are independent random variables having the same Poisson distribution with parameter λ, but where λ is also random, being exponentially distributed with parameter θ. What is the conditional distribution for X given that $X + Y = n$?

7. Let X and Y be jointly distributed random variables whose joint probability mass function is given in the following table:

		x		
		-1	0	1
y	-1	$\frac{1}{9}$	$\frac{2}{9}$	0
	0	0	$\frac{1}{9}$	$\frac{2}{9}$
	1	$\frac{2}{9}$	0	$\frac{1}{9}$

$$p(x, y) = \Pr\{X = x, Y = y\}$$

Show that the covariance between X and Y is zero even though X and Y are not independent.

8. Let X_0, X_1, X_2, . . . be independent identically distributed nonnegative random variables having a continuous distribution. Let N be the first index k for which $X_k > X_0$. That is, $N = 1$ if $X_1 > X_0$; $N = 2$ if $X_1 \leq X_0$ and $X_2 > X_0$; etc. Determine the probability mass function for N and the mean $E[N]$. (Interpretation: X_0, X_1, . . . are successive offers or bids on a car that you are trying to sell. Then N is the index of the first bid that is better than the initial bid.)

Chapter 3 | Markov Chains: Introduction

3.1 Definitions

A *Markov process* $\{X_t\}$ is a stochastic process with the property that, given the value of X_t, the values of X_s for $s > t$ are not influenced by the values of X_u for $u < t$. In words, the probability of any particular future behavior of the process, when its current state is known exactly, is not altered by additional knowledge concerning its past behavior. A *discrete time Markov chain* is a Markov process whose state space is a finite or countable set, and whose (time) index set is $T = (0, 1, 2, \ldots)$. In formal terms, the Markov property is that

$$\Pr\{X_{n+1} = j | X_0 = i_0, \ldots, X_{n-1} = i_{n-1}, X_n = i\}$$
$$= \Pr\{X_{n+1} = j | X_n = i\} \tag{3.1}$$

for all time points n and all states $i_0, \ldots, i_{n-1}, i, j$.

It is frequently convenient to label the state space of the Markov chain by the nonnegative integers $\{0, 1, 2, \ldots\}$, which we will do unless the contrary is explicitly stated, and it is customary to speak of X_n as being in state i if $X_n = i$.

The probability of X_{n+1} being in state j given that X_n is in state i is called the *one-step transition probability* and is denoted by $P_{ij}^{n,n+1}$. That is,

$$P_{ij}^{n,n+1} = \Pr\{X_{n+1} = j | X_n = i\}. \tag{3.2}$$

The notation emphasizes that in general the transition probabilities are functions not only of the initial and final states, but also of the time of transition as well. When the one-step transition probabilities are independent of the time variable n, we say that the Markov chain has *stationary transition probabilities*. Since the vast majority of Markov chains that we shall encounter have stationary transition probabilities, we limit our discussion to this case. Then $P_{ij}^{n,n+1} = P_{ij}$ is independent of n and P_{ij} is the conditional probability that the state value undergoes a transition from i to j in one trial. It is customary to arrange these numbers P_{ij} in a *matrix*, in the infinite square array

$$
\mathbf{P} = \left\|
\begin{array}{ccccc}
P_{00} & P_{01} & P_{02} & P_{03} & \cdots \\
P_{10} & P_{11} & P_{12} & P_{13} & \cdots \\
P_{20} & P_{21} & P_{22} & P_{23} & \cdots \\
\cdot & \cdot & \cdot & \cdot & \\
\cdot & \cdot & \cdot & \cdot & \\
\cdot & \cdot & \cdot & \cdot & \\
P_{i0} & P_{i1} & P_{i2} & P_{i3} & \cdots \\
\cdot & \cdot & \cdot & \cdot & \\
\cdot & \cdot & \cdot & \cdot & \\
\cdot & \cdot & \cdot & \cdot &
\end{array}
\right\|
$$

and refer to $\mathbf{P} = \|P_{ij}\|$ as the Markov matrix or *transition probability matrix* of the process.

The $(i + 1)$st row of $\mathbf{P}$ is the probability distribution of the values of X_{n+1} under the condition that $X_n = i$. If the number of states is finite, then $\mathbf{P}$ is a finite square matrix whose order (the number of rows) is equal to the number of states. Clearly the quantities P_{ij} satisfy the conditions

$$P_{ij} \geq 0 \qquad \text{for} \quad i, j = 0, 1, 2, \ldots, \tag{3.3}$$

$$\sum_{j=0}^{\infty} P_{ij} = 1 \qquad \text{for} \quad i = 0, 1, 2, \ldots. \tag{3.4}$$

The condition (3.4) merely expresses the fact that some transition occurs at each trial. (For convenience, one says that a transition has occurred even if the state remains unchanged.)

A Markov process is completely defined once its transition probability matrix and initial state X_0 (or, more generally, the probability distribution of X_0) are specified. We shall now prove this fact.

Let $\Pr\{X_0 = i\} = p_i$. It is enough to show how to compute the quantities

$$\Pr\{X_0 = i_0, X_1 = i_1, X_2 = i_2, \ldots, X_n = i_n\} \tag{3.5}$$

since any probability involving $X_{j_1}, \ldots, X_{j_k}$, for $j_1 < \ldots < j_k$, can be obtained, according to the axiom of total probability, by summing terms of the form (3.5).

By the definition of conditional probabilities we obtain

$$\Pr\{X_0 = i_0, X_1 = i_1, X_2 = i_2, \ldots, X_n = i_n\}$$
$$= \Pr\{X_0 = i_0, X_1 = i_1, \ldots, X_{n-1} = i_{n-1}\} \quad (3.6)$$
$$\times \Pr\{X_n = i_n | X_0 = i_0, X_1 = i_1, \ldots, X_{n-1} = i_{n-1}\}.$$

Now by the definition of a Markov process,

$$\Pr\{X_n = i_n | X_0 = i_0, X_1 = i_1, \ldots, X_{n-1} = i_{n-1}\}$$
$$= \Pr\{X_n = i_n | X_{n-1} = i_{n-1}\} = P_{i_{n-1}, i_n}. \quad (3.7)$$

Substituting (3.7) into (3.6) gives

$$\Pr\{X_0 = i_0, X_1 = i_1, \ldots, X_n = i_n\}$$
$$= \Pr\{X_0 = i_0, X_1 = i_1, \ldots, X_{n-1} = i_{n-1}\} P_{i_{n-1}, i_n}.$$

Then, by induction, (3.5) becomes

$$\Pr\{X_0 = i_0, X_1 = i_1, \ldots, X_n = i_n\}$$
$$= p_{i_0} P_{i_0, i_1} \ldots P_{i_{n-2}, i_{n-1}} P_{i_{n-1}, i_n}. \quad (3.8)$$

This shows that all finite dimensional probabilities are specified once the transition probabilities and initial distribution are given, and in this sense the process is defined by these quantities.

Related computations show that (3.1) is equivalent to the Markov property in the form

$$\Pr\{X_{n+1} = j_1, \ldots, X_{n+m} = j_m | X_0 = i_0, \ldots, X_n = i_n\}$$
$$= \Pr\{X_{n+1} = j_1, \ldots, X_{n+m} = j_m | X_n = i_n\} \quad (3.9)$$

for all time points n, m and all states $i_0, \ldots, i_n, j_1, \ldots, j_m$. In other words, once (3.9) is established for the value $m = 1$, it holds for all $m \geq 1$ as well.

Problems 3.1

1. A Markov chain $X_0, X_1, \ldots$ on states 0, 1, 2 has the transition probability matrix

$$\mathbf{P} = \begin{array}{c} 0 \\ 1 \\ 2 \end{array} \begin{Vmatrix} \begin{array}{ccc} 0 & 1 & 2 \\ .1 & .2 & .7 \\ .9 & .1 & 0 \\ .1 & .8 & .1 \end{array} \end{Vmatrix}$$

and initial distribution $p_0 = \Pr\{X_0 = 0\} = .3$, $p_1 = \Pr\{X_0 = 1\} = .4$, and $p_2 = \Pr\{X_0 = 2\} = .3$. Determine $\Pr\{X_0 = 0, X_1 = 1, X_2 = 2\}$.

2. Consider the problem of sending a binary message, 0 or 1, through a signal channel consisting of several stages, where transmission through each stage is subject to a fixed probability of error α. Suppose that

$X_0 = 0$ is the signal that is sent and let X_n be the signal that is received at the nth stage. Assume that $\{X_n\}$ is a Markov chain with transition probabilities $P_{00} = P_{11} = 1 - \alpha$ and $P_{01} = P_{10} = \alpha$ where $0 < \alpha < 1$.
(a) Determine $\Pr\{X_0 = 0, X_1 = 0, X_2 = 0\}$, the probability that no error occurs up to stage $n = 2$.
(b) Determine the probability that a correct signal is received at stage 2. *Hint:* This is $\Pr\{X_0 = 0, X_1 = 0, X_2 = 0\} + \Pr\{X_0 = 0, X_1 = 1, X_2 = 0\}$.

3.2 Transition Probability Matrices of a Markov Chain

A Markov chain is completely defined by its one-step transition probability matrix and the specification of a probability distribution on the state of the process at time 0. The analysis of a Markov chain concerns mainly the calculation of the probabilities of the possible realizations of the process. Central in these calculations are the n-step transition probability matrices $\mathbf{P}^{(n)} = \|P_{ij}^{(n)}\|$. Here $P_{ij}^{(n)}$ denotes the probability that the process goes from state i to state j in n transitions. Formally,

$$P_{ij}^{(n)} = \Pr\{X_{m+n} = j | X_m = i\}. \qquad (3.10)$$

Observe that we are dealing only with temporally homogeneous processes having stationary transition probabilities, since otherwise the left side of (3.10) would also depend on m.

The Markov property allows us to express (3.10) in terms of $\|P_{ij}\|$ as stated in the following theorem.

Theorem 3.1 *The n-step transition probabilities of a Markov chain satisfy*

$$P_{ij}^{(n)} = \sum_{k=0}^{\infty} P_{ik} P_{kj}^{(n-1)}, \qquad (3.11)$$

where we define

$$P_{ij}^{(0)} = \begin{cases} 1 & \text{if } i = j, \\ 0 & \text{if } i \neq j. \end{cases}$$

From the theory of matrices we recognize the relation (3.11) as the formula for matrix multiplication, so that $\mathbf{P}^{(n)} = \mathbf{P} \times \mathbf{P}^{(n-1)}$. By iterating this formula, we obtain

$$\mathbf{P}^{(n)} = \underbrace{\mathbf{P} \times \mathbf{P} \times \cdots \times \mathbf{P}}_{n \text{ factors}} = \mathbf{P}^n; \qquad (3.12)$$

in other words, the n-step transition probabilities $P_{ij}^{(n)}$ are the entries in the matrix $\mathbf{P}^n$, the nth power of $\mathbf{P}$.

Proof The proof proceeds via a *first step analysis,* a breaking down or analysis of the possible transitions on the first step, followed by an application of the Markov property. The event of going from state i to state j in n transitions can be realized in the mutually exclusive ways of going to some intermediate state k, $(k = 0, 1, . . .)$, in the first transition, and then going from state k to state j in the remaining $(n - 1)$ transitions. Because of the Markov property, the probability of the second transition is $P_{kj}^{(n-1)}$, and that of the first is clearly P_{ik}. If we use the law of total probability, then (3.11) follows. The steps are

$$P_{ij}^{(n)} = \Pr\{X_n = j | X_0 = i\} = \sum_{k=0}^{\infty} \Pr\{X_n = j, X_1 = k | X_0 = i\}$$

$$= \sum_{k=0}^{\infty} \Pr\{X_1 = k | X_0 = i\} \Pr\{X_n = j | X_0 = i, X_1 = k\}$$

$$= \sum_{k=0}^{\infty} P_{ik} P_{kj}^{(n-1)}.$$

If the probability of the process initially being in state j is p_j, i.e., the distribution law of X_0 is $\Pr\{X_0 = j\} = p_j$, then the probability of the process being in state k at time n is

$$p_k^{(n)} = \sum_{j=0}^{\infty} p_j P_{jk}^{(n)} = \Pr\{X_n = k\}. \tag{3.13}$$

Problems 3.2

1. A Markov chain $\{X_n\}$ on the states 0, 1, 2 has the transition probability matrix

$$\mathbf{P} = \begin{matrix} & \begin{matrix} 0 & \quad 1 & \quad 2 \end{matrix} \\ \begin{matrix} 0 \\ 1 \\ 2 \end{matrix} & \left\| \begin{matrix} .1 & .2 & .7 \\ .2 & .2 & .6 \\ .6 & .1 & .3 \end{matrix} \right\| \end{matrix}.$$

   (a) Compute the two step transition matrix P^2.
   (b) What is $\Pr\{X_3 = 1 | X_1 = 0\}$?
   (c) What is $\Pr\{X_3 = 1 | X_0 = 0\}$?

2. A particle moves among the states 0, 1, 2 according to a Markov process whose transition probability matrix is

$$\mathbf{P} = \begin{matrix} & \begin{matrix} 0 & \quad 1 & \quad 2 \end{matrix} \\ \begin{matrix} 0 \\ 1 \\ 2 \end{matrix} & \left\| \begin{matrix} 0 & \frac{1}{2} & \frac{1}{2} \\ \frac{1}{2} & 0 & \frac{1}{2} \\ \frac{1}{2} & \frac{1}{2} & 0 \end{matrix} \right\| \end{matrix}.$$

   Let X_n denote the position of the particle at the nth move. Calculate $\Pr\{X_n = 0 | X_0 = 0\}$ for $n = 0, 1, 2, 3, 4$.

3. Consider the problem of sending a binary message, 0 or 1, through a signal channel consisting of several stages, where transmission through each stage is subject to a fixed probability of error α. Let X_0 be the signal that is sent and let X_n be the signal that is received at the nth stage. Suppose X_n is a Markov chain with transition probabilities $P_{00} = P_{11} = 1 - \alpha$ and $P_{01} = P_{10} = \alpha$, $(0 < \alpha < 1)$. Determine $\Pr\{X_5 = 0 | X_0 = 0\}$, the probability of correct transmission through five stages.

4. Let X_n denote the quality of the nth item produced by a production system with $X_n = 0$ meaning "Good" and $X_n = 1$ meaning "Defective." Suppose that X_n evolves as a Markov chain whose transition probability matrix is

$$\mathbf{P} = \begin{array}{c} 0 \\ 1 \end{array} \left\| \begin{array}{cc} 0 & 1 \\ .99 & .01 \\ .12 & .88 \end{array} \right\|.$$

   What is the probability that the fourth item is defective given that the first item is defective?

5. Suppose X_n is a two state Markov chain whose transition probability matrix is

$$\mathbf{P} = \begin{array}{c} 0 \\ 1 \end{array} \left\| \begin{array}{cc} 0 & 1 \\ \alpha & 1 - \alpha \\ 1 - \beta & \beta \end{array} \right\|.$$

   Then $Z_n = (X_{n-1}, X_n)$ is a Markov chain having the four states $(0, 0)$, $(0, 1)$, $(1, 0)$, and $(1, 1)$. Determine the transition probability matrix.

3.3 Some Markov Chain Models

The importance of Markov chains lies in the large number of natural physical, biological and economic phenomena that can be described by them and is enhanced by the amenability of Markov chains to quantitative manipulation. In this section we give several examples of Markov chain models that arise in various arenas of science. General methods for computing certain functionals on Markov chains are derived in the following section.

3.3.1 An Inventory Model

Consider a situation in which a commodity is stocked in order to satisfy a continuing demand. We assume that the replenishment of stock takes place at the end of periods labeled $n = 0, 1, 2, \ldots$, and we assume that the total aggregate demand for the commodity during period n is a random variable ξ_n whose distribution function is independent of the time period,

$$\Pr\{\xi_n = k\} = a_k \quad \text{for} \quad k = 0, 1, 2, \ldots \qquad (3.14)$$

where $a_k \geq 0$ and $\sum_{k=0}^{\infty} a_k = 1$. The stock level is examined at the end of each period. A replenishment policy is prescribed by specifying two nonnegative critical numbers s and $S > s$ whose interpretation is: If the end-of-period stock quantity is not greater than s, then an amount sufficient to increase the quantity of stock on hand up to the level S is immediately procured. If, however, the available stock is in excess of s, then no replenishment of stock is undertaken. Let X_n denote the quantity on hand at the end of period n just prior to restocking. The states of the process $\{X_n\}$ consist of the possible values of stock size

$$S, S - 1, \ldots, +1, 0, -1, -2, \ldots,$$

where a negative value is interpreted as an unfilled demand that will be satisfied immediately upon restocking.

The process $\{X_n\}$ is depicted in Figure 3.1.

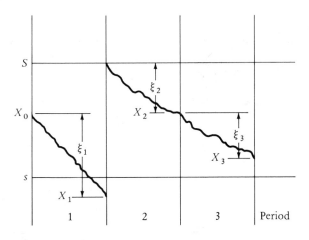

Figure 3.1   The inventory process

According to the rules of the inventory policy, the stock levels at two consecutive periods are connected by the relation

$$X_{n+1} = \begin{cases} X_n - \xi_{n+1} & \text{if} \quad s < X_n \leq S, \\ \\ S - \xi_{n+1} & \text{if} \quad X_n \leq s, \end{cases} \qquad (3.15)$$

where ξ_n is the quantity demanded in the nth period, stipulated to follow the probability law (3.14). If we assume that the successive demands ξ_1, ξ_2, . . . are independent random variables, then the stock values X_0, X_1,

$X_2, \ldots$ constitute a Markov chain whose transition probability matrix can be calculated in accordance with relation (3.15). Explicitly,

$$P_{ij} = \Pr\{X_{n+1} = j | X_n = i\}$$

$$= \begin{cases} \Pr\{\xi_{n+1} = i - j\} & \text{if } s < i \leq S, \\ \\ \Pr\{\xi_{n+1} = S - j\} & \text{if } i \leq s. \end{cases}$$

Consider a spare parts inventory model as a numerical example in which either 0, 1, or 2 repair parts are demanded in any period, with

$$\Pr\{\xi_n = 0\} = .5, \Pr\{\xi_n = 1\} = .4, \Pr\{\xi_n = 2\} = .1,$$

and suppose $s = 0$ while $S = 2$. The possible values for X_n are $S = 2, 1, 0,$ and -1. To illustrate the transition probability calculations, we will consider first the determination of $P_{10} = \Pr\{X_{n+1} = 0 | X_n = 1\}$. When $X_n = 1$, then no replenishment takes place and the next state $X_{n+1} = 0$ results when the demand $\xi_{n+1} = 1$, and this occurs with probability $P_{10} = .4$. To illustrate another case, if $X_n = 0$, then instantaneous replenishment to $S = 2$ ensues, and a next period level of $X_{n+1} = 0$ results from the demand quantity $\xi_{n+1} = 2$. The corresponding probability of this outcome yields $P_{00} = .1$. Continuing in this manner, we obtain the transition probability matrix

$$\mathbf{P} = \begin{array}{c} \\ -1 \\ 0 \\ +1 \\ +2 \end{array} \begin{array}{cccc} -1 & 0 & +1 & +2 \\ \left\| \begin{array}{cccc} 0 & .1 & .4 & .5 \\ 0 & .1 & .4 & .5 \\ .1 & .4 & .5 & 0 \\ 0 & .1 & .4 & .5 \end{array} \right\| \end{array}.$$

Important quantities of interest in inventory models of this type are the long run fraction of periods in which demand is not met ($X_n < 0$) and long run average inventory level. Using the notation $p_j^{(n)} = \Pr\{X_n = j\}$, we give these quantities respectively, as $\lim_{n \to \infty} \Sigma_{j<0} p_j^{(n)}$ and $\lim_{n \to \infty} \Sigma_{j>0} j p_j^{(n)}$. This illustrates the importance of determining conditions under which the probabilities $p_j^{(n)}$ stabilize and approach limiting probabilities π_j as $n \to \infty$, and of determining methods for calculating the limiting probabilities π_j when they exist. These topics are the subject of Chapter 4.

3.3.2 The Ehrenfest Urn Model

A classical mathematical description of diffusion through a membrane is the famous Ehrenfest urn model. Imagine two containers containing a total of $2a$ balls (molecules). Suppose the first container, labeled A, holds k balls and the second container, B, holds the remaining $2a - k$ balls. A ball is selected at random (all selections are equally likely) from the totality of the $2a$ balls and moved to the other container. (A molecule diffuses at random through the membrane.) Each selection generates a transition of the process. Clearly the balls fluctuate between the two containers with an average drift from the urn with the excess numbers to the one with the smaller concentration.

Let Y_n be the number of balls in Urn A at the nth stage and define $X_n = Y_n - a$. Then $\{X_n\}$ is a Markov chain on the states $i = -a$, $-a + 1, \ldots, -1, 0, +1, \ldots, a$ with transition probabilities

$$
P_{ij} = \begin{cases} \dfrac{a - i}{2a} & \text{if } j = i + 1, \\ \dfrac{a + i}{2a} & \text{if } j = i - 1, \\ 0 & \text{otherwise.} \end{cases}
$$

An important quantity in the Ehrenfest urn model is the long run or equilibrium distribution of the number of balls in each urn.

3.3.3 Markov Chains in Genetics

The following idealized genetics model was introduced by S. Wright to investigate the fluctuation of gene frequency under the influence of mutation and selection. We begin by describing the so-called simple haploid model of random reproduction, disregarding mutation pressures and selective forces. We assume that we are dealing with a fixed population size of $2N$ genes composed of type-a and type-A individuals. The makeup of the next generation is determined by $2N$ independent binomial trials as follows: If the parent population consists of j a-genes and $2N - j$ A-genes, then each trial results in a or A with probabilities

$$
p_j = \frac{j}{2N}, \qquad q_j = 1 - \frac{j}{2N},
$$

respectively. Repeated selections are done with replacement. By this procedure we generate a Markov chain $\{X_n\}$ where X_n is the number of a-genes in the nth generation among a constant population size of $2N$ individuals. The state space contains the $2N + 1$ values $\{0, 1, 2, \ldots, 2N\}$. The transition probability matrix is computed according to the binomial distribution as

$$
\Pr\{X_{n+1} = k | X_n = j\} = P_{jk} = \binom{2N}{k} p_j^k q_j^{2N-k}
$$

$$
(j, k = 0, 1, \ldots, 2N). \quad (3.16)
$$

For some discussion of the biological justification of these postulates we refer the reader to Fisher.

Notice that states 0 and $2N$ are completely absorbing in the sense that once $X_n = 0$ (or $2N$) then $X_{n+k} = 0$ (or $2N$, respectively) for all $k \geq 0$. One of the questions of interest is to determine the probability under the condition $X_0 = i$ that the population will attain fixation, i.e., that it will become a pure population composed only of a-genes or A-genes. It is also pertinent to determine the rate of approach to fixation. We will examine such questions in our general analysis of absorption probabilities.

A more complete model takes account of mutation pressures. We assume that prior to the formation of the new generation each gene has the possibility to mutate, that is, to change into a gene of the other kind. Specifically, we assume that for each gene the mutation a $\to$ A occurs with probability x_1, and A $\to$ a occurs with probability x_2. Again we assume that the composition of the next generation is determined by 2N independent binomial trials. The relevant values of p_j and q_j when the parent population consists of j a-genes are now taken to be

$$p_j = \frac{j}{2N}(1 - x_1) + \left(1 - \frac{j}{2N}\right)x_2$$

and   (3.17)

$$q_j = \frac{j}{2N}x_1 + \left(1 - \frac{j}{2N}\right)(1 - x_2).$$

The rationale is as follows: We assume that the mutation pressures operate first, after which a new gene is chosen by random selection from the population. Now the probability of selecting an a-gene after the mutation forces have acted is just 1/2N times the number of a-genes present; hence the average probability (averaged with respect to the possible mutations) is simply 1/2N times the average number of a-genes after mutation. But this average number is clearly $j(1 - x_1) + (2N - j)x_2$, which leads at once to (3.17).

The transition probabilities of the associated Markov chain are calculated by (3.16) using the values of p_j and q_j given in (3.17).

If $x_1x_2 > 0$ then fixation will not occur in any state. Instead, as $n \to \infty$, the distribution function of X_n will approach a steady state distribution of a random variable ξ where $\Pr\{\xi = k\} = \pi_k$ ($k = 0, 1, 2, \ldots, 2N$) ($\sum_{k=0}^{n} \pi_k = 1$, $\pi_k > 0$). The distribution function of ξ is called the steady state gene frequency distribution.

We return to the simple random mating model and discuss the concept of a selection force operating in favor of, say, a-genes. Suppose we wish to impose a selective advantage for a-genes over A-genes so that the relative number of offspring have expectations proportional to $1 + s$ and 1, respectively, where s is small and positive. We replace $p_j = j/2N$ and $q_j = 1 - j/2N$ by

$$p_j = \frac{(1 + s)j}{2N + sj}, \qquad q_j = 1 - p_j$$

and build the next generation by binomial sampling as before. If the parent population consisted of j a-genes, then in the next generation the expected population sizes of a-genes and A-genes, respectively, are

$$2N \frac{(1 + s)j}{2N + sj}, \qquad 2N \frac{(2N - j)}{2N + sj}.$$

The ratio of expected population size of a-genes to A-genes at the $(n + 1)$th generation is

$$\frac{1 + s}{1} \times \frac{j}{2N - j} = \left(\frac{1 + s}{1} \right) \left(\frac{\text{number of a-genes in the } n\text{th generation}}{\text{number of A-genes in the } n\text{th generation}} \right)$$

which explains the meaning of selection.

3.3.4 A Discrete Queueing Markov Chain

Customers arrive for service and take their place in a waiting line. During each period of time, a single customer is served, provided that at least one customer is present. If no customer awaits service, then during this period no service is performed. (We can imagine, for example, a taxi stand at which a cab arrives at fixed time intervals to give service. If no one is present, the cab immediately departs.) During a service period new customers may arrive. We suppose that the actual number of customers that arrive during the nth period is a random variable ξ_n whose distribution is independent of the period and is given by

$$\text{Pr}\{k \text{ customers arrive in a service period}\} = \text{Pr}\{\xi_n = k\} = a_k$$

for $k = 0, 1, \ldots$ where $a_k \geq 0$ and $\sum_{k=0}^{\infty} a_k = 1$.

We also assume that $\xi_1, \xi_2, \ldots$ are independent random variables. The state of the system at the start of each period is defined to be the number of customers waiting in line for service. If the present state is i, then after the lapse of one period the state is

$$j = \begin{cases} i - 1 + \xi & \text{if } i \geq 1, \\ \xi & \text{if } i = 0, \end{cases} \qquad (3.18)$$

where ξ is the number of new customers having arrived in this period while a single customer was served. In terms of the random variables of the process we can express (3.18) formally as

$$X_{n+1} = (X_n - 1)^+ + \xi_n,$$

where $Y^+ = \max\{Y, 0\}$. In view of (3.18), the transition probability matrix may be calculated easily, and we obtain

$$P = \begin{Vmatrix} a_0 & a_1 & a_2 & a_3 & a_4 & \cdots \\ a_0 & a_1 & a_2 & a_3 & a_4 & \cdots \\ 0 & a_0 & a_1 & a_2 & a_3 & \cdots \\ 0 & 0 & a_0 & a_1 & a_2 & \cdots \\ 0 & 0 & 0 & a_0 & a_1 & \cdots \\ \cdot & \cdot & \cdot & \cdot & \cdot & \\ \cdot & \cdot & \cdot & \cdot & \cdot & \\ \cdot & \cdot & \cdot & \cdot & \cdot & \end{Vmatrix}.$$

It is intuitively clear that if the expected number of new customers, $\sum_{k=0}^{\infty} k a_k$, that arrive during a service period exceeds one, then with the passage of time the length of the waiting line increases without limit. On the other hand, if $\sum_{k=0}^{\infty} k a_k < 1$, then the length of the waiting line approaches a statistical equilibrium which is described by a limiting distribution

$$\lim_{n \to \infty} \Pr\{X_n = k | X_0 = j\} = \pi_k > 0 \qquad \text{for} \quad k = 0, 1, \ldots$$

where $\sum_{k=0}^{\infty} \pi_k = 1$. Important quantities to be determined by this model include the long run fraction of time that the service facility is idle, given by π_0, the long run mean time that a customer spends in the system, given by $\sum_{k=0}^{\infty}(1 + k)\pi_k$.

Problems 3.3

1. Consider a spare parts inventory model in which either 0, 1, or 2 repair parts are demanded in any period, with

$$\Pr\{\xi_n = 0\} = .4, \ \Pr\{\xi_n = 1\} = .3, \ \Pr\{\xi_n = 2\} = .3,$$

and suppose $s = 0$ and $S = 3$. Determine the transition probability matrix for the Markov chain $\{X_n\}$ where X_n is defined to be the quantity on hand at the end of period n.

2. Consider two urns A and B containing a total of N balls. An experiment is performed in which a ball is selected at random (all selections equally likely) at time t ($t = 1, 2, \ldots$) from among the totality of N balls. Then an urn is selected at random (A is chosen with probability p and B is chosen with probability q) and the ball previously drawn is placed in this urn. The state of the system at each trial is represented by the number of balls in A. Determine the transition matrix for this Markov chain.

3. Two urns A and B contain a total of N balls. Assume that at time t there were exactly k balls in A. At time $t + 1$ an urn is selected at random in proportion to its contents (i.e., A is chosen with probability k/N and B is chosen with probability $(N - k)/N$). Then a ball is selected from A with probability p or from B with probability q and placed in the previously chosen urn. Determine the transition matrix for this Markov chain.

4. Suppose that two urns A and B contain a total of N balls. Assume that at time t there are exactly k balls in A. At time $t + 1$ a ball and an urn are chosen with probability depending on the contents of the urn (i.e., a ball is chosen from A with probability k/N or from B with probability $(N - k)/N$. Then the ball is placed into one of the urns, where urn A is chosen with probability k/N or urn B is chosen with probability $(N - k)/N$. Determine the transition matrix of the Markov chain with states represented by the contents of A.

5. Consider a discrete time, periodic review inventory model and let ξ_n be the total demand in period n, and let X_n be the inventory quantity on hand at the end of period n. An (s, S) inventory policy is used: If the end of period stock is not greater than s, then a quantity is instantly procured to bring the level up to S. If the end of period stock exceeds s, then no replenishment takes place.
   (a) Suppose that $s = 1$, $S = 4$, and $X_0 = S = 4$. If the period demands turn out to be $\xi_1 = 2$, $\xi_2 = 3$, $\xi_3 = 4$, $\xi_4 = 0$, $\xi_5 = 2$, $\xi_6 = 1$, $\xi_7 = 2$, $\xi_8 = 2$, what are the end of period stock levels X_n for periods $n = 1$, 2, . . . , 8?
   (b) Suppose ξ_1, ξ_2, . . . are independent random variables where $\Pr\{\xi_n = 0\} = .1$, $\Pr\{\xi_n = 1\} = .3$, $\Pr\{\xi_n = 2\} = .3$, $\Pr\{\xi_n = 3\} = .2$ and $\Pr\{\xi_n = 4\} = .1$. Then X_0, X_1, . . . is a Markov chain. Determine P_{41} and P_{04}.

3.4 First Step Analysis

A surprising number of functionals on a Markov chain can be evaluated by a technique that we call *first step analysis*. This method proceeds by analyzing, or breaking down, the possibilities that can arise at the end of the first transition, and then invoking the law of total probability coupled with the Markov property to establish a characterizing relationship among the unknown variables. We first applied this technique in Theorem 3.1. In this section we develop a series of applications of the technique.

3.4.1 Simple First Step Analyses

Consider the Markov chain $\{X_n\}$ whose transition probability matrix is

$$\mathbf{P} = \begin{array}{c} \\ 0 \\ 1 \\ 2 \end{array} \begin{array}{ccc} 0 & 1 & 2 \\ \left\| \begin{array}{ccc} 1 & 0 & 0 \\ \alpha & \beta & \gamma \\ 0 & 0 & 1 \end{array} \right\|, \end{array}$$

where $\alpha > 0$, $\beta > 0$, $\gamma > 0$ and $\alpha + \beta + \gamma = 1$. If the Markov chain begins in state 1, it remains there for a random duration, and then proceeds either to state 0 or to state 2 where it is trapped or absorbed. That is, once in state 0 the process remains there forever after, as it also does in state 2. Two questions arise: In which state, 0 or 2, is the process ultimately trapped, and how long, on the average, does it take to reach one of these states? Both questions are easily answered by instituting a first step analysis.

We begin by more precisely defining the questions. Let

$$T = \min\{n \ge 0; X_n = 0 \text{ or } X_n = 2\}$$

be the time of absorption of the process. In terms of this random absorption time, the two questions ask us to find

$$u = \Pr\{X_T = 0 | X_0 = 1\}$$

and

$$v = E[T | X_0 = 1].$$

We proceed to institute a first step analysis, considering separately the three contingencies $X_1 = 0$, $X_1 = 1$ and $X_1 = 2$, with respective probabilities α, β, and γ. Consider $u = \Pr\{X_T = 0 | X_0 = 1\}$. If $X_1 = 0$, which occurs with probability α, then $T = 1$ and $X_T = 0$. If $X_1 = 2$, which occurs with probability γ, then again $T = 1$ but $X_T = 2$. Finally, if $X_1 = 1$, which occurs with probability β, then the process returns to state 1 and the problem repeats from the same state as before. In symbols, we claim

$$\Pr\{X_T = 0 | X_1 = 0\} = 1,$$

$$\Pr\{X_T = 0 | X_1 = 2\} = 0,$$

$$\Pr\{X_T = 0 | X_1 = 1\} = u,$$

which inserted into the law of total probability gives

$$u = \Pr\{X_T = 0 | X_0 = 1\}$$

$$= \sum_{k=0}^{2} \Pr\{X_T = 0 | X_0 = 1, X_1 = k\} \Pr\{X_1 = k | X_0 = 1\}$$

$$= \sum_{k=0}^{2} \Pr\{X_T = 0 | X_1 = k\} \Pr\{X_1 = k | X_0 = 1\}$$

(by the Markov property)

$$= 1(\alpha) + u(\beta) + 0(\gamma).$$

Thus we obtain the equation

$$u = \alpha + \beta u \tag{3.19}$$

which solves to give

$$u = \frac{\alpha}{1 - \beta} = \frac{\alpha}{\alpha + \gamma}.$$

Observe that this quantity is the conditional probability of a transition to 0, given that a transition to 0 or 2 occurred. That is, the answer makes sense.

We turn to determining the mean time to absorption, again analyzing the possibilities arising on the first step. The absorption time T is always at least 1. If either $X_1 = 0$ or $X_1 = 2$, then no further steps are required. If, on the other hand, $X_1 = 1$, then the process is back at its starting point, and, on the average, $v = E[T|X_0 = 1]$ *additional* steps are required for absorption. Weighing these contingencies by their respective probabilities, we obtain for $v = E[T|X_0 = 1]$,

$$\begin{aligned} v &= 1 + \alpha(0) + \beta(v) + \gamma(0) \\ &= 1 + \beta v \end{aligned} \tag{3.20}$$

which solves to give

$$v = \frac{1}{1 - \beta}.$$

In the example just studied, the reader is invited to verify that T has the geometric distribution in which

$$\Pr\{T > k|X_0 = 1\} = \beta^k \quad \text{for} \quad k = 0, 1, \ldots.$$

and therefore

$$E[T|X_0 = 1] = \sum_{k=0}^{\infty} \Pr\{T > k|X_0 = 1\} = \frac{1}{1 - \beta}.$$

That is, a direct calculation verifies the result of the first step analysis. Unfortunately, in more general Markov chains a direct calculation is rarely possible, and first step analysis provides the only solution technique.

A significant extension occurs when we move up to the four state Markov chain whose transition probability matrix is

$$\mathbf{P} = \begin{array}{c} \\ 0 \\ 1 \\ 2 \\ 3 \end{array} \begin{array}{cccc} 0 & 1 & 2 & 3 \\ \left\| \begin{array}{cccc} 1 & 0 & 0 & 0 \\ P_{10} & P_{11} & P_{12} & P_{13} \\ P_{20} & P_{21} & P_{22} & P_{23} \\ 0 & 0 & 0 & 1 \end{array} \right\| \end{array}.$$

Absorption now occurs in states 0 and 3, and states 1 and 2 are "transient." The probability of ultimate absorption in state 0, say, now depends on the transient state in which the process began. Accordingly, we must extend our notation to include the starting state. Let

$$T = \min\{n \geq 0; X_n = 0 \text{ or } X_n = 3\},$$
$$u_i = \Pr\{X_T = 0 | X_0 = i\} \qquad \text{for} \quad i = 1, 2,$$

and

$$v_i = E[T | X_0 = i] \qquad \text{for} \quad i = 1, 2.$$

We may extend the definitions for u_i and v_i in a consistent and common sense manner by prescribing $u_0 = 1$, $u_3 = 0$, and $v_0 = v_3 = 0$.

The first step analysis now requires us to consider the two possible starting states $X_0 = 1$ and $X_0 = 2$ separately. Considering $X_0 = 1$ and applying a first step analysis to $u_1 = \Pr\{X_T = 0 | X_0 = 1\}$, we obtain

$$u_1 = P_{10} + P_{11} u_1 + P_{12} u_2. \tag{3.21}$$

The three terms on the right correspond to the contingencies $X_1 = 0$, $X_1 = 1$ and $X_1 = 2$, respectively, with the conditional probabilities

$$\Pr\{X_T = 0 | X_1 = 0\} = 1,$$
$$\Pr\{X_T = 0 | X_1 = 1\} = u_1,$$

and

$$\Pr\{X_T = 0 | X_1 = 2\} = u_2.$$

The law of total probability then applies to give (3.21) just as it was used in obtaining (3.19). A similar equation is obtained for u_2:

$$u_2 = P_{20} + P_{21} u_1 + P_{22} u_2. \tag{3.22}$$

The two equations in u_1 and u_2 are now solved simultaneously. To give a numerical example, we will suppose

$$
\mathbf{P} = \begin{array}{c} \\ 0 \\ 1 \\ 2 \\ 3 \end{array}
\begin{array}{cccc}
0 & 1 & 2 & 3 \\
\left\|\begin{array}{cccc}
1 & 0 & 0 & 0 \\
.4 & .3 & .2 & .1 \\
.1 & .3 & .3 & .3 \\
0 & 0 & 0 & 1
\end{array}\right\|
\end{array}. \tag{3.23}
$$

The first step analysis equations (3.21) and (3.22) for u_1 and u_2 are

$$u_1 = .4 + .3\, u_1 + .2\, u_2$$
$$u_2 = .1 + .3\, u_1 + .3\, u_2$$

or

$$.7\, u_1 - .2\, u_2 = .4,$$
$$-.3\, u_1 + .7\, u_2 = .1.$$

The solution is $u_1 = \frac{30}{43}$ and $u_2 = \frac{19}{43}$. Note that one cannot, in general, solve for u_1 without bringing in u_2, and vice versa. The result $u_2 = \frac{19}{43}$ tells us that, once begun in state $X_0 = 2$, the Markov chain $\{X_n\}$ described by (3.23) will ultimately end up in state 0 with probability $u_2 = \frac{19}{43}$, and, alternatively, will be absorbed in state 3 with probability $1 - u_2 = \frac{24}{43}$.

The mean time to absorption also depends on the starting state. The first step analysis equations for $v_i = E[T|X_0 = i]$ are

$$v_1 = 1 + P_{11} v_1 + P_{12} v_2$$
$$v_2 = 1 + P_{21} v_1 + P_{22} v_2. \tag{3.24}$$

The right side of (3.24) asserts that at least one step is always taken. If the first move is to either $X_1 = 1$ or $X_1 = 2$, then additional steps are needed, and, on the average, these are v_1 and v_2 respectively. Weighing the contingencies $X_1 = 1$ and $X_1 = 2$ by their respective probabilities and summing according to the law of total probability results in (3.24).

For the transition matrix given in (3.23), the equations are

$$v_1 = 1 + .3\, v_1 + .2\, v_2$$
$$v_2 = 1 + .3\, v_1 + .3\, v_2,$$

and their solutions are $v_1 = \frac{90}{43}$ and $v_2 = \frac{100}{43}$. Again, v_1 cannot be obtained without also considering v_2, and vice versa. For a process that begins in state $X_0 = 2$, on the average $v_2 = \frac{100}{43} = 2.33$ steps will transpire prior to absorption.

To study the method in a more general context, let $\{X_n\}$ be a finite state Markov chain whose states are labeled $0, 1, \ldots, N$. Suppose that states 0, $1, \ldots, r - 1$ are *transient*[*] in that $P_{ij}^{(n)} \to 0$ as $n \to \infty$ for $0 \leq i, j < r$ while states $r, \ldots, N$ are *absorbing* ($P_{ii} = 1$ for $r \leq i \leq N$). The transition matrix has the form

$$\mathbf{P} = \left\| \begin{matrix} \mathbf{Q} & \mathbf{R} \\ \mathbf{O} & \mathbf{I} \end{matrix} \right\| \tag{3.25}$$

where $\mathbf{O}$ is an $(N - r + 1) \times r$ matrix all of whose entries are zero, $\mathbf{I}$ is an $(N - r + 1) \times (N - r + 1)$ identity matrix, and $Q_{ij} = P_{ij}$ for $0 \leq i, j < r$.

Started at one of the transient states $X_0 = i$, where $0 \leq i < r$, such a process will remain in the transient states for some random duration, but ultimately the process gets trapped in one of the absorbing states $i = r, \ldots, N$. Functionals of importance are the mean duration until absorption and the probability distribution over the states in which absorption takes place.

Let us consider the second question first and fix a state k among the absorbing states ($r \leq k \leq N$). The probability of ultimate absorption in state

---

[*]The definition of a transient state is different for an infinite state Markov chain. See Section 4.3.

k, as opposed to some other absorbing state, depends on the initial state $X_0 = i$. Let $U_{ik} = u_i$ denote this probability, where we suppress the target state k in the notation for typographical convenience.

We begin a first step analysis by enumerating the possibilities in the first transition. Starting from state i, then with probability P_{ik} the process immediately goes to state k, thereafter to remain, and this is the first possibility considered. Alternatively the process could move on its first step to an absorbing state $j \neq k$, where $r \leq j \leq N$, in which case ultimate absorption in state k is precluded. Finally the process could move to a transient state $j < r$. Because of the Markov property, once in state j, then the probability of ultimate absorption in state k is $u_j = U_{jk}$ by definition. Weighing the enumerated possibilities by their respective probabilities via the law of total probability, we obtain the relation

$$u_i = \Pr\{\text{Absorption in } k | X_0 = i\}$$

$$= \sum_{j=0}^{N} \Pr\{\text{Absorption in } k | X_0 = i, X_1 = j\} P_{ij}$$

$$= P_{ik} + \sum_{\substack{j=r \\ j \neq k}}^{N} P_{ij} \times 0 + \sum_{j=0}^{r-1} P_{ij} u_j.$$

To summarize, for a fixed absorbing state k, the quantities

$$u_i = U_{ik} = \Pr\{\text{Absorption in } k | X_0 = i\} \quad \text{for} \quad 0 \leq i < r$$

satisfy the nonhomogeneous system of linear equations

$$U_{ik} = P_{ik} + \sum_{j=0}^{r-1} P_{ij} U_{jk}, \quad i = 0, 1, \ldots, r - 1. \tag{3.26}$$

Example *A Maze* A white rat is put into the maze shown:

In the absence of learning, one might hypothesize that the rat would move through the maze at random, i.e., if there are k ways to leave a compartment, then the rat would choose each of these with probability $1/k$. Assume that the rat makes one change to some adjacent compartment at each unit of time and let X_n denote the compartment occupied at stage n. We suppose that compartment 7 contains food and compartment 8 contains an electrical shocking mechanism, and we ask the probability that the rat, moving at random, encounters the food before being shocked. The appropriate transition probability matrix is

	0	1	2	3	4	5	6	7	8
0		$\frac{1}{2}$	$\frac{1}{2}$						
1	$\frac{1}{3}$			$\frac{1}{3}$				$\frac{1}{3}$	
2	$\frac{1}{3}$			$\frac{1}{3}$					$\frac{1}{3}$
3		$\frac{1}{4}$	$\frac{1}{4}$		$\frac{1}{4}$	$\frac{1}{4}$			
4				$\frac{1}{3}$			$\frac{1}{3}$	$\frac{1}{3}$	
5				$\frac{1}{3}$			$\frac{1}{3}$		$\frac{1}{3}$
6					$\frac{1}{2}$	$\frac{1}{2}$			
7								1	
8									1

$$\mathbf{P} = $$

Let $u_i = u_i(7)$ denote the probability of absorption in the food compartment 7, given that the rat is dropped initially in compartment i. Then equations (3.26) become, in this particular instance,

$$u_0 = \tfrac{1}{2}\,u_1 + \tfrac{1}{2}\,u_2$$
$$u_1 = \tfrac{1}{3} + \tfrac{1}{3}\,u_0 \qquad\qquad + \tfrac{1}{3}\,u_3$$
$$u_2 = \qquad \tfrac{1}{3}\,u_0 \qquad\qquad + \tfrac{1}{3}\,u_3$$
$$u_3 = \tfrac{1}{4}\,u_1 + \tfrac{1}{4}\,u_2 \qquad\qquad + \tfrac{1}{4}\,u_4 + \tfrac{1}{4}\,u_5$$
$$u_4 = \tfrac{1}{3} \qquad\qquad + \tfrac{1}{3}\,u_3 \qquad\qquad + \tfrac{1}{3}\,u_6$$
$$u_5 = \qquad\qquad \tfrac{1}{3}\,u_3 \qquad\qquad + \tfrac{1}{3}\,u_6$$
$$u_6 = \qquad\qquad \tfrac{1}{2}\,u_4 + \tfrac{1}{2}\,u_5$$

Turning to the solution, we see that the symmetry of the maze implies that $u_0 = u_6$, $u_2 = u_5$ and $u_1 = u_4$. We also must have $u_3 = \tfrac{1}{2}$. With these simplifications the equations for u_0, u_1, and u_2 become

$$u_0 = \qquad \tfrac{1}{2}\,u_1 + \tfrac{1}{2}\,u_2$$
$$u_1 = \tfrac{1}{2} + \tfrac{1}{3}\,u_0$$
$$u_2 = \tfrac{1}{6} + \tfrac{1}{3}\,u_0,$$

and the natural substitutions give $u_0 = \tfrac{1}{2}(\tfrac{1}{2} + \tfrac{1}{3}\,u_0) + \tfrac{1}{2}(\tfrac{1}{6} + \tfrac{1}{3}\,u_0)$ or $u_0 = \tfrac{1}{2}$, $u_1 = \tfrac{2}{3}$, and $u_2 = \tfrac{1}{3}$.

One might compare these theoretical values under random moves with actual observations as an indication of whether or not learning is taking place.

We turn to a more general form of the first question by introducing the random absorption time T. Formally we define

$$T = \min\{n \geq 0;\ X_n \geq r\}.$$

Let us suppose that associated with each transient state i is a rate $g(i)$ and that we wish to determine the mean total rate that is accumulated up to absorption. Let w_i be this mean total amount, where the subscript i denotes the starting position $X_0 = i$. To be precise, let

$$w_i = E[\sum_{n=0}^{T-1} g(X_n)|X_0 = i].$$

The choice $g(i) = 1$ for all i yields $\Sigma_{n=0}^{T-1} g(X_n) = \Sigma_{n=0}^{T-1} 1 = T$, and then w_i is identical to $v_i \equiv E[T|X_0 = i]$, the mean time until absorption. For a transient state k, the choice

$$g(i) = \begin{cases} 1 & \text{if } i = k \\ 0 & \text{if } i \neq k \end{cases}$$

gives $w_i = W_{ik}$, the mean number of visits to state k $(0 \leq k < r)$ prior to absorption.

We again proceed via a first step analysis. The sum $\Sigma_{n=0}^{T-1} g(X_n)$ always includes the first term $g(X_0) = g(i)$. In addition, if a transition is made from i to a transient state j, then the sum includes future terms as well. By invoking the Markov property we deduce that this future sum proceeding from state j has an expected value equal to w_j. Weighing this by the transition probability P_{ij} and then summing all contributions in accordance with the law of total probabilities, we obtain the joint relations

$$w_i = g(i) + \sum_{j=0}^{r-1} P_{ij}w_j \quad \text{for} \quad i = 0, \ldots, r - 1. \qquad (3.27)$$

The special case in which $g(i) = 1$ for all i determines $v_i = E[T|X_0 = i]$ as solving

$$v_i = 1 + \sum_{j=0}^{r-1} P_{ij}v_j \quad \text{for} \quad i = 0, 1, \ldots, r - 1. \qquad (3.28)$$

The case in which

$$g(i) = \delta_{ik} = \begin{cases} 1 & \text{if } i = k \\ 0 & \text{if } i \neq k \end{cases}$$

determines W_{ik}, the number of visits to state k prior to absorption starting from state i, as solving

$$W_{ik} = \delta_{ik} + \sum_{j=0}^{r-1} P_{ij}W_{jk} \quad \text{for} \quad i = 0, 1, \ldots, r - 1. \qquad (3.29)$$

Example *A Model of Fecundity*   Changes in sociological patterns such as increase in age at marriage, more remarriages after widowhood, increased divorce rates, have profound effects on overall population growth rates. Here we attempt to model the life span of a female in a population in order to provide a framework for analyzing the effect of social changes on average fecundity.

The general model we propose has a large number of states delimiting the age and status of a typical female in the population. For example, we begin with the twelve age groups 0–4 years, 5–9 years, . . . , 50–54 years, 55 years and over. In addition, each of these age groups might be further subdivided according to marital status: single, married, separated or divorced, or widowed, and might also be subdivided according to the number of children. Each female would begin in the (0–4, single) category and end in a distinguished state Δ corresponding to death or emigration from the population. However, the duration spent in the various other states might differ between different females. Of interest is the mean duration spent in the categories of maximum fertility, or more generally a mean sum of durations weighted by appropriate fecundity rates.

When there are a large number of states in the model, as just sketched, the relevant calculations require a computer. We turn to a simpler model which, while less realistic, will serve to illustrate the concepts and approach. We introduce the states:

E_0: Prepuberty E_3: Divorced
E_1: Single E_4: Widowed
E_2: Married E_5: Δ

and we are interested in the mean duration spent in state E_2: Married, since this corresponds to the state of maximum fecundity. To illustrate the computations, we will suppose the transition probability matrix is

$$\mathbf{P} = \begin{array}{c c} & \begin{array}{c c c c c c} E_0 & E_1 & E_2 & E_3 & E_4 & E_5 \end{array} \\ \begin{array}{c} E_0 \\ E_1 \\ E_2 \\ E_3 \\ E_4 \\ E_5 \end{array} & \left\| \begin{array}{c c c c c c} 0 & .9 & 0 & 0 & 0 & .1 \\ 0 & .5 & .4 & 0 & 0 & .1 \\ 0 & 0 & .6 & .2 & .1 & .1 \\ 0 & 0 & .4 & .5 & 0 & .1 \\ 0 & 0 & .4 & 0 & .5 & .1 \\ 0 & 0 & 0 & 0 & 0 & 1.0 \end{array} \right\| \end{array}$$

In practice, such a matrix would be estimated from demographic data.

Every person begins in state E_0 and ends in state E_5, but a variety of intervening states may be visited. We wish to determine the mean duration spent in state E_2: Married. The powerful approach of *first step analysis* begins by considering the slightly more general problem in which the initial state is varied. Let $w_i = W_{i2}$ be the mean duration in state E_2 given the initial state $X_0 = E_i$ for $i = 0, 1, \ldots, 5$. We are interested in w_0, the mean duration corresponding to the initial state E_0.

First step analysis breaks down, or analyzes, the possibilities arising in the first transition, and using the Markov property, an equation that relates $w_0, \ldots, w_5$ results.

We begin by considering w_0. From state E_0, a transition to one of the states E_1 or E_5 occurs, and the mean duration spent in E_2 starting from E_0 must be the appropriately weighted average of w_1 and w_5. That is

$$w_0 = .9w_1 + .1w_5.$$

Proceeding in a similar manner we obtain

$$w_1 = .5w_1 + .4w_2 + .1w_5.$$

The situation changes when the process begins in state E_2, because in counting the mean duration spent in E_2, we must count this initial visit plus any subsequent visits that may occur. Thus for E_2 we have

$$w_2 = 1 + .6w_2 + .2w_3 + .1w_4 + .1w_5.$$

The other states give us

$$w_3 = .4w_2 + .5w_3 + .1w_5$$
$$w_4 = .4w_2 + .5w_4 + .1w_5$$
$$w_5 = w_5.$$

Since state E_5 corresponds to death, it is clear that we must have $w_5 = 0$. With this prescription, the reduced equations become, after elementary simplification,

$$-1.0w_0 + .9w_1 \qquad\qquad\qquad = 0$$
$$- .5w_1 + .4w_2 \qquad\qquad = 0$$
$$- .4w_2 + .2w_3 + .1w_4 = -1$$
$$.4w_2 - .5w_3 \qquad = 0$$
$$.4w_2 \qquad - .5w_4 = 0.$$

The unique solution is

$$w_0 = 4.5, \quad w_1 = 5.00, \quad w_2 = 6.25, \quad w_3 = w_4 = 5.00.$$

Each female, on the average, spends $w_0 = W_{02} = 4.5$ periods in the childbearing state E_2 during her lifetime.

Problems 3.4

1. Find the mean time to reach state 3 starting from state 0 for the Markov chain whose transition probability matrix is

$$\mathbf{P} = \begin{array}{c c} & \begin{array}{cccc} 0 & 1 & 2 & 3 \end{array} \\ \begin{array}{c} 0 \\ 1 \\ 2 \\ 3 \end{array} & \left\| \begin{array}{cccc} .4 & .3 & .2 & .1 \\ 0 & .7 & .2 & .1 \\ 0 & 0 & .9 & .1 \\ 0 & 0 & 0 & 1 \end{array} \right\| \end{array}.$$

2. A white rat is put into compartment 4 of the maze shown here:

```
 ┌──────┬──────┬──────┐
 │      │      │  3   │
 │  1   │  2   │ food │
 │      │      │      │
 ├──  ──┼──  ──┤      │
 │      │      │      │
 │  4   │  5   │  6   │
 │      │      │      │
 ├──  ──┴──────┴──────┘
 │      │
 │  7   │
 │shock │
 └──────┘
```

He moves through the compartments at random, i.e., if there are k ways to leave a compartment, he chooses each of these with probability $1/k$. What is the probability that the rat finds the food in compartment 3 before feeling the electric shock in compartment 7?

3. A coin is tossed repeatedly until two successive heads appear. Find the mean number of tosses required.

   Hint: Let X_n be the cumulative number of successive heads. The state space is 0, 1, 2 and the transition probability matrix is

$$
\mathbf{P} = \begin{array}{c} 0 \\ 1 \\ 2 \end{array}
\begin{array}{ccc}
0 & 1 & 2 \\
\left\|\begin{array}{ccc}
\frac{1}{2} & \frac{1}{2} & 0 \\
\frac{1}{2} & 0 & \frac{1}{2} \\
0 & 0 & 1
\end{array}\right\|.
\end{array}
$$

   Determine the mean time to reach state 2 starting from state 0 by invoking a first step analysis.

4. A coin is tossed repeatedly until either two successive heads appear or two successive tails appear. Suppose the first coin toss results in a head. Find the probability that the game ends with two successive tails.

5. Consider the Markov chain whose transition matrix is

$$
\mathbf{P} = \begin{array}{c} 0 \\ 1 \\ 2 \\ 3 \\ 4 \end{array}
\begin{array}{ccccc}
0 & 1 & 2 & 3 & 4 \\
\left\|\begin{array}{ccccc}
q & p & 0 & 0 & 0 \\
q & 0 & p & 0 & 0 \\
q & 0 & 0 & p & 0 \\
q & 0 & 0 & 0 & p \\
0 & 0 & 0 & 0 & 1
\end{array}\right\|
\end{array}
$$

   where $p + q = 1$. Determine the mean time to reach state 4 starting from state 0. That is, find $E[T|X_0 = 0]$ where $T = \min\{n \geq 0: X_n = 4\}$.

   Hint: Let $v_i = E[T|X_0 = i]$ for $i = 0, 1, \ldots, 4$. Establish equations for $v_0, v_1, \ldots, v_4$ by using a first step analysis and the boundary condition $v_4 = 0$. Then solve for v_0.

6. Let X_n be a Markov chain with transition probabilities P_{ij}. We are given a "discount factor" β with $0 < \beta < 1$ and a cost function $c(i)$, and we wish to determine the total expected discounted cost starting from state i, defined by

$$h_i = E\left[\sum_{n=0}^{\infty} \beta^n c(X_n) | X_0 = i\right].$$

Using a first step analysis, show that h_i satisfies the linear equation set

$$h_i = c(i) + \beta \sum_j P_{ij} h_j \qquad \text{for all states } i.$$

3.5 Some Special Markov Chains

We introduce several particular Markov chains that arise in a variety of applications.

3.5.1 The Two State Markov Chain

Let

$$\mathbf{P} = \begin{array}{c} \\ 0 \\ 1 \end{array}\begin{array}{c} 0 \qquad\qquad 1 \\ \left\| \begin{array}{cc} 1 - a & a \\ b & 1 - b \end{array} \right\| \end{array} \qquad \text{where} \quad 0 < a, b < 1, \qquad (3.30)$$

be the transition matrix of a two state Markov chain.

When $a = 1 - b$ so that the rows of $\mathbf{P}$ are the same, then the states $X_1, X_2, \ldots$ are independent identically distributed random variables with $\Pr\{X_n = 0\} = b$ and $\Pr\{X_n = 1\} = a$. When $a \neq 1 - b$, the probability distribution for X_n varies depending on the outcome X_{n-1} at the previous stage.

For the two state Markov chain, it is readily verified by induction that the n-step transition matrix is given by

$$\mathbf{P}^n = \frac{1}{a+b}\left\| \begin{array}{cc} b & a \\ b & a \end{array} \right\| + \frac{(1 - a - b)^n}{a+b}\left\| \begin{array}{cc} a & -a \\ -b & b \end{array} \right\|. \qquad (3.31)$$

To verify this general formula, introduce the abbreviations

$$\mathbf{A} = \left\| \begin{array}{cc} b & a \\ b & a \end{array} \right\| \quad \text{and} \quad \mathbf{B} = \left\| \begin{array}{cc} a & -a \\ -b & b \end{array} \right\|$$

so that (3.31) can be written

$$\mathbf{P}^n = (a+b)^{-1}[\mathbf{A} + (1 - a - b)^n \mathbf{B}].$$

Next, check the multiplications

$$\mathbf{AP} = \begin{Vmatrix} b & a \\ b & a \end{Vmatrix} \times \begin{Vmatrix} 1-a & a \\ b & 1-b \end{Vmatrix} = \begin{Vmatrix} b & a \\ b & a \end{Vmatrix} = \mathbf{A}$$

and

$$\mathbf{BP} = \begin{Vmatrix} a & -a \\ -b & b \end{Vmatrix} \times \begin{Vmatrix} 1-a & a \\ b & 1-b \end{Vmatrix}$$

$$= \begin{Vmatrix} a - a^2 - ab & a^2 - a + ab \\ -b + ab + b^2 & -ab + b - b^2 \end{Vmatrix} = (1 - a - b)\mathbf{B}.$$

Now (3.31) is easily seen to be true when $n = 1$, since then

$$\mathbf{P}^1 = \frac{1}{a+b} \begin{Vmatrix} b & a \\ b & a \end{Vmatrix} + \frac{(1-a-b)}{a+b} \begin{Vmatrix} a & -a \\ -b & b \end{Vmatrix}$$

$$= \frac{1}{a+b} \begin{Vmatrix} b + a - a^2 - ab & a - a + a^2 + ab \\ b - b + ab + b^2 & a + b - ab - b^2 \end{Vmatrix}$$

$$= \begin{Vmatrix} 1-a & a \\ b & 1-b \end{Vmatrix} = \mathbf{P}.$$

To complete an induction proof, assume the formula is true for n. Then

$$\begin{aligned} \mathbf{P}^n\mathbf{P} &= (a+b)^{-1}[\mathbf{A} + (1-a-b)^n\mathbf{B}]\mathbf{P} \\ &= (a+b)^{-1}[\mathbf{AP} + (1-a-b)^n\mathbf{BP}] \\ &= (a+b)^{-1}[\mathbf{A} + (1-a-b)^{n+1}\mathbf{B}] = \mathbf{P}^{n+1}. \end{aligned}$$

We have verified that the formula holds for $n + 1$. It therefore is established for all n.

Note that $|1 - a - b| < 1$ when $0 < a, b < 1$, and thus $|1 - a - b|^n \to 0$ as $n \to \infty$ and

$$\lim_{n \to \infty} \mathbf{P}^n = \begin{Vmatrix} \dfrac{b}{a+b} & \dfrac{a}{a+b} \\ \dfrac{b}{a+b} & \dfrac{a}{a+b} \end{Vmatrix}. \tag{3.32}$$

This tells us that such a system, in the long run, will be in state 0 with probability $b/(a + b)$ and in state 1 with probability $a/(a + b)$, irrespective of the initial state in which the system started.

For a numerical example, suppose that the items produced by a certain worker are graded as defective or not, and that due to trends in raw material quality, whether or not a particular item is defective depends in part on whether or not the previous item was defective. Let X_n denote the quality of the nth item with $X_n = 0$ meaning "Good" and $X_n = 1$ meaning

"Defective." Suppose that $\{X_n\}$ evolves as a Markov chain whose transition matrix is

$$\mathbf{P} = \begin{array}{c} \\ 0 \\ 1 \end{array} \begin{array}{cc} 0 & 1 \\ \left\| \begin{array}{cc} .99 & .01 \\ .12 & .88 \end{array} \right\| \end{array}.$$

Defective items would tend to appear in bunches in the output of such a system.

In the long run, the probability that an item produced by this system is defective is given by $a/(a + b) = .01/(.01 + .12) = .077$.

3.5.2 Markov Chains Associated with iid Random Variables

Let ξ denote a discrete valued random variable whose possible values are the nonnegative integers and where $\Pr\{\xi = i\} = a_i \geq 0$ for $i = 0, 1, \ldots$ and $\sum_{i=0}^{\infty} a_i = 1$. Let $\xi_1, \xi_2, \ldots, \xi_n, \ldots$ represent independent observations of ξ.

We shall now describe three different Markov chains connected with the sequence $\xi_1, \xi_2, \ldots$. In each case the state space of the process coincides with the set of nonnegative integers.

Example *Independent Random Variables* Consider the process X_n, $n = 0, 1, 2, \ldots$, defined by $X_n = \xi_n$, ($X_0 = \xi_0$ prescribed). Its Markov matrix has the form

$$\mathbf{P} = \left\| \begin{array}{cccc} a_0 & a_1 & a_2 & \cdots \\ a_0 & a_1 & a_2 & \cdots \\ a_0 & a_1 & a_2 & \cdots \\ \cdot & \cdot & \cdot & \\ \cdot & \cdot & \cdot & \\ \cdot & \cdot & \cdot & \end{array} \right\|. \tag{3.33}$$

Each row being identical plainly expresses the fact that the random variable X_{n+1} is independent of X_n.

Example *Successive Maxima* The partial maxima of $\xi_1, \xi_2, \ldots$ define a second important Markov chain. Let

$$\theta_n = \max\{\xi_1, \ldots, \xi_n\} \quad \text{for} \quad n = 1, 2, \ldots,$$

with $\theta_0 = 0$. The process defined by $X_n = \theta_n$ is readily seen to be a Markov chain, and the relation $X_{n+1} = \max\{X_n, \xi_{n+1}\}$ allows the transition probabilities to be computed to be

$$\mathbf{P} = \left\| \begin{array}{ccccc} A_0 & a_1 & a_2 & a_3 & \cdots \\ 0 & A_1 & a_2 & a_3 & \cdots \\ 0 & 0 & A_2 & a_3 & \cdots \\ 0 & 0 & 0 & A_3 & \cdots \\ \cdot & \cdot & \cdot & \cdot & \\ \cdot & \cdot & \cdot & \cdot & \\ \cdot & \cdot & \cdot & \cdot & \end{array} \right\| \qquad (3.34)$$

where $A_k = a_0 + \ldots + a_k$ for $k = 0, 1, \ldots$.

Suppose $\xi_1, \xi_2, \ldots$ represent successive bids on a certain asset that is offered for sale. Then $X_n = \max\{\xi_1, \ldots, \xi_n\}$ is the maximum that is bid up to stage n. Suppose that the bid that is accepted is the first bid that equals or exceeds a prescribed level M. The time of sale is the random variable $T = \min\{n \geq 1; X_n \geq M\}$. A first step analysis shows that the mean $\mu = E[T]$ satisfies

$$\mu = 1 + \mu \Pr\{\xi_1 < M\} \qquad (3.35)$$

or $\mu = 1/\Pr\{\xi_1 \geq M\} = 1/(a_M + a_{M+1} + \ldots)$. The first step analysis invoked in establishing (3.35) considers the two possibilities $\{\xi_1 < M\}$ and $\{\xi_1 \geq M\}$. With this breakdown, the law of total probabilities justifies the sum

$$E[T] = E[T|\xi_1 \geq M]\Pr\{\xi_1 \geq M\} + E[T|\xi_1 < M]\Pr\{\xi_1 < M\}. \quad (3.36)$$

Clearly $E[T|\xi_1 \geq M] = 1$, since no further bids are examined in this case. On the other hand, when $\xi_1 < M$ we have the first bid, which was not accepted, plus some future bids. The future bids $\xi_2, \xi_3, \ldots$ have the same probabilistic properties as in the original problem, and they are examined until the first acceptable bid appears. This reasoning leads to $E[T|\xi_1 < M] = 1 + \mu$. Substitution into (3.36) then yields (3.35) as follows:

$$\begin{aligned} E[T] &= 1 \times \Pr\{\xi_1 \geq M\} + (1 + \mu)\Pr\{\xi_1 < M\} \\ &= 1 + \mu \Pr\{\xi_1 < M\}. \end{aligned}$$

To restate the argument somewhat differently, one always examines the first bid ξ_1. If $\xi_1 < M$, then further bids are examined in a future that is probabilistically similar to the original problem. That is, when $\xi_1 < M$, then on the average μ bids in addition to ξ_1 must be examined before an acceptable bid appears. Equation (3.35) results.

Example *Partial Sums*   Another important Markov chain arises from consideration of the successive partial sums η_n of the ξ_i, i.e.,

$$\eta_n = \xi_1 + \cdots + \xi_n, \qquad n = 1, 2, \ldots$$

and, by definition, $\eta_0 = 0$. The process $X_n = \eta_n$ is readily seen to be a Markov chain via

$$\Pr\{X_{n+1} = j | X_1 = i_1, \ldots, X_{n-1} = i_{n-1}, X_n = i\}$$
$$= \Pr\{\xi_{n+1} = j - i | \xi_1 = i_1, \xi_2 = i_2 - i_1, \ldots, \xi_n = i - i_{n-1}\}$$
$$= \Pr\{\xi_{n+1} = j - i\} \qquad \text{(independence of } \xi_1, \xi_2, \ldots)$$
$$= \Pr\{X_{n+1} = j | X_n = i\}.$$

The transition probability matrix is determined by

$$\Pr\{X_{n+1} = j | X_n = i\} = \Pr\{\xi_1 + \cdots + \xi_{n+1} = j | \xi_1 + \cdots + \xi_n = i\}$$
$$= \Pr\{\xi_{n+1} = j - i\}$$
$$= \begin{cases} a_{j-i} & \text{for } j \geq i, \\ \\ 0 & \text{for } j < i, \end{cases}$$

where we have used the independence of the ξ_i.

Schematically, we have

$$\mathbf{P} = \begin{Vmatrix} a_0 & a_1 & a_2 & a_3 & \cdots \\ 0 & a_0 & a_1 & a_2 & \cdots \\ 0 & 0 & a_0 & a_1 & \cdots \\ \cdot & \cdot & \cdot & \cdot \\ \cdot & \cdot & \cdot & \cdot \\ \cdot & \cdot & \cdot & \cdot \end{Vmatrix}. \tag{3.37}$$

If the possible values of the random variable ξ are permitted to be the positive and negative integers, then the possible values of η_n for each n will be contained among the totality of all integers. Instead of labeling the states conventionally by means of the nonnegative integers, it is more convenient to identify the state space with the totality of integers, since the transition probability matrix will then appear in a more symmetric form. The state space consists then of the values . . . $-2, -1, 0, 1, 2, \ldots$. The transition probability matrix becomes

$$\mathbf{P} = \begin{Vmatrix} \cdot & \cdot & \cdot & \cdot & \cdot \\ \cdot & \cdot & \cdot & \cdot & \cdot \\ \cdots & a_{-1} & a_0 & a_1 & a_2 & a_3 & \cdots \\ \cdots & a_{-2} & a_{-1} & a_0 & a_1 & a_2 & \cdots \\ \cdots & a_{-3} & a_{-2} & a_{-1} & a_0 & a_1 & \cdots \\ \cdot & \cdot & \cdot & \cdot & \cdot \\ \cdot & \cdot & \cdot & \cdot & \cdot \end{Vmatrix},$$

where $\Pr\{\xi = k\} = a_k$ for $k = 0, \pm 1, \pm 2, \ldots$, and $a_k \geq 0$, $\sum_{k=-\infty}^{+\infty} a_k = 1$.

3.5.3 One-Dimensional Random Walks

When we discuss random walks, it is an aid to intuition to speak about the state of the system as the position of a moving "particle."

A one-dimensional random walk is a Markov chain whose state space is a finite or infinite subset $a, a + 1, \ldots, b$ of the integers, in which the particle, if it is in state i, can in a single transition either stay in i or move to one of the neighboring states $i - 1, i + 1$. If the state space is taken as the nonnegative integers, the transition matrix of a random walk has the form

$$
\mathbf{P} =
\begin{matrix}
 & 0 & 1 & 2 & i-1 & i & i+1 \\
0 & r_0 & p_0 & 0 \cdots & 0 & \cdots & \\
1 & q_1 & r_1 & p_1 \cdots & 0 & \cdots & \\
2 & 0 & q_2 & r_2 \cdots & 0 & \cdots & \\
\vdots & & & & & \\
i & & & 0 & q_i & r_i & p_i & 0 \\
\end{matrix}
, \quad (3.38)
$$

where $p_i > 0$, $q_i > 0$, $r_i \geq 0$, and $q_i + r_i + p_i = 1$, $i = 1, 2, \ldots$ $(i \geq 1)$, $p_0 \geq 0$, $r_0 \geq 0$, $r_0 + p_0 = 1$. Specifically, if $X_n = i$ then, for $i \geq 1$,

$$\Pr\{X_{n+1} = i + 1 | X_n = i\} = p_i, \qquad \Pr\{X_{n+1} = i - 1 | X_n = i\} = q_i,$$

and

$$\Pr\{X_{n+1} = i | X_n = i\} = r_i,$$

with the obvious modifications holding for $i = 0$.

The designation "random walk" seems apt since a realization of the process describes the path of a person (suitably intoxicated) moving randomly one step forward or backward.

The fortune of a player engaged in a series of contests is often depicted by a random walk process. Specifically, suppose an individual (player A) with fortune k plays a game against an infinitely rich adversary and has probability p_k of winning one unit and probability $q_k = 1 - p_k$ $(k \geq 1)$ of losing one unit in the next contest (the choice of the contest at each stage may depend on his fortune), and $r_0 = 1$. The process X_n, where X_n represents his fortune after n contests, is clearly a random walk. Note that once the state 0 is reached (i.e., player A is wiped out), the process remains in that state. The event of reaching state $k = 0$ is commonly known as the "gambler's ruin."

If the adversary, player B, also starts with a limited fortune l and player A has an initial fortune k $(k + l = N)$, then we may again consider the Markov chain process X_n representing player A's fortune. However, the states of the process are now restricted to the values $0, 1, 2, \ldots, N$. At any trial, $N - X_n$ is interpreted as player B's fortune. If we allow the possibility of neither player winning in a contest, the transition probability matrix takes the form

$$
\mathbf{P} =
\begin{matrix}
 & \begin{matrix} 0 & \ 1 & \ 2 & \ 3 & & N \end{matrix} \\
\begin{matrix} 0 \\ 1 \\ 2 \\ \\ \\ \\ N \end{matrix} &
\left\|
\begin{matrix}
1 & 0 & 0 & 0 & \cdot\ \cdot\ \cdot & \\
q_1 & r_1 & p_1 & 0 & \cdot\ \cdot\ \cdot & \\
0 & q_2 & r_2 & p_2 & \cdot\ \cdot\ \cdot & \\
 & & \cdot & & & \\
 & & \cdot & & & \\
 & & & q_{N-1} & r_{N-1} & p_{N-1} \\
0 & \cdot\ \cdot\ \cdot & & \cdot\ \cdot\ \cdot & 0 & 0 & 1
\end{matrix}
\right\|
\end{matrix}
\quad . \quad (3.39)
$$

Again $p_i(q_i)$, $i = 1, 2, \ldots, N - 1$, denotes the probability of player A's fortune increasing (decreasing) by 1 at the subsequent trial when his present fortune is i, and r_i may be interpreted as the probability of a draw. Note that, in accordance with the Markov chain given in (3.39), when player A's fortune (the state of the process) reaches 0 or N it remains in this same state forever. We say player A is ruined when the state of the process reaches 0 and player B is ruined when the state of the process reaches N.

The probability of gambler's ruin (for Player A) is derived in the next section by solving a first step analysis. Some more complex functionals on random walk processes are also derived there.

The random walk corresponding to $p_k = p$, $q_k = 1 - p = q$ for all $k \geq 1$ and $r_0 = 1$ describes the situation of identical contests. There is a definite advantage to Player A in each individual trial if $p > q$, and conversely, an advantage to Player B if $p < q$. A "fair" contest corresponds to $p = q = \frac{1}{2}$. Suppose the total fortunes of both players is N. Then the corresponding walk, where X_n is Player A's fortune at stage n, has the transition probability matrix

$$
\mathbf{P} =
\begin{matrix}
 & \begin{matrix} 0 & 1 & 2 & 3 & & N-1 & N \end{matrix} \\
\begin{matrix} 0 \\ 1 \\ 2 \\ \\ \\ \\ N-1 \\ N \end{matrix} &
\left\|
\begin{matrix}
1 & 0 & 0 & 0 & \cdot\ \cdot\ \cdot & 0 & 0 \\
q & 0 & p & 0 & \cdot\ \cdot\ \cdot & 0 & 0 \\
0 & q & 0 & p & \cdot\ \cdot\ \cdot & 0 & 0 \\
\cdot & \cdot & \cdot & \cdot & \cdot & \cdot & \cdot \\
\cdot & \cdot & \cdot & \cdot & \cdot & \cdot & \cdot \\
0 & 0 & 0 & 0 & \cdot\ \cdot\ \cdot & 0 & p \\
0 & 0 & 0 & 0 & \cdot\ \cdot\ \cdot & 0 & 1
\end{matrix}
\right\|
\end{matrix}
\quad . \quad (3.40)
$$

Let $u_i = U_{i0}$ be the probability of gambler's ruin starting with the initial fortune i. Then u_i is the probability that the random walk reaches state 0 before reaching state N, starting from $X_0 = i$. The first step analysis of Section 3.4, as used in deriving Equation (3.26), shows that these ruin probabilities satisfy

$$u_i = pu_{i+1} + qu_{i-1} \quad \text{for} \quad i = 1, \ldots, N-1 \qquad (3.41)$$

together with the obvious boundary conditions

$$u_0 = 1 \quad \text{and} \quad u_N = 0.$$

These equations are solved in the next section following a straightforward but arduous method. There it is shown that the gambler's ruin probabilities corresponding to the transition probability matrix given in (3.40) are

$$
u_i = \Pr\{X_n \text{ reaches state } 0 \text{ before state } N | X_0 = i\}
$$

$$
= \begin{cases}
\dfrac{N-i}{N} & \text{when} \quad p = q = \dfrac{1}{2}, \\[3mm]
\dfrac{(q/p)^i - (q/p)^N}{1 - (q/p)^N} & \text{when} \quad p \neq q.
\end{cases} \qquad (3.42)
$$

The ruin probabilities u_i given by (3.42) have the following interpretation. In a game in which Player A begins with an initial fortune of i units, and Player B begins with $N - i$ units, then the probability that Player A loses all his money before Player B goes broke is given by u_i where p is the probability that Player A wins in a single contest. If Player B is infinitely rich ($N \to \infty$) then passing to the limit in (3.42) and using $(q/p)^N \to \infty$ as $N \to \infty$ if $p < q$ while $(q/p)^N \to 0$ if $p > q$, we see that the ruin probabilities become

$$
u_i = \begin{cases}
1 & \text{if} \quad p \leq q \\[3mm]
\left(\dfrac{q}{p}\right)^i & \text{if} \quad p > q.
\end{cases} \qquad (3.43)
$$

(In passing to the limit, the case $p = q = \frac{1}{2}$ must be treated separately.) We see that ruin is certain ($u_i = 1$) against an infinitely rich adversary when the game is unfavorable ($p < q$), and even when the game is fair ($p = q$). In a favorable game ($p > q$), starting with initial fortune i, then ruin occurs (Player A goes broke) with probability $(q/p)^i$. This ruin probability decreases as the initial fortune i increases. In a favorable game against an infinitely rich opponent, with probability $1 - (q/p)^i$ Player A's fortune increases, in the long run, without limit.

More complex gambler's ruin type problems find practical relevance in certain models describing the fluctuation of insurance company assets over time.

Random walks are not only useful in simulating situations of gambling but frequently serve as reasonable discrete approximations to physical processes describing the motion of diffusing particles. If a particle is subjected to collisions and random impulses, then its position fluctuates randomly, although the particle describes a continuous path. If the future position (i.e., its probability distribution) of the particle depends only on the present position, then the process X_t, where X_t is the position at time t, is Markov. A discrete approximation to such a continuous motion corresponds to a random walk. A classical discrete version of Brownian motion is provided by the symmetric random walk. By a symmetric random walk on the integers (say all the integers) we mean a Markov chain with state space the totality of all integers and whose transition probability matrix has the elements

$$P_{ij} = \begin{cases} p & \text{if } j = i + 1, \\ p & \text{if } j = i - 1, \\ r & \text{if } j = i, \\ 0 & \text{otherwise,} \end{cases} \quad i, j = 0, 1, 2, \ldots,$$

where $p > 0$, $r \geq 0$, and $2p + r = 1$. Conventionally, "symmetric random walk" refers only to the case $r = 0$, $p = \frac{1}{2}$.

The classical symmetric random walk in n dimensions admits the following formulation. The state space is identified with the set of all integral lattice points in E^n (Euclidean n space): that is, a state is an n-tuple $k = (k_1, k_2, \ldots, k_n)$ of integers. The transition probability matrix is defined by

$$\mathbf{P_{kl}} = \begin{cases} \dfrac{1}{2n} & \text{if } \sum_{i=1}^{n} |l_i - k_i| = 1, \\ \\ 0 & \text{otherwise.} \end{cases}$$

Analogous to the one-dimensional case, the symmetric random walk in E^n represents a discrete version of n-dimensional Brownian motion.

3.5.4 Success Runs

Consider a Markov chain on the nonnegative integers with transition probability matrix of the form

$$\mathbf{P} = \begin{array}{c} 0 \\ 1 \\ 2 \\ 3 \\ {} \\ {} \\ {} \end{array} \begin{Vmatrix} \begin{array}{ccccc} p_0 & q_0 & 0 & 0 & 0 \\ p_1 & r_1 & q_1 & 0 & 0 \\ p_2 & 0 & r_2 & q_2 & 0 \\ p_3 & 0 & 0 & r_3 & q_3 \\ \cdot & \cdot & \cdot & \cdot & \cdot \\ \cdot & \cdot & \cdot & \cdot & \cdot \\ \cdot & \cdot & \cdot & \cdot & \cdot \end{array} \end{Vmatrix} \begin{array}{c} 0 \ 1 \ 2 \ 3 \ 4 \\ \cdot \\ \cdot \\ \cdot \\ \cdot \end{array} \tag{3.44}$$

where $q_i > 0$, $p_i > 0$ and $p_i + q_i + r_i = 1$ for $i = 0, 1, 2, \ldots$. The zero state plays a distinguished role in that it can be reached in one transition from any other state, while state $i + 1$ can be reached only from state i.

This example arises surprisingly often in applications and, at the same time, is very easy to compute with. We will frequently illustrate concepts and results in terms of it.

A special case of this transition matrix arises when one is dealing with success runs resulting from repeated trials each of which admits two possible outcomes, success S or failure F. More explicitly, consider a sequence of trials with two possible outcomes S or F. Moreover, suppose that in each trial, the probability of S is α and the probability of F is $\beta = 1 - \alpha$. We say a success run of length r happened at trial n if the outcomes in the preceding $r + 1$ trials, including the present trial as the last, were respectively, F, S, $S, \ldots , S$. Let us now label the present state of the process by the length of the success run currently under way. In particular, if the last trial resulted in a failure then the state is zero. Similarly, when the preceding $r + 1$ trials in order have the outcomes F, S, $S, \ldots , S$, the state variable would carry the label r. The process is clearly Markov (since the individual trials were independent of each other), and its transition matrix has the form (3.44) where

$$ p_n = \beta, \quad r_n = 0 \quad \text{and} \quad q_n = \alpha \quad \text{for} \quad n = 0, 1, 2, \ldots . $$

A second example is furnished by the *current age* in a *renewal process*. Consider a light bulb whose lifetime, measured in discrete units, is a random variable ξ, where

$$ \Pr\{\xi = k\} = a_k > 0 \quad \text{for} \quad k = 1, 2, \ldots , \sum_{k=0}^{\infty} a_k = 1. $$

Let each bulb be replaced by a new one when it burns out. Suppose the first bulb lasts until time ξ_1, the second bulb until time $\xi_1 + \xi_2$, and the nth bulb until time $\xi_1 + \ldots + \xi_n$, where the individual lifetimes $\xi_1, \xi_2, \ldots$ are independent random variables each having the same distribution as ξ. Let X_n be the age of the bulb in service at time n. This current age process is depicted in Figure 3.2.

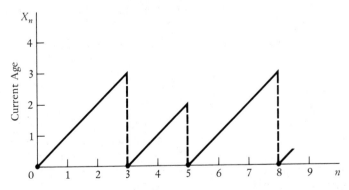

Figure 3.2 The current age X_n in a renewal process. Here $\xi_1 = 3$, $\xi_2 = 2$, and $\xi_3 = 3$.

By convention we set $X_n = 0$ at the time of a failure.
The current age is a success runs Markov process for which

$$p_k = \frac{a_{k+1}}{a_{k+1} + a_{k+2} + \cdots}, \qquad r_k = 0, \; q_k = 1 - p_k$$

$$\text{for} \quad k = 0, 1, \ldots \quad (3.45)$$

We reason as follows: The age process reverts to zero upon failure of the item in service. Given that the age of the item in current service is k, then failure occurs in the next time period with *conditional probability* $p_k = a_{k+1}/(a_{k+1} + a_{k+2} + \cdots)$. Given that the item has survived k periods, it survives at least to the next period with the remaining probability $q_k = 1 - p_k$.

Renewal processes are extensively discussed in Chapter 7.

Problems 3.5

1. The probability of the thrower winning in the dice game called "craps" is $p = .4929$. Suppose Player A is the thrower and begins the game with \$5 and Player B, his opponent, begins with \$10. What is the probability that Player A goes bankrupt before Player B? Assume that the bet is \$1 per round.
   Hint: Use Equation (3.42).

2. Determine the gambler's ruin probability for Player A when both players begin with \$50, bet \$1 on each play, and where the win probability for Player A in each game is
   (a) $p = .49292929$
   (b) $p = .5029237$
   (See Section 2.2.)

   What are the gambler's ruin probabilities when each player begins with \$500?

3. Determine $\mathbf{P}^n$ for $n = 2, 3, 4, 5$ for the Markov chain whose transition probability matrix is

$$\mathbf{P} = \left\| \begin{matrix} .4 & .6 \\ .7 & .3 \end{matrix} \right\|.$$

4. A coin is tossed repeatedly until three heads in a row appear. Let X_n record the current number of successive heads that have appeared. That is, $X_n = 0$ if the nth toss resulted in tails; $X_n = 1$ if the nth toss was heads and the $(n-1)$st toss was tails and so on. Model X_n as a success runs Markov chain by specifying the probabilities p_i and q_i.

5. A component of a computer has an active life, measured in discrete units, that is a random variable T where $\Pr\{T = k\} = a_k$ for $k = 1, 2, \ldots$

Suppose one starts with a fresh component and each component is replaced by a new component upon failure. Let X_n be the age of the component in service at time n. Then $\{X_n\}$ is a success runs Markov chain.

(a) Specify the probabilities p_i and q_i.

(b) A "planned replacement" policy calls for replacing the component upon its failure or upon its reaching age N, whichever occurs first. Specify the success runs probabilities p_i and q_i under the planned replacement policy.

6. **A Batch Processing Model.** Customers arrive at a facility and wait there until a total number of K customers have accumulated. Upon the arrival of the Kth customer, all are instantaneously served, and the process repeats. Let $\xi_0, \xi_1, \ldots$ denote the arrivals in successive periods, assumed to be independent random variables whose distribution is given by

$$\Pr\{\xi_k = 0\} = \alpha, \qquad \Pr\{\xi_k = 1\} = 1 - \alpha$$

where $0 < \alpha < 1$. Let X_n denote the number of customers in the system at time n. Then $\{X_n\}$ is a Markov chain on the states $0, 1, \ldots,$ $K - 1$. With $K = 3$, give the transition probability matrix for $\{X_n\}$. Be explicit about any assumptions you make.

3.6 Functionals of Random Walks and Success Runs

Consider first the random walk on $N + 1$ states whose transition probability matrix is given by

$$\mathbf{P} = \begin{array}{c} \\ 0 \\ 1 \\ 2 \\ \cdot \\ \cdot \\ \cdot \\ N \end{array} \begin{array}{c} \begin{matrix} 0 & 1 & 2 & 3 & \cdots & N \end{matrix} \\ \left\| \begin{matrix} 1 & 0 & 0 & 0 & \cdots & 0 \\ q & 0 & p & 0 & \cdots & 0 \\ 0 & q & 0 & p & \cdots & 0 \\ \cdot & & & & & \cdot \\ \cdot & & & & & \cdot \\ \cdot & & & & & \cdot \\ 0 & 0 & 0 & 0 & \cdots & 1 \end{matrix} \right\| \end{array}.$$

"Gambler's ruin" is the event that the process reaches state 0 before reaching state N. This event can be stated more formally if we introduce the concept of *hitting time*. Let T be the (random) time that the process first reaches or hits state 0 or N. In symbols

$$T = \min\{n \geq 0; X_n = 0 \text{ or } X_n = N\}.$$

The random time T is shown in Figure 3.3 in a typical case.

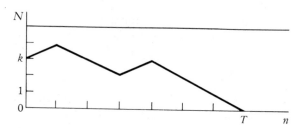

Figure 3.3   The hitting time to 0 or N. As depicted here, state 0 was reached first.

In terms of T, the event written as $X_T = 0$ is the event of gambler's ruin, and the probability of this event starting from the initial state k is

$$u_k = \Pr\{X_T = 0 | X_0 = k\}.$$

Figure 3.4 shows the first step analysis that leads to the equations

$$u_k = p u_{k+1} + q u_{k-1}, \qquad \text{for} \quad k = 1, \ldots, N - 1, \qquad (3.46)$$

with the obvious boundary conditions

$$u_0 = 1, u_N = 0.$$

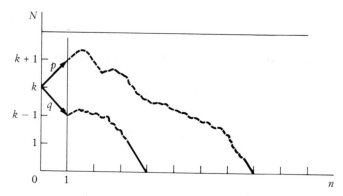

Figure 3.4   First step analysis for the gambler's ruin problem

Equations (3.46) yield to straightforward but tedious manipulations. Because the approach has considerable generality and arises frequently, it is well worth pursuing in this simplest case.

We begin the solution by introducing the differences $x_k = u_k - u_{k-1}$ for $k = 1, \ldots, N$. Using $p + q = 1$ to write $u_k = (p + q)u_k = p u_k + q u_k$, then Equations (3.46) become

$$k = 1; \quad 0 = p(u_2 - u_1) - q(u_1 - u_0) = px_2 - qx_1$$
$$k = 2; \quad 0 = p(u_3 - u_2) - q(u_2 - u_1) = px_3 - qx_2$$
$$k = 3; \quad 0 = p(u_4 - u_3) - q(u_3 - u_2) = px_4 - qx_3$$

$$\cdot$$
$$\cdot$$
$$\cdot$$

$$k = N - 1; \quad 0 = p(u_N - u_{N-1}) - q(u_{N-1} - u_{N-2}) = px_N - qx_{N-1}$$

or

$$x_2 = (q/p)x_1$$
$$x_3 = (q/p)x_2 = (q/p)^2 x_1$$
$$x_4 = (q/p)x_3 = (q/p)^3 x_1$$

$$\cdot$$
$$\cdot$$
$$\cdot$$

$$x_k = (q/p)x_{k-1} = (q/p)^{k-1} x_1$$

$$\cdot$$
$$\cdot$$
$$\cdot$$

$$x_N = (q/p)x_{N-1} = (q/p)^{N-1} x_1.$$

We now recover $u_0, u_1, \ldots, u_N$ by invoking the conditions $u_0 = 1$, $u_N = 0$ and summing the x_k's:

$$x_1 = u_1 - u_0 = u_1 - 1$$
$$x_2 = u_2 - u_1 \qquad\qquad x_1 + x_2 = u_2 - 1$$
$$x_3 = u_3 - u_2 \qquad\qquad x_1 + x_2 + x_3 = u_3 - 1$$

$$\cdot \qquad\qquad\qquad\qquad \cdot$$
$$\cdot \qquad\qquad\qquad\qquad \cdot$$
$$\cdot \qquad\qquad\qquad\qquad \cdot$$

$$x_k = u_k - u_{k-1} \qquad\qquad x_1 + \cdots + x_k = u_k - 1$$

$$\cdot \qquad\qquad\qquad\qquad \cdot$$
$$\cdot \qquad\qquad\qquad\qquad \cdot$$
$$\cdot \qquad\qquad\qquad\qquad \cdot$$

$$x_N = u_N - u_{N-1} = -u_{N-1} \qquad\qquad x_1 + \cdots + x_N = u_N - 1 = -1.$$

The equation for general k gives

$$\begin{aligned} u_k &= 1 + x_1 + x_2 + \cdots + x_k \\ &= 1 + x_1 + (q/p)x_1 + \cdots + (q/p)^{k-1} x_1 \\ &= 1 + [1 + (q/p) + \cdots + (q/p)^{k-1}]x_1, \end{aligned} \qquad (3.47)$$

which expresses u_k in terms of the as yet undetermined x_1. But $u_N = 0$ gives

$$0 = 1 + [1 + (q/p) + \cdots + (q/p)^{N-1}]x_1$$

or

$$x_1 = -\frac{1}{1 + (q/p) + \cdots + (q/p)^{N-1}},$$

which substituted into (3.47) gives

$$u_k = 1 - \frac{1 + (q/p) + \cdots + (q/p)^{k-1}}{1 + (q/p) + \cdots + (q/p)^{N-1}}.$$

The geometric series sums to

$$1 + (q/p) + \cdots + (q/p)^{k-1} = \begin{cases} k & \text{if } p = q = \tfrac{1}{2}; \\[2mm] \dfrac{1 - (q/p)^k}{1 - (q/p)} & \text{if } p \neq q, \end{cases}$$

whence

$$u_k = \begin{cases} 1 - (k/N) = (N - k)/N & \text{when } p = q = \tfrac{1}{2}, \\[2mm] 1 - \dfrac{1 - (q/p)^k}{1 - (q/p)^N} = \dfrac{(q/p)^k - (q/p)^N}{1 - (q/p)^N} & \text{when } p \neq q. \end{cases} \tag{3.48}$$

A similar approach works to evaluate the mean duration

$$v_i = E[T|X_0 = i]. \tag{3.49}$$

The time T is comprised of a first step plus the remaining steps. With probability p the first step is to state $i + 1$ and then the remainder, on the average, is v_{i+1} additional steps. With probability q the first step is to $i - 1$ and then, on the average, there are v_{i-1} further steps. Thus for the mean duration a first step analysis leads to the equation

$$v_i = 1 + pv_{i+1} + qv_{i-1} \qquad \text{for } i = 1, \ldots, N - 1. \tag{3.50}$$

Of course the game ends in states 0 and N and thus

$$v_0 = 0, \qquad v_N = 0.$$

We will solve equations (3.50) when $p = q = \tfrac{1}{2}$. The solution for other values of p proceeds in a similar manner, and the solution for a general random walk is given later in this section.

Again we introduce the differences $x_k = v_k - v_{k-1}$ for $k = 1, \ldots,$ N, writing (3.50) in the form

$$
\begin{aligned}
k = 1; &\quad -1 = \tfrac{1}{2}(v_2 - v_1) - \tfrac{1}{2}(v_1 - v_0) = \tfrac{1}{2}x_2 - \tfrac{1}{2}x_1; \\
k = 2; &\quad -1 = \tfrac{1}{2}(v_3 - v_2) - \tfrac{1}{2}(v_2 - v_1) = \tfrac{1}{2}x_3 - \tfrac{1}{2}x_2; \\
k = 3; &\quad -1 = \tfrac{1}{2}(v_4 - v_3) - \tfrac{1}{2}(v_3 - v_2) = \tfrac{1}{2}x_4 - \tfrac{1}{2}x_3;
\end{aligned}
$$

$$\vdots$$

$$k = N - 1; \quad -1 = \tfrac{1}{2}(v_N - v_{N-1}) - \tfrac{1}{2}(v_{N-1} - v_{N-2}) = \tfrac{1}{2}x_N - \tfrac{1}{2}x_{N-1}.$$

The right side forms a collapsing sum. Upon adding we obtain

$$k = 1; \qquad -1 = \tfrac{1}{2}x_2 - \tfrac{1}{2}x_1;$$
$$k = 2; \qquad -2 = \tfrac{1}{2}x_3 - \tfrac{1}{2}x_1;$$
$$k = 3; \qquad -3 = \tfrac{1}{2}x_4 - \tfrac{1}{2}x_1;$$

$$k = N - 1; \qquad -(N - 1) = \tfrac{1}{2}x_N - \tfrac{1}{2}x_1.$$

The general line gives $x_k = x_1 - 2(k - 1)$ for $k = 2, 3, \ldots, N$. We return to the v_k's by means of

$$x_1 = v_1 - v_0 = v_1;$$
$$x_2 = v_2 - v_1; \qquad\qquad x_1 + x_2 = v_2;$$
$$x_3 = v_3 - v_2; \qquad\qquad x_1 + x_2 + x_3 = v_3;$$

$$x_k = v_k - v_{k-1}; \qquad\qquad x_1 + \cdots + x_k = v_k,$$

or

$$v_k = kv_1 - 2[1 + 2 + \cdots + (k - 1)] = kv_1 - k(k - 1), \quad (3.51)$$

which gives v_k in terms of the as yet unknown v_1. We impose the boundary condition $v_N = 0$ to obtain $0 = Nv_1 - N(N - 1)$ or $v_1 = (N - 1)$. Substituting this into (3.51) we obtain

$$v_k = k(N - k) \qquad k = 0, 1, \ldots, N, \qquad (3.52)$$

for the mean duration of the game. Note that the mean duration is greatest for initial fortunes k that are midway between the boundaries 0 and N, as we would expect.

3.6.1 The General Random Walk

We give the results of similar derivations on the random walk whose transition matrix is

$$\mathbf{P} = \begin{array}{c} \\ 0 \\ 1 \\ 2 \\ \cdot \\ \cdot \\ \cdot \\ N \end{array}
\begin{array}{ccccccc}
0 & 1 & 2 & 3 & \cdots & N \\
1 & 0 & 0 & 0 & \cdots & 0 \\
q_1 & r_1 & p_1 & 0 & \cdots & 0 \\
0 & q_2 & r_2 & p_2 & \cdots & 0 \\
\cdot & \cdot & \cdot & \cdot & & \cdot \\
\cdot & \cdot & \cdot & \cdot & & \cdot \\
\cdot & \cdot & \cdot & \cdot & & \cdot \\
0 & 0 & 0 & 0 & \cdots & 1
\end{array}$$

where $q_k > 0$ and $p_k > 0$ for $k = 1, \ldots, N - 1$. Let $T = \min\{n \geq 0; X_n = 0 \text{ or } X_n = N\}$ be the hitting time to states 0 and N.

Problem 1   The probability of gambler's ruin

$$u_i = \Pr\{X_T = 0 | X_0 = i\} \tag{3.53}$$

satisfies the first step analysis equation

$$u_i = q_i u_{i-1} + r_i u_i + p_i u_{i+1} \quad \text{for} \quad i = 1, \ldots, N - 1,$$

and

$$u_0 = 1, \quad u_N = 0.$$

The solution is

$$u_i = \frac{\rho_i + \cdots + \rho_{N-1}}{1 + \rho_1 + \rho_2 + \cdots + \rho_{N-1}}, \quad i = 1, \ldots, N - 1. \tag{3.54}$$

where

$$\rho_k = \frac{q_1 q_2 \cdots q_k}{p_1 p_2 \cdots p_k}, \quad k = 1, \ldots, N - 1. \tag{3.55}$$

Problem 2   The mean hitting time

$$v_k = E[T | X_0 = k] \tag{3.56}$$

satisfies the equation

$$v_k = 1 + q_k v_{k-1} + r_k v_k + p_k v_{k+1} \quad \text{and} \quad v_0 = v_N = 0. \tag{3.57}$$

The solution is

$$v_k = \left(\frac{\Phi_1 + \cdots + \Phi_{N-1}}{1 + \rho_1 + \cdots + \rho_{N-1}} \right)(1 + \rho_1 + \cdots + \rho_{k-1})$$

$$- (\Phi_1 + \cdots + \Phi_{k-1}) \quad \text{for} \quad k = 1, \ldots, N - 1 \tag{3.58}$$

where ρ_i is given in (3.55) and

$$\Phi_i = \left(\frac{1}{q_1} + \frac{1}{q_2 \rho_1} + \cdots + \frac{1}{q_i \rho_{i-1}} \right) \rho_i \tag{3.59}$$

$$= \frac{q_2 \cdots q_i}{p_1 \cdots p_i} + \frac{q_3 \cdots q_i}{p_2 \cdots p_i} + \cdots + \frac{q_i}{p_{i-1} p_i} + \frac{1}{p_i}$$

$$\text{for} \quad i = 1, \ldots, N - 1.$$

Problem 3 Fix a state k, where $0 < k < N$, and let W_{ik} be the mean total visits to state k starting from i. Formally the definition is

$$W_{ik} = E\left[\sum_{n=0}^{T-1}\mathbf{1}\{X_n = k\}|X_0 = i\right] \tag{3.60}$$

where

$$\mathbf{1}\{X_n = k\} = \begin{cases} 1 & \text{if } X_n = k, \\ 0 & \text{if } X_n \neq k. \end{cases}$$

Then W_{ik} satisfies the equation

$$W_{ik} = \delta_{ik} + q_i W_{i-1,k} + r_i W_{ik} + p_i W_{i+1,k} \qquad \text{for} \quad i = 1, \ldots, N-1$$

and

$$W_{0k} = W_{Nk} = 0,$$

where

$$\delta_{ik} = \begin{cases} 1 & \text{if } i = k \\ 0 & \text{if } i \neq k. \end{cases}$$

The solution is

$$W_{ik} = \begin{cases} \dfrac{(1 + \cdots + \rho_{i-1})(\rho_k + \cdots + \rho_{N-1})}{1 + \cdots + \rho_{N-1}}\left(\dfrac{1}{q_k\rho_{k-1}}\right) & \text{for } i \leq k \\[4mm] \left[\dfrac{(1 + \cdots + \rho_{i-1})(\rho_k + \cdots + \rho_{N-1})}{1 + \cdots + \rho_{N-1}} \right. \\[4mm] \left. \qquad\qquad - (\rho_k + \cdots + \rho_{i-1})\right]\left(\dfrac{1}{q_k\rho_{k-1}}\right) & \text{for } i \geq k. \end{cases} \tag{3.61}$$

Example As a sample calculation of these functionals, we consider the special case in which the transition probabilities are the same from row to row. That is, we study the random walk whose transition probability matrix is

$$\mathbf{P} = \begin{matrix} & \begin{matrix} 0 & 1 & 2 & 3 & \cdots & N \end{matrix} \\ \begin{matrix} 0 \\ 1 \\ 2 \\ \cdot \\ \cdot \\ \cdot \\ N \end{matrix} & \left\|\begin{matrix} 1 & 0 & 0 & 0 & \cdots & 0 \\ q & r & p & 0 & \cdots & 0 \\ 0 & q & r & p & \cdots & 0 \\ \cdot & \cdot & \cdot & \cdot & & \cdot \\ \cdot & \cdot & \cdot & \cdot & & \cdot \\ \cdot & \cdot & \cdot & \cdot & & \cdot \\ 0 & 0 & 0 & 0 & & 1 \end{matrix}\right\|, \end{matrix}$$

with $p > 0$, $q > 0$ and $p + q + r = 1$. Let us abbreviate by setting $\theta = (q/p)$, and then ρ_k, as defined in (3.55), simplifies according to

$$\rho_k = \frac{q_1 q_2 \cdots q_k}{p_1 p_2 \cdots p_k} = \left(\frac{q}{p}\right)^k = \theta^k \qquad \text{for} \quad k = 1, \ldots, N - 1.$$

The probability of gambler's ruin, as defined in (3.53) and evaluated in (3.54), becomes

$$
\begin{aligned}
u_k &= \Pr\{X_T = 0 | X_0 = k\} \\[6pt]
&= \frac{\theta^k + \cdots + \theta^{N-1}}{1 + \theta + \cdots + \theta^{N-1}} \\[6pt]
&= \begin{cases} \dfrac{\theta^k - \theta^N}{1 - \theta^N} & \text{if} \quad \theta \equiv (q/p) \neq 1, \\[12pt] \dfrac{N - k}{N} & \text{if} \quad \theta \equiv (q/p) = 1. \end{cases}
\end{aligned}
$$

This, of course, agrees with the answer given in (3.48).

We turn to evaluating the mean time

$$v_k = E[T | X_0 = k] \qquad \text{for} \quad k = 1, \ldots, N - 1,$$

by first substituting $\rho_i = \theta^i$ into (3.59) to obtain

$$
\begin{aligned}
\Phi_i &= \left(\frac{1}{q} + \frac{1}{q\theta} + \cdots + \frac{1}{q\theta^{i-1}}\right)\theta^i \\[6pt]
&= \frac{1}{q}(\theta^i + \theta^{i-1} + \cdots + \theta) \\[6pt]
&= \frac{1}{p}(1 + \theta + \cdots + \theta^{i-1}) \\[6pt]
&= \begin{cases} \dfrac{i}{p} & \text{when} \quad p = q \ (\theta = 1) \\[12pt] \dfrac{1}{p}\left(\dfrac{1 - \theta^i}{1 - \theta}\right) & \text{when} \quad p \neq q \ (\theta \neq 1). \end{cases}
\end{aligned}
$$

Now observe that

$$
\begin{aligned}
1 + \rho_1 + \cdots + \rho_{i-1} &= 1 + \theta + \cdots + \theta^{i-1} \\
&= p\Phi_i
\end{aligned}
$$

so that (3.58) reduces to

$$v_k = \frac{\Phi_k}{\Phi_N}(\Phi_1 + \cdots + \Phi_{N-1}) - (\Phi_1 + \cdots + \Phi_{k-1}). \qquad (3.62)$$

In order to continue, we need to simplify the terms of the form $\Phi_1 + \cdots + \Phi_{j-1}$. We consider the two cases $\theta \equiv (q/p) = 1$ and $\theta \equiv (q/p) \neq 1$ separately.

When $p = q$, or equivalently, $\theta = 1$, then $\Phi_i = i/p$ whence

$$\Phi_1 + \cdots + \Phi_{j-1} = \frac{1 + \cdots + (j-1)}{p} = \frac{j(j-1)}{2p},$$

which inserted into (3.62) gives

$$\begin{aligned} v_i &\equiv E[T|X_0 = i] \\ &= \frac{i}{N}\left[\frac{N(N-1)}{2p}\right] - \frac{i(i-1)}{2p} \\ &= \frac{i(N-i)}{2p} \quad \text{if} \quad p = q. \end{aligned} \tag{3.63}$$

When $p = \frac{1}{2}$, then $v_i = i(N-i)$ in agreement with (3.52).
When $p \neq q$ so that $\theta \equiv q/p \neq 1$, then

$$\Phi_i = \frac{1}{p}\left(\frac{1 - \theta^i}{1 - \theta}\right)$$

whence

$$\begin{aligned} \Phi_1 + \cdots + \Phi_{j-1} &= \frac{1}{p(1-\theta)}[(j-1) - (\theta + \theta^2 + \cdots + \theta^{j-1})] \\ &= \frac{1}{p(1-\theta)}\left[(j-1) - \theta\left(\frac{1-\theta^{j-1}}{1-\theta}\right)\right], \end{aligned}$$

and

$$\begin{aligned} v_i &= E[T|X_0 = i] \\ &= \left(\frac{1-\theta^i}{1-\theta^N}\right)\frac{1}{p(1-\theta)}\left[N - \left(\frac{1-\theta^N}{1-\theta}\right)\right] - \frac{1}{p(1-\theta)}\left[i - \left(\frac{1-\theta^i}{1-\theta}\right)\right] \\ &= \frac{1}{p(1-\theta)}\left[N\left(\frac{1-\theta^i}{1-\theta^N}\right) - i\right] \end{aligned}$$

when $\theta \equiv (q/p) \neq 1$.

Finally we evaluate W_{ik}, expressed verbally as the mean number of visits to state k starting from $X_0 = i$, and defined formally in (3.60). Again we consider the two cases $\theta \equiv (q/p) = 1$ and $\theta \equiv (q/p) \neq 1$.

When $\theta = 1$, then $\rho_j = \theta^j = 1$ and $1 + \ldots + \rho_{i-1} = i$, $\rho_k + \ldots + \rho_{N-1} = N - k$, and (3.61) simplifies to

$$\begin{aligned} W_{ik} &= \begin{cases} \dfrac{i(N-k)}{qN} & \text{for} \quad 0 < i \leq k < N \\[3mm] \dfrac{1}{q}\left[\dfrac{i(N-k)}{N} - (i-k)\right] = \dfrac{k(N-i)}{qN} & \text{for} \quad 0 < k < i < N, \end{cases} \\[3mm] &= \frac{i(N-k)}{qN} - \frac{\max\{0, i-k\}}{q}. \end{aligned} \tag{3.64}$$

When $\theta = (q/p) \neq 1$, then $\rho_j = \theta^j$ and

$$1 + \cdot\cdot\cdot + \rho_{i-1} = \frac{1 - \theta^i}{1 - \theta},$$

$$\rho_k + \cdot\cdot\cdot + \rho_{N-1} = \frac{\theta^k - \theta^N}{1 - \theta},$$

and

$$q\rho_{k-1} = p\rho_k = p\theta^k.$$

In this case, (3.61) simplifies to

$$W_{ik} = \frac{(1 - \theta^i)(\theta^k - \theta^N)}{(1 - \theta)(1 - \theta^N)}\left(\frac{1}{p\theta^k}\right) \qquad \text{for} \quad 0 < i \leq k < N,$$

and

$$\begin{aligned} W_{ik} &= \left[\frac{(1 - \theta^i)(\theta^k - \theta^N)}{(1 - \theta)(1 - \theta^N)} - \frac{\theta^k - \theta^i}{1 - \theta}\right]\left(\frac{1}{p\theta^k}\right) \\ &= \frac{(1 - \theta^k)(\theta^i - \theta^N)}{(1 - \theta)(1 - \theta^N)}\left(\frac{1}{p\theta^k}\right) \qquad \text{for} \quad 0 < k < i < N. \end{aligned}$$

We may write the expression for W_{ik} in a single line by introducing the notation $(i - k)^+ = \max\{0, i - k\}$. Then

$$W_{ik} = \frac{(1 - \theta^i)(1 - \theta^{N-k})}{p(1 - \theta)(1 - \theta^N)} - \frac{1 - \theta^{(i-k)^+}}{p(1 - \theta)}. \qquad (3.65)$$

3.6.2 Cash Management

Short term cash management is the review and control of a corporation's cash balances, short term loan balances, and short term marketable security holdings. The objective is to maintain the smallest cash balances that are adequate to meet future disbursements. The corporation cashier tries to eliminate idle cash balances (by reducing short term loans or buying treasury bills, for example), but to cover potential cash shortages (by selling treasury bills or increasing short terms loans). The analogous problem for an individual is to maintain an optimal balance between a checking and a savings account.

In the absence of intervention, the corporation's cash level fluctuates randomly as the result of many relatively small transactions. We model this by dividing time into successive, equal length periods, each of short duration, and by assuming that from period to period, the cash level moves up or down one unit, each with probability one-half. Let X_n be the cash on hand in period n. We are assuming that $\{X_n\}$ is the random walk in which

$$\Pr\{X_{n+1} = k \pm 1 | X_n = k\} = \tfrac{1}{2}.$$

The cashier's job is to intervene if the cash level ever gets too low or too high. We consider cash management strategies that are specified by two parameters, s and $\mathcal{S}$, where $0 < s < \mathcal{S}$. The policy is as follows: If the cash level ever drops to zero, then sell sufficient treasury bills to replenish the cash level up to s. If the cash level ever increases up to $\mathcal{S}$, then invest in treasury bills in order to reduce the cash level to s. A typical sequence of cash levels $\{X_n\}$ when $s = 2$ and $\mathcal{S} = 5$ is depicted in Figure 3.5.

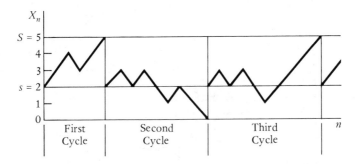

Figure 3.5   Several typical cycles in a cash inventory model

We see that the cash level fluctuates in a series of statistically similar cycles, each cycle beginning with s units of cash on hand and ending at the next intervention, whether a replenishment or reduction in cash. We begin our study by evaluating the mean length of a cycle and the mean total unit-periods of cash on hand during a cycle. Later we use these quantities to evaluate the long run performance of the model.

Let T denote the random time at which the cash on hand first reaches the level $\mathcal{S}$ or 0. That is, T is the time of the first transaction. Let $v_s = E[T | X_0 = s]$ be the mean time to the first transaction, or the mean cycle length. From (3.52) we have

$$v_s = s(\mathcal{S} - s). \tag{3.66}$$

Next, fix an arbitrary state k $(0 < k < \mathcal{S})$ and let W_{sk} be the mean number of visits to k up to time T for a process starting at $X_0 = s$. From (3.64) we have

$$W_{sk} = 2\left[\frac{s}{\mathcal{S}}(\mathcal{S} - k) - (s - k)^+\right]. \tag{3.67}$$

Using this we obtain the mean total unit-periods of cash on hand up to time T starting from $X_0 = s$ by weighting W_{sk} by k and summing according to

$$W_s = \sum_{k=1}^{\mathscr{S}-1} k\, W_{sk}$$

$$= 2\left\{ \frac{s}{\mathscr{S}} \sum_{k=1}^{\mathscr{S}-1} k(\mathscr{S} - k) - \sum_{k=1}^{s-1} k(s - k) \right\}$$

$$= 2\left\{ \frac{s}{\mathscr{S}} \left[\frac{\mathscr{S}(\mathscr{S} - 1)(\mathscr{S} + 1)}{6} \right] - \frac{s(s - 1)(s + 1)}{6} \right\}_\star$$

$$= \frac{s}{3}[\mathscr{S}^2 - s^2]. \tag{3.68}$$

Having obtained these single cycle results, we will use them to evaluate the long run behavior of the model. Note that each cycle starts from the cash level s, and thus the cycles are statistically independent. Let K be the fixed cost of each transaction. Let T_i be the duration of the ith cycle and let R_i be the total opportunity cost of holding cash on hand during that time. Over n cycles the average cost per unit-time is

$$\text{Average cost} = \frac{nK + R_1 + \cdots + R_n}{T_1 + \cdots + T_n}.$$

Next, divide the numerator and denominator by n, let $n \to \infty$, and invoke the law of large numbers to obtain

$$\text{Long run average cost} = \frac{K + E[R_i]}{E[T_i]}.$$

Let r denote the opportunity cost per unit-time of cash on hand. Then $E[R_i] = r\, W_s$, while $E[T_i] = v_s$. Since these quantities were determined in (3.66) and (3.68) we have

$$\text{Long run average cost} = \frac{K + (1/3)rs(\mathscr{S}^2 - s^2)}{s(\mathscr{S} - s)}. \tag{3.69}$$

In order to use calculus to determine the cost minimizing values for $\mathscr{S}$ and s, it simplifies matters if we introduce the new variable $x = s/\mathscr{S}$. Then (3.69) becomes

$$\text{Long run average cost} = \frac{K + (1/3)r\mathscr{S}^3\, x(1 - x^2)}{\mathscr{S}^2\, x(1 - x)},$$

whence

$$\frac{d(\text{average cost})}{dx} = 0 = -\frac{K(1 - 2x)}{\mathscr{S}^2\, x^2(1 - x)^2} + \frac{1}{3}\, r\mathscr{S},$$

$$\frac{d(\text{average cost})}{d\mathscr{S}} = 0 = -\frac{2K}{\mathscr{S}^3\, x(1 - x)} + \frac{r(1 + x)}{3}$$

$\star$Use the sum $\sum_{k=1}^{a-1} k(a - k) = \frac{1}{6}a(a + 1)(a - 1)$.

which solve to give

$$x_{\text{opt}} = \frac{1}{3} \quad \text{and} \quad \mathcal{S}_{\text{opt}} = 3 s_{\text{opt}} = 3 \sqrt[3]{\frac{3K}{4r}}$$

Implementing the cash management strategy with the values s_{opt} and $\mathcal{S}_{\text{opt}}$ results in the optimal balance between transaction costs and the opportunity cost of holding cash on hand.

3.6.3 The Success Runs Markov Chain

Consider the success runs Markov chain on $N + 1$ states whose transition matrix is

$$
\mathbf{P} =
\begin{array}{c c}
 & \begin{array}{c c c c c c c}
0 & 1 & 2 & 3 & \cdots & & N
\end{array} \\
\begin{array}{c}
0 \\ 1 \\ 2 \\ \\ \\ \\ N-1 \\ N
\end{array} &
\left\| \begin{array}{c c c c c c c}
1 & 0 & 0 & 0 & \cdots & & 0 \\
p_1 & r_1 & q_1 & 0 & \cdots & & 0 \\
p_2 & 0 & r_2 & q_2 & \cdots & & 0 \\
\cdot & \cdot & \cdot & \cdot & & & \cdot \\
\cdot & \cdot & \cdot & \cdot & & & \cdot \\
\cdot & \cdot & \cdot & \cdot & & & \cdot \\
p_{N-1} & 0 & 0 & 0 & \cdots & & q_{N-1} \\
0 & 0 & 0 & 0 & \cdots & & 1
\end{array} \right\|
\end{array}.
$$

Note that states 0 and N are absorbing; once the process reaches one of these two states it remains there.

Let T be the hitting time to states 0 or N,

$$T = \min\{n \geq 0; \ X_n = 0 \text{ or } X_n = N\}.$$

Problem 1   The probability of absorption at 0 starting from state k

$$u_k = \Pr\{X_T = 0 | X_0 = k\} \qquad (3.70)$$

satisfies the equation

$$u_k = p_k + r_k u_k + q_k u_{k+1},$$

$$k = 1, \ldots, N-1 \quad \text{and} \quad u_0 = 1, \quad u_N = 0.$$

The solution is

$$u_k = 1 - \left(\frac{q_k}{p_k + q_k} \right) \cdots \left(\frac{q_{N-1}}{p_{N-1} + q_{N-1}} \right)$$

$$\text{for} \quad k = 1, \ldots, N-1. \quad (3.71)$$

Problem 2 The mean hitting time

$$v_k = E[T|X_0 = k] \tag{3.72}$$

satisfies the equation

$$v_k = 1 + r_k v_k + q_k v_{k+1}$$

$$\text{for} \quad k = 1, \ldots, N - 1 \quad \text{and} \quad v_0 = v_N = 0.$$

The solution is

$$v_k = \frac{1}{p_k + q_k} + \frac{\pi_{k,k+1}}{p_{k+1} + q_{k+1}} + \cdots + \frac{\pi_{k,N-1}}{p_{N-1} + q_{N-1}}, \tag{3.73}$$

where

$$\pi_{kj} = \left(\frac{q_k}{p_k + q_k}\right)\left(\frac{q_{k+1}}{p_{k+1} + q_{k+1}}\right) \cdots \left(\frac{q_{j-1}}{p_{j-1} + q_{j-1}}\right)$$

$$\text{for} \quad k < j. \tag{3.74}$$

Problem 3   Fix a state j $(0 < j < N)$ and let W_{ij} be the mean total visits to state j starting from state i [see Equation (3.60)]. Then

$$W_{ij} = \begin{cases} \dfrac{1}{p_i + q_i} & \text{for} \quad j = i, \\[2ex] \left(\dfrac{q_i}{p_i + q_i}\right) \cdots \left(\dfrac{q_{j-1}}{p_{j-1} + q_{j-1}}\right)\dfrac{1}{p_j + q_j} & \text{for} \quad i < j, \tag{3.75} \\[2ex] 0 & \text{for } i > j. \end{cases}$$

Problems 3.6

1. A rat is put into the linear maze as shown

(a) Assume that the rat is equally likely to move right or left at each step. What is the probability that the rat finds the food before getting shocked?

(b) As a result of learning, at each step the rat moves to the right with probability $p > \frac{1}{2}$, and to the left with probability $q = 1 - p < \frac{1}{2}$. What is the probability that the rat finds the food before getting shocked?

2. Customer accounts receivable at Smith Company are classified each month according to

> 0: Current
> 1: 30 to 60 days past due
> 2: 60 to 90 days past due
> 3: Over 90 days past due

Consider a particular customer account and suppose that it evolves month-to-month as a Markov chain $\{X_n\}$ whose transition probability matrix is

$$
\mathbf{P} = \begin{array}{c} 0 \\ 1 \\ 2 \\ 3 \end{array} \begin{Vmatrix} .9 & .1 & 0 & 0 \\ .5 & 0 & .5 & 0 \\ .3 & 0 & 0 & .7 \\ .2 & 0 & 0 & .8 \end{Vmatrix}.
$$

Suppose that a certain customer's account is now in state 1: 30 to 60 days past due. What is the probability that this account will be paid (and thereby enter state 0: Current) before it becomes over 90 days past due? That is, let $T = \min\{n \geq 0: X_n = 0 \text{ or } X_n = 3\}$. Determine $\Pr\{X_T = 0 | X_0 = 1\}$.

3. Players A and B each have $50 at the beginning of a game in which each player bets $1 at each play, and the game continues until one player is broke. Suppose there is a constant probability $p = .492929 \ldots$ that Player A wins on any given bet. What is the mean duration of the game?

4. Consider the Markov chain $\{X_n\}$ whose transition matrix is

$$
\mathbf{P} = \begin{array}{c} 0 \\ 1 \\ 2 \\ 3 \end{array} \begin{Vmatrix} \alpha & \beta & 0 & 0 \\ \alpha & 0 & 0 & \beta \\ \alpha & \beta & 0 & 0 \\ 0 & 0 & 0 & 1 \end{Vmatrix},
$$

where $\alpha > 0$, $\beta > 0$ and $\alpha + \beta = 1$. Determine the mean time to reach state 3 starting from state 0. That is, find $E[T | X_0 = 0]$ where $T = \min\{n \geq 0; X_n = 3\}$.

3.7 Another Look at First Step Analysis★

In this section we provide an alternative approach to evaluating the functionals treated in Section 3.4. The nth power of a transition probability matrix having both transient and absorbing states is directly evaluated. From these nth powers it is possible to extract the mean number of visits to a transient state j prior to absorption, the mean time until absorption, and the probability of absorption in any particular absorbing state k. These functionals all depend on the initial state $X_0 = i$, and as a by-product of the derivation, we show that, as functions of this initial state i, these functionals satisfy their appropriate first step analysis equations.

Consider a Markov chain whose states are labeled $0, 1, \ldots, N$. States $0, 1, \ldots, r - 1$ are transient in that $P_{ij}^{(n)} \to 0$ as $n \to \infty$ for $0 \le i, j < r$, while states $r, \ldots, N$ are absorbing, or trap, and here $P_{ii} = 1$ for $r \le i \le N$. The transition matrix has the form

$$\mathbf{P} = \begin{Vmatrix} \mathbf{Q} & \mathbf{R} \\ \mathbf{O} & \mathbf{I} \end{Vmatrix} \tag{3.76}$$

where $\mathbf{0}$ is an $(N - r + 1) \times r$ matrix all of whose components are zero, $\mathbf{I}$ is an $(N - r + 1) \times (N - r + 1)$ identity matrix and $Q_{ij} = P_{ij}$ for $0 \le i, j < r$.

To illustrate the calculations, begin with the four state transition matrix

$$\mathbf{P} = \begin{array}{c} 0 \\ 1 \\ 2 \\ 3 \end{array} \begin{Vmatrix} \begin{array}{cccc} 0 & 1 & 2 & 3 \end{array} \\ Q_{00} & Q_{01} & R_{02} & R_{03} \\ Q_{10} & Q_{11} & R_{12} & R_{13} \\ 0 & 0 & 1 & 0 \\ 0 & 0 & 0 & 1 \end{Vmatrix}. \tag{3.77}$$

Straightforward matrix multiplication shows the square of $\mathbf{P}$ to be

$$\mathbf{P}^2 = \begin{Vmatrix} \mathbf{Q}^2 & \mathbf{R} + \mathbf{QR} \\ \mathbf{O} & \mathbf{I} \end{Vmatrix}. \tag{3.78}$$

Continuing on to the third power, we have

$$\mathbf{P}^3 = \begin{Vmatrix} \mathbf{Q} & \mathbf{R} \\ \mathbf{O} & \mathbf{I} \end{Vmatrix} \times \begin{Vmatrix} \mathbf{Q}^2 & \mathbf{R} + \mathbf{QR} \\ \mathbf{O} & \mathbf{I} \end{Vmatrix} = \begin{Vmatrix} \mathbf{Q}^3 & \mathbf{R} + \mathbf{QR} + \mathbf{Q}^2\mathbf{R} \\ \mathbf{O} & \mathbf{I} \end{Vmatrix}$$

and for higher values of n,

$$\mathbf{P}^n = \begin{Vmatrix} \mathbf{Q}^n & (\mathbf{I} + \mathbf{Q} + \cdots + \mathbf{Q}^{n-1})\mathbf{R} \\ \mathbf{O} & \mathbf{I} \end{Vmatrix}. \tag{3.79}$$

★This section contains material at a more difficult level. It is not prerequisite to what follows.

The consideration of four states was for typographical convenience only. It is straightforward to verify that the nth power of $\mathbf{P}$ is given by (3.79) for the general $(N + 1)$ state transition matrix of (3.76) in which states $0, 1, \ldots, r - 1$ are *transient* $(P_{ij}^{(n)} \to 0$ as $n \to \infty$ for $0 \le i, j < r)$ while states $r, \ldots, N$ are *absorbing* $(P_{ii} = 1$ for $r \le i \le N)$.

We turn to the interpretation of (3.79). Let $W_{ij}^{(n)}$ be the mean number of visits to state j up to stage n for a Markov chain starting in state i. Formally,

$$W_{ij}^{(n)} = E\left[\sum_{l=0}^{n} \mathbf{1}\{X_l = j\} | X_0 = i\right] \tag{3.80}$$

where

$$\mathbf{1}\{X_l = j\} = \begin{cases} 1 & \text{if } X_l = j, \\ 0 & \text{if } X_l \ne j. \end{cases} \tag{3.81}$$

Now $E[\mathbf{1}\{X_l = j\} | X_0 = i] = \Pr\{X_l = j | X_0 = i\} = P_{ij}^{(l)}$ and since the expected value of a sum is the sum of the expected values, we obtain from (3.80) that

$$W_{ij}^{(n)} = \sum_{l=0}^{n} E[\mathbf{1}\{X_l = j\} | X_0 = i]$$

$$= \sum_{l=0}^{n} P_{ij}^{(l)}. \tag{3.82}$$

Equation (3.82) holds for all states i, j, but it has the most meaning when i and j are transient. Because (3.79) asserts that $P_{ij}^{(l)} = Q_{ij}^{(l)}$ when $0 \le i, j < r$, then

$$W_{ij}^{(n)} = Q_{ij}^{(0)} + Q_{ij}^{(1)} + \cdots + Q_{ij}^{(n)}, \qquad 0 \le i, j < r$$

where

$$Q_{ij}^{(0)} = \begin{cases} 1 & \text{if } i = j \\ 0 & \text{if } i \ne j. \end{cases}$$

In matrix notation, $\mathbf{Q}^{(0)} = \mathbf{I}$ and because $\mathbf{Q}^{(n)} = \mathbf{Q}^n$, the nth power of $\mathbf{Q}$, then

$$\begin{aligned}\mathbf{W}^{(n)} &= \mathbf{I} + \mathbf{Q} + \mathbf{Q}^2 + \cdots + \mathbf{Q}^n \\ &= \mathbf{I} + \mathbf{Q}(\mathbf{I} + \mathbf{Q} + \cdots + \mathbf{Q}^{n-1}) \\ &= \mathbf{I} + \mathbf{Q}\mathbf{W}^{(n-1)}. \end{aligned} \tag{3.83}$$

Upon writing out the matrix equation (3.83) in terms of the matrix entries, we recognize the results of a first step analysis. We have

$$W_{ij}^{(n)} = \delta_{ij} + \sum_{k=0}^{r-1} Q_{ik} W_{kj}^{(n-1)}$$

$$= \delta_{ij} + \sum_{k=0}^{r-1} P_{ik} W_{kj}^{(n-1)}.$$

In words, the equation asserts that the mean number of visits to state j in the first n stages starting from the initial state i includes the initial visit if $i = j$

(δ_{ij}) plus the future visits during the $n - 1$ remaining stages weighted by the appropriate transition probabilities.

We pass to the limit in (3.83) and obtain for

$$W_{ij} = \lim_{n \to \infty} W_{ij}^{(n)} = E[\text{Total visits to } j | X_0 = i] \qquad 0 \le i, j < r,$$

the matrix equations

$$W = I + Q + Q^2 + \cdots$$

and

$$W = I + QW. \qquad (3.84)$$

In terms of its entries, (3.84) is

$$W_{ij} = \delta_{ij} + \sum_{l=0}^{r-1} P_{il} W_{lj} \qquad \text{for} \quad i, j = 0, \ldots, r - 1. \qquad (3.85)$$

Equation (3.85) is the same as Equation (3.29) which was derived by a first step analysis.

Rewriting Equation (3.84) in the form

$$\mathbf{W} - \mathbf{QW} = (\mathbf{I} - \mathbf{Q})\mathbf{W} = \mathbf{I} \qquad (3.86)$$

we see that $\mathbf{W} = (\mathbf{I} - \mathbf{Q})^{-1}$, the inverse matrix to $\mathbf{I} - \mathbf{Q}$. The matrix $\mathbf{W}$ is often called the *fundamental* matrix associated with $\mathbf{Q}$.

Let T be the time of absorption. Formally, since states $r, r + 1, \ldots,$ N are the absorbing ones, the definition is

$$T = \min\{n \ge 0 : r \le X_n \le N\}.$$

Then the $i - j$th element W_{ij} of the fundamental matrix W evaluates

$$W_{ij} = E\left[\sum_{n=0}^{T-1} \mathbf{1}\{X = j\} | X_0 = i \right] \qquad \text{for} \quad 0 \le i, j < r. \qquad (3.87)$$

Let $v_i = E[T | X_0 = i]$ be the mean time to absorption starting from state i. The time to absorption is comprised of sojourns in the transient states. Formally,

$$\sum_{j=0}^{r-1} \sum_{n=0}^{T-1} \mathbf{1}\{X_n = j\} = \sum_{n=0}^{T-1} \sum_{j=0}^{r-1} \mathbf{1}\{X_n = j\}$$

$$= \sum_{n=0}^{T-1} 1 = T.$$

It follows from (3.87), then, that

$$\sum_{j=0}^{r-1} W_{ij} = \sum_{j=0}^{r-1} E\left[\sum_{n=0}^{T-1} \mathbf{1}\{X_n = j\} | X_0 = i \right]$$

$$= E[T | X_0 = i] = v_i \qquad \text{for} \quad 0 \le i < r. \qquad (3.88)$$

Summing Equation (3.85) over transient states j as follows,

$$\sum_{j=0}^{r-1} W_{ij} = \sum_{j=0}^{r-1} \delta_{ij} + \sum_{j=0}^{r-1}\sum_{k=0}^{r-1} P_{ik}W_{kj} \qquad \text{for} \quad i = 0, 1, \ldots, r-1,$$

and using the equivalence $v_i = \sum_{j=0}^{r-1} W_{ij}$ leads to

$$v_i = 1 + \sum_{k=0}^{r-1} P_{ik}v_k \qquad \text{for} \quad i = 0, 1, \ldots, r-1. \tag{3.89}$$

This equation is identical with that derived by first step analysis in (3.22). We turn to the hitting probabilities. Recall that states $k = r, \ldots, N$ are absorbing. Since such a state cannot be left once entered, the probability of absorption in a particular absorbing state k up to time n, starting from initial state i, is simply

$$\begin{aligned} P_{ik}^{(n)} &= \Pr\{X_n = k | X_0 = i\} \\ &= \Pr\{T \le n \text{ and } X_T = k | X_0 = i\} \end{aligned} \tag{3.90}$$

$$\text{for} \quad i = 0, \ldots, r-1; \quad k = r, \ldots, N,$$

where $T = \min\{n \ge 0 : r \le X_n \le N\}$ is the time of absorption. Let

$$U_{ik}^{(n)} = \Pr\{T \le n \text{ and } X_T = k | X_0 = i\}$$

$$\text{for} \quad 0 \le i < r \text{ and } r \le k \le N. \tag{3.91}$$

Referring to (3.79) and (3.90), we give the matrix $\mathbf{U}^{(n)}$ by

$$\begin{aligned} \mathbf{U}^{(n)} &= (\mathbf{I} + \mathbf{Q} + \cdots + \mathbf{Q}^{n-1})\mathbf{R} \\ &= \mathbf{W}^{(n-1)}\mathbf{R} \qquad [\text{by } (3.83)]. \end{aligned} \tag{3.92}$$

If we pass to the limit in n, we obtain the hitting probabilities

$$U_{ik} = \lim_{n \to \infty} U_{ik}^{(n)} = \Pr\{X_T = k | X_0 = i\} \qquad \text{for} \quad 0 \le i < r \text{ and } r \le k \le N.$$

Equation (3.92) then leads to an expression of the hitting probability matrix $\mathbf{U}$ in terms of the fundamental matrix $\mathbf{W}$ as simply $\mathbf{U} = \mathbf{WR}$, or

$$U_{ik} = \sum_{j=0}^{r-1} W_{ij}R_{jk} \qquad \text{for} \quad 0 \le i < r \text{ and } r \le k \le N. \tag{3.93}$$

Equation (3.93) may be used in conjunction with (3.85) to verify the first step analysis equation for U_{ik}. We multiply (3.85) by R_{jk} and sum, obtaining thereby,

$$\sum_{j=0}^{r-1} W_{ij}R_{jk} = \sum_{j=0}^{r-1} \delta_{ij}R_{jk} + \sum_{j=0}^{r-1}\sum_{l=0}^{r-1} P_{il}W_{lj}R_{jk},$$

which with (3.93) gives

$$\begin{aligned} U_{ik} &= R_{ik} + \sum_{l=0}^{r-1} P_{il}U_{lk} \\ &= P_{ik} + \sum_{l=0}^{r-1} P_{il}U_{lk} \qquad \text{for} \quad 0 \le i < r \text{ and } r \le k \le N. \end{aligned}$$

This equation was derived earlier by first step analysis in (3.20).

Chapter 4 | The Long Run Behavior of Markov Chains

4.1 Regular Transition Probability Matrices

Suppose that a transition probability matrix $\mathbf{P} = \|P_{ij}\|$ on a finite number of states labeled 0, 1, . . . , N, has the property that, when raised to some power k, the matrix $\mathbf{P}^k$ has all of its elements strictly positive. Such a transition probability matrix, or the corresponding Markov chain, is called *regular*. The most important fact concerning a regular Markov chain is the existence of a *limiting probability distribution* $\boldsymbol{\pi} = (\pi_0, \pi_1, . . . , \pi_N)$ where $\pi_j > 0$ for $j = 0, 1, . . . , N$ and $\Sigma_j \pi_j = 1$, and this distribution is independent of the initial state. Formally, for a regular transition probability matrix $\mathbf{P} = \|P_{ij}\|$ we have the convergence

$$\lim_{n \to \infty} P_{ij}^{(n)} = \pi_j > 0 \qquad \text{for} \quad j = 0, 1, . . . , N,$$

or, in terms of the Markov chain $\{X_n\}$,

$$\lim_{n \to \infty} \Pr\{X_n = j | X_0 = i\} = \pi_j > 0 \qquad \text{for} \quad j = 0, 1, . . . , N.$$

This convergence means that, in the long run ($n \to \infty$), the probability of finding the Markov chain in state j is approximately π_j no matter in which state the chain began at time 0.

Example The Markov chain whose transition probability matrix is

$$\mathbf{P} = \begin{matrix} & 0 & 1 \\ 0 \\ 1 \end{matrix} \left\| \begin{matrix} 1 - a & a \\ b & 1 - b \end{matrix} \right\| \tag{4.1}$$

is regular when $0 < a, b < 1$, and in this case the limiting distribution is $\pi = (b/(a + b), a/(a + b))$. To give a numerical example, we will suppose

$$\mathbf{P} = \left\| \begin{matrix} .33 & .67 \\ .75 & .25 \end{matrix} \right\|.$$

The first several powers of P are given here:

$$\mathbf{P}^2 = \left\| \begin{matrix} .6114 & .3886 \\ .4350 & .5650 \end{matrix} \right\|, \quad \mathbf{P}^3 = \left\| \begin{matrix} .4932 & .5068 \\ .5673 & .4327 \end{matrix} \right\|$$

$$\mathbf{P}^4 = \left\| \begin{matrix} .5428 & .4572 \\ .5117 & .4883 \end{matrix} \right\|, \quad \mathbf{P}^5 = \left\| \begin{matrix} .5220 & .4780 \\ .5350 & .4560 \end{matrix} \right\|$$

$$\mathbf{P}^6 = \left\| \begin{matrix} .5307 & .4693 \\ .5253 & .4747 \end{matrix} \right\|, \quad \mathbf{P}^7 = \left\| \begin{matrix} .5271 & .4729 \\ .5294 & .4706 \end{matrix} \right\|.$$

By $n = 7$, the entries agree row-to-row to two decimal places. The limiting probabilities are $b/(a + b) = .5282$ and $a/(a + b) = .4718$.

Example Sociologists often assume that the social classes of successive generations in a family can be regarded as a Markov chain. Thus, the occupation of a son is assumed to depend only on his father's occupation and not on his grandfather's. Suppose that such a model is appropriate and that the transition probability matrix is given by

Son's Class

		Lower	Middle	Upper
Father's Class	Lower	.40	.50	.10
	Middle	.05	.70	.25
	Upper	.05	.50	.45

For such a population, what fraction of people are middle class in the long run?

For the time being, we will answer the question by computing sufficiently high powers of $\mathbf{P}^n$. A better method for determining the limiting distribution will be presented later in this section.

We compute

$$\mathbf{P}^2 = \mathbf{P} \times \mathbf{P} = \left\| \begin{matrix} .1900 & .6000 & .2100 \\ .0675 & .6400 & .2925 \\ .0675 & .6000 & .3325 \end{matrix} \right\|,$$

$$\mathbf{P}^4 = \mathbf{P}^2 \times \mathbf{P}^2 = \left\| \begin{matrix} .0908 & .6240 & .2852 \\ .0758 & .6256 & .2986 \\ .0758 & .6240 & .3002 \end{matrix} \right\|,$$

$$\mathbf{P}^8 = \mathbf{P}^4 \times \mathbf{P}^4 = \left\| \begin{matrix} .0772 & .6250 & .2978 \\ .0769 & .6250 & .2981 \\ .0769 & .6250 & .2981 \end{matrix} \right\|.$$

Note that we have not computed $\mathbf{P}^n$ for consecutive values of n but have speeded up the calculations by evaluating the successive squares $\mathbf{P}^2$, $\mathbf{P}^4$, $\mathbf{P}^8$.

In the long run, approximately 62.5 percent of the population are middle class under the assumptions of the model.

Computing the limiting distribution by raising the transition probability matrix to a high power suffers from being inexact, since $n = \infty$ is never attained, and it also requires more computational effort than is necessary. Theorem 4.1 provides an alternative computational approach by asserting that the limiting distribution is the unique solution to a set of linear equations. For this social class example, the exact limiting distribution, computed using the method of Theorem 4.1, is $\pi_0 = \frac{1}{13} = .0769$, $\pi_1 = \frac{5}{8} = .6250$, and $\pi_2 = \frac{31}{104} = .2981$.

If a transition probability matrix $\mathbf{P}$ on N states is regular, then $\mathbf{P}^{N^2}$ will have no zero elements. Conversely, if $\mathbf{P}^{N^2}$ is not strictly positive, then the Markov chain is not regular. Furthermore, once it happens that $\mathbf{P}^k$ has no zero entries, then every higher power $\mathbf{P}^{k+n}$, $n = 1, 2, \ldots$ will have no zero entries. Thus it suffices to check the successive squares $\mathbf{P}$, $\mathbf{P}^2$, $\mathbf{P}^4$, $\mathbf{P}^8$, $\ldots$. Finally, to determine whether or not the square of a transition probability matrix has only strictly positive entries, it is not necessary to perform the actual multiplication, but only to record whether or not the product is nonzero.

Example Consider the transition probability matrix

$$
\mathbf{P} = \begin{Vmatrix}
.9 & .1 & 0 & 0 & 0 & 0 & 0 \\
.9 & 0 & .1 & 0 & 0 & 0 & 0 \\
.9 & 0 & 0 & .1 & 0 & 0 & 0 \\
.9 & 0 & 0 & 0 & .1 & 0 & 0 \\
.9 & 0 & 0 & 0 & 0 & .1 & 0 \\
.9 & 0 & 0 & 0 & 0 & 0 & .1 \\
.9 & 0 & 0 & 0 & 0 & 0 & .1
\end{Vmatrix}.
$$

We recognize this as a success runs Markov chain. We record the nonzero entries as $+$ and write $\mathbf{P} \times \mathbf{P}$ in the form

$$
\mathbf{P} \times \mathbf{P} = \begin{Vmatrix}
+ & + & 0 & 0 & 0 & 0 & 0 \\
+ & 0 & + & 0 & 0 & 0 & 0 \\
+ & 0 & 0 & + & 0 & 0 & 0 \\
+ & 0 & 0 & 0 & + & 0 & 0 \\
+ & 0 & 0 & 0 & 0 & + & 0 \\
+ & 0 & 0 & 0 & 0 & 0 & + \\
+ & 0 & 0 & 0 & 0 & 0 & +
\end{Vmatrix}
$$

$$
\times
\begin{Vmatrix}
+ & + & 0 & 0 & 0 & 0 & 0 \\
+ & 0 & + & 0 & 0 & 0 & 0 \\
+ & 0 & 0 & + & 0 & 0 & 0 \\
+ & 0 & 0 & 0 & + & 0 & 0 \\
+ & 0 & 0 & 0 & 0 & + & 0 \\
+ & 0 & 0 & 0 & 0 & 0 & + \\
+ & 0 & 0 & 0 & 0 & 0 & +
\end{Vmatrix}
$$

$$
=
\begin{Vmatrix}
+ & + & + & 0 & 0 & 0 & 0 \\
+ & + & 0 & + & 0 & 0 & 0 \\
+ & + & 0 & 0 & + & 0 & 0 \\
+ & + & 0 & 0 & 0 & + & 0 \\
+ & + & 0 & 0 & 0 & 0 & + \\
+ & + & 0 & 0 & 0 & 0 & + \\
+ & + & 0 & 0 & 0 & 0 & +
\end{Vmatrix}
= \mathbf{P}^2,
$$

$$
\mathbf{P}^4 =
\begin{Vmatrix}
+ & + & + & + & + & 0 & 0 \\
+ & + & + & + & 0 & + & 0 \\
+ & + & + & + & 0 & 0 & + \\
+ & + & + & + & 0 & 0 & + \\
+ & + & + & + & 0 & 0 & + \\
+ & + & + & + & 0 & 0 & + \\
+ & + & + & + & 0 & 0 & +
\end{Vmatrix},
$$

$$
\mathbf{P}^8 =
\begin{Vmatrix}
+ & + & + & + & + & + & + \\
+ & + & + & + & + & + & + \\
+ & + & + & + & + & + & + \\
+ & + & + & + & + & + & + \\
+ & + & + & + & + & + & + \\
+ & + & + & + & + & + & + \\
+ & + & + & + & + & + & +
\end{Vmatrix}.
$$

We see that $\mathbf{P}^8$ has all strictly positive entries, and therefore $\mathbf{P}$ is regular. The limiting distribution for a similar matrix is computed in Section 4.2.2.

Every transition probability matrix on the states $0, 1, \ldots, N$ that satisfies the following two conditions is regular:

(1) For every pair of states i, j there is a path $k_1, \ldots, k_r$ for which $P_{ik_1} P_{k_1 k_2} \cdots P_{k_r j} > 0$.
(2) There is at least one state i for which $P_{ii} > 0$.

Theorem 4.1 *Let $\mathbf{P}$ be a regular transition probability matrix on the states $0, 1, \ldots, N$. Then the limiting distribution $\boldsymbol{\pi} = (\pi_0, \pi_1, \ldots, \pi_N)$ is the unique nonnegative solution of the equations*

$$\pi_j = \sum_{k=0}^{N} \pi_k P_{kj}, \qquad j = 0, 1, \ldots, N, \tag{4.2}$$

$$\sum_{k=0}^{N} \pi_k = 1. \tag{4.3}$$

Proof   Because the Markov chain is regular, we have a limiting distribution, $\lim_{n \to \infty} P_{ij}^{(n)} = \pi_j$, for which $\sum_{k=0}^{N} \pi_k = 1$. Write $\mathbf{P}^n$ as the matrix product $\mathbf{P}^{n-1}\mathbf{P}$ in the form

$$P_{ij}^{(n)} = \sum_{k=0}^{N} P_{ik}^{(n-1)} P_{kj}, \qquad j = 0, \ldots, N, \tag{4.4}$$

and now let $n \to \infty$. Then $P_{ij}^{(n)} \to \pi_j$ while $P_{ik}^{(n-1)} \to \pi_k$ and (4.4) passes into $\pi_j = \sum_{k=0}^{N} \pi_k P_{kj}$ as claimed.

It remains to show that the solution is unique. Suppose that $x_0, x_1,$ $\ldots, x_N$ solves

$$x_j = \sum_{k=0}^{N} x_k P_{kj} \qquad \text{for } j = 0, \ldots, N \tag{4.5}$$

and

$$\sum_{k=0}^{N} x_k = 1. \tag{4.6}$$

We wish to show that $x_j = \pi_j$, the limiting probability. Begin by multiplying (4.5) on the right by P_{jl} and then sum over j to get

$$\sum_{j=0}^{N} x_j P_{jl} = \sum_{j=0}^{N}\sum_{k=0}^{N} x_k P_{kj} P_{jl} = \sum_{k=0}^{N} x_k P_{kl}^{(2)}. \tag{4.7}$$

But by (4.5) we have $x_l = \sum_{j=0}^{N} x_j P_{jl}$ whence (4.7) becomes

$$x_l = \sum_{k=0}^{N} x_k P_{kl}^{(2)} \qquad \text{for } l = 0, \ldots, N.$$

Repeating this argument n times we deduce that

$$x_l = \sum_{k=0}^{N} x_k P_{kl}^{(n)} \qquad \text{for } l = 0, \ldots, N,$$

and then passing to the limit in n and using that $P_{kl}^{(n)} \to \pi_l$ we see that

$$x_l = \sum_{k=0}^{N} x_k \pi_l, \qquad l = 0, \ldots, N.$$

But by (4.6) we have $\sum_k x_k = 1$, whence $x_l = \pi_l$ as claimed.   □

Example   For the social class matrix

$$\mathbf{P} = \begin{array}{c} \\ 0 \\ 1 \\ 2 \end{array} \begin{array}{ccc} 0 & 1 & 2 \\ \left\| \begin{array}{ccc} .40 & .50 & .10 \\ .05 & .70 & .25 \\ .05 & .50 & .45 \end{array} \right\| \end{array}$$

the equations determining the limiting distribution (π_0, π_1, π_2) are

$$.40\pi_0 + .05\pi_1 + .05\pi_2 = \pi_0 \tag{4.8}$$

$$.50\pi_0 + .70\pi_1 + .50\pi_2 = \pi_1 \tag{4.9}$$

$$.10\pi_0 + .25\pi_1 + .45\pi_2 = \pi_2 \tag{4.10}$$

$$\pi_0 + \quad \pi_1 + \quad \pi_2 = 1. \tag{4.11}$$

One of the equations (4.8), (4.9), and (4.10) is redundant, because of the linear constraint $\Sigma_k P_{ik} = 1$. We arbitrarily strike out (4.10) and simplify the remaining equations to get

$$-60\pi_0 + 5\pi_1 + 5\pi_2 = 0 \tag{4.12}$$

$$5\pi_0 - 3\pi_1 + 5\pi_2 = 0 \tag{4.13}$$

$$\pi_0 + \quad \pi_1 + \quad \pi_2 = 1. \tag{4.14}$$

We eliminate π_2 by subtracting (4.12) from (4.13) and five times (4.14) to reduce the system to

$$65\pi_0 - 8\pi_1 = 0$$
$$65\pi_0 \quad\quad = 5.$$

Then $\pi_0 = \frac{5}{65} = \frac{1}{13}$, $\pi_1 = \frac{5}{8}$, and then $\pi_2 = 1 - \pi_0 - \pi_1 = \frac{31}{104}$ as given earlier.

Doubly Stochastic Matrices

A transition probability matrix is called doubly stochastic if the columns sum to one as well as the rows. Formally $\mathbf{P} = \|P_{ij}\|$ is doubly stochastic if

$$P_{ij} \geq 0 \quad \text{and} \quad \sum_k P_{ik} = \sum_k P_{kj} = 1 \quad\quad \text{for all} \quad i, j.$$

Consider a doubly stochastic transition probability matrix on the N states $0, 1, \ldots, N - 1$. If the matrix is regular, then the unique limiting distribution is the uniform distribution $\boldsymbol{\pi} = (1/N, \ldots, 1/N)$. Because there is only one solution to $\pi_j = \Sigma_k \pi_k P_{kj}$ and $\Sigma_k \pi_k = 1$ when P is regular, we need only check that $\boldsymbol{\pi} = (1/N, \ldots, 1/N)$ is a solution when $\mathbf{P}$ is doubly stochastic in order to establish the claim. By using the doubly stochastic feature $\Sigma_j P_{jk} = 1$ we verify that

$$\frac{1}{N} = \sum_j \frac{1}{N} P_{jk} = \frac{1}{N}.$$

As an example, let Y_n be the sum of n independent rolls of a fair die and consider the problem of determining with what probability Y_n is a multiple of 7 in the long run. Let X_n be the remainder when Y_n is divided by 7. Then X_n is a Markov chain on the states $0, 1, \ldots, 6$ with transition probability matrix

$$
\mathbf{P} = \begin{array}{c} \\ 0 \\ 1 \\ 2 \\ 3 \\ 4 \\ 5 \\ 6 \end{array}
\begin{array}{ccccccc}
0 & 1 & 2 & 3 & 4 & 5 & 6 \\
\left\| \begin{array}{ccccccc}
0 & \frac{1}{6} & \frac{1}{6} & \frac{1}{6} & \frac{1}{6} & \frac{1}{6} & \frac{1}{6} \\
\frac{1}{6} & 0 & \frac{1}{6} & \frac{1}{6} & \frac{1}{6} & \frac{1}{6} & \frac{1}{6} \\
\frac{1}{6} & \frac{1}{6} & 0 & \frac{1}{6} & \frac{1}{6} & \frac{1}{6} & \frac{1}{6} \\
\frac{1}{6} & \frac{1}{6} & \frac{1}{6} & 0 & \frac{1}{6} & \frac{1}{6} & \frac{1}{6} \\
\frac{1}{6} & \frac{1}{6} & \frac{1}{6} & \frac{1}{6} & 0 & \frac{1}{6} & \frac{1}{6} \\
\frac{1}{6} & \frac{1}{6} & \frac{1}{6} & \frac{1}{6} & \frac{1}{6} & 0 & \frac{1}{6} \\
\frac{1}{6} & \frac{1}{6} & \frac{1}{6} & \frac{1}{6} & \frac{1}{6} & \frac{1}{6} & 0
\end{array} \right\|
\end{array}.
$$

The matrix is doubly stochastic, and it is regular ($\mathbf{P}^2$ has only strictly positive entries), hence the limiting distribution is $\boldsymbol{\pi} = (\frac{1}{7}, \ldots, \frac{1}{7})$. Furthermore, Y_n is a multiple of 7 if and only if $X_n = 0$. Thus the limiting probability that Y_n is a multiple of 7 is $\frac{1}{7}$.

Interpretation of the Limiting Distribution

Given a regular transition matrix $\mathbf{P}$ for a Markov process $\{X_n\}$ on the $N + 1$ states $0, 1, \ldots, N$, we solve the linear equations

$$
\pi_i = \sum_{k=0}^{N} \pi_k P_{ki} \quad \text{for} \quad i = 0, 1, \ldots, N
$$

and

$$
\pi_0 + \pi_1 + \cdots + \pi_N = 1.
$$

The primary interpretation of the solution $(\pi_0, \ldots, \pi_N)$ is as the limiting distribution

$$
\pi_j = \lim_{n \to \infty} P_{ij}^{(n)} = \lim_{n \to \infty} \Pr\{X_n = j \mid X_0 = i\}.
$$

In words, after the process has been in operation for a long duration, the probability of finding the process in state j is π_j, irrespective of the starting state.

There is a second interpretation of the limiting distribution $\boldsymbol{\pi} = (\pi_0, \pi_1, \ldots, \pi_N)$ that plays a major role in many models. We claim that π_j also gives the long run mean fraction of time that the process $\{X_n\}$ is in state j. Thus if each visit to state j incurs a "cost" of c_j, then the long run mean cost per unit time associated with this Markov chain is

$$
\text{Long run mean cost per unit time} = \sum_{j=0}^{N} \pi_j c_j.
$$

To verify this interpretation, recall that if a sequence $a_0, a_1, \ldots$ of real numbers converges to a limit a, then the averages of these numbers also converge in the manner

$$
\lim_{n \to \infty} \frac{1}{m} \sum_{k=0}^{m-1} a_k = a.
$$

We apply this result to the convergence $\lim_{n \to \infty} P_{ij}^{(n)} = \pi_j$ to conclude that

$$\lim_{m \to \infty} \frac{1}{m} \sum_{k=0}^{m-1} P_{ij}^{(k)} = \pi_j.$$

Now $(1/m) \sum_{k=0}^{m-1} P_{ij}^{(k)}$ is exactly the mean fraction of time during steps $0, 1, \ldots, m - 1$ that the process spends in state j. Indeed, the actual (random) fraction of time in state j is

$$\frac{1}{m} \sum_{k=0}^{m-1} 1\{X_k = j\}$$

where

$$1\{X_k = j\} = \begin{cases} 1 & \text{if } X_k = j, \\ 0 & \text{if } X_k \neq j. \end{cases}$$

Therefore the *mean* fraction of visits is obtained by taking expected values according to

$$E\left[\frac{1}{m} \sum_{k=0}^{m-1} 1\{X_k = j\} \Big| X_0 = i \right] = \frac{1}{m} \sum_{k=0}^{m-1} E[1\{X_k = j\} | X_0 = i\}$$

$$= \frac{1}{m} \sum_{k=0}^{m-1} \Pr\{X_k = j | X_0 = i\}$$

$$= \frac{1}{m} \sum_{k=0}^{m-1} P_{ij}^{(k)}.$$

Because $\lim_{n \to \infty} P_{ij}^{(n)} = \pi_j$, the long run mean fraction of time that the process spends in state j is

$$\lim_{m \to \infty} E\left[\frac{1}{m} \sum_{k=0}^{m-1} 1\{X_k = j\} \Big| X_0 = i \right] = \lim_{m \to \infty} \frac{1}{m} \sum_{k=0}^{m-1} P_{ij}^{(k)} = \pi_j,$$

independent of the starting state i.

Problems 4.1

1. Compute the limiting distribution for the transition probability matrix

$$\mathbf{P} = \begin{array}{c} \\ 0 \\ 1 \\ 2 \end{array} \begin{array}{ccc} 0 & 1 & 2 \\ \left\| \begin{array}{ccc} \frac{1}{2} & \frac{1}{2} & 0 \\ \frac{1}{3} & \frac{1}{3} & \frac{1}{3} \\ \frac{1}{6} & \frac{1}{2} & \frac{1}{3} \end{array} \right\| \end{array}.$$

2. A Markov chain on the states 0, 1, 2, 3 has the transition probability matrix

$$\mathbf{P} = \begin{array}{c} \\ 0 \\ 1 \\ 2 \\ 3 \end{array}\begin{array}{c}0\\\left|\left|\begin{array}{c}.1\\0\\0\\1\end{array}\right.\end{array}\begin{array}{c}1\\.2\\.3\\0\\0\end{array}\begin{array}{c}2\\.3\\.3\\.6\\0\end{array}\begin{array}{c}3\\\left.\begin{array}{c}.4\\.4\\.4\\0\end{array}\right|\right|\end{array}.$$

Determine the corresponding limiting distribution.

3. Suppose that the social classes of successive generations in a family follow a Markov chain with transition probability matrix given by

Son's Class

		Lower	Middle	Upper
Father's Class	Lower	.7	.2	.1
	Middle	.2	.6	.2
	Upper	.1	.4	.5

What fraction of families are upper class in the long run?

4. Show that the transition probability matrix

$$\mathbf{P} = \begin{array}{c} \\ 0 \\ 1 \\ 2 \\ 3 \\ 4 \end{array}\left|\left|\begin{array}{ccccc} 0 & \frac{1}{2} & \frac{1}{2} & 0 & 0 \\ \frac{1}{2} & 0 & \frac{1}{2} & 0 & 0 \\ \frac{1}{3} & \frac{1}{3} & 0 & \frac{1}{3} & 0 \\ 0 & 0 & \frac{1}{2} & 0 & \frac{1}{2} \\ \frac{1}{2} & 0 & 0 & \frac{1}{2} & 0 \end{array}\right|\right|$$

with columns labeled $0\ 1\ 2\ 3\ 4$

is regular and compute the limiting distribution.

5. Consider a Markov chain with transition probability matrix

$$\mathbf{P} = \left|\left|\begin{array}{ccccc} p_0 & p_1 & p_2 & \cdots & p_N \\ p_N & p_0 & p_1 & \cdots & p_{N-1} \\ p_{N-1} & p_N & p_0 & \cdots & p_{N-2} \\ \cdot & \cdot & \cdot & & \cdot \\ \cdot & \cdot & \cdot & & \cdot \\ \cdot & \cdot & \cdot & & \cdot \\ p_1 & p_2 & p_3 & \cdots & p_0 \end{array}\right|\right|$$

where $0 < p_0 < 1$ and $p_0 + p_1 + \ldots + p_N = 1$. Determine the limiting distribution.

6. Determine the following limits in terms of the transition probability matrix $\mathbf{P} = \|P_{ij}\|$ and limiting distribution $\boldsymbol{\pi} = \|\pi_j\|$ of a finite state regular Markov chain $\{X_n\}$:

   (a) $\lim\limits_{n\to\infty} \Pr\{X_{n+1} = j | X_0 = i\}$,

   (b) $\lim\limits_{n\to\infty} \Pr\{X_n = k, X_{n+1} = j | X_0 = i\}$.

7. Determine the limiting distribution for the Markov chain whose transition probability matrix is

$$
\mathbf{P} = \begin{array}{c} \\ 0 \\ 1 \\ 2 \\ 3 \end{array}
\begin{array}{cccc}
0 & 1 & 2 & 3 \\
\left\| \begin{matrix} \frac{1}{2} & 0 & 0 & \frac{1}{2} \\ 1 & 0 & 0 & 0 \\ 0 & \frac{1}{2} & \frac{1}{3} & \frac{1}{6} \\ 0 & 0 & 1 & 0 \end{matrix} \right\|
\end{array}.
$$

8. Determine the limiting distribution for the Markov chain whose transition probability matrix is

$$
\mathbf{P} = \begin{array}{c} \\ 0 \\ 1 \\ 2 \end{array}
\begin{array}{ccc}
0 & 1 & 2 \\
\left\| \begin{matrix} \frac{1}{2} & \frac{1}{2} & 0 \\ \frac{1}{3} & \frac{1}{2} & \frac{1}{6} \\ 0 & \frac{1}{4} & \frac{3}{4} \end{matrix} \right\|
\end{array}.
$$

9. Determine the long run or limiting distribution for the Markov chain whose transition matrix is

$$
\mathbf{P} = \begin{array}{c} \\ 0 \\ 1 \\ 2 \\ 3 \end{array}
\begin{array}{cccc}
0 & 1 & 2 & 3 \\
\left\| \begin{matrix} 0 & 0 & 1 & 0 \\ 0 & 0 & 0 & 1 \\ \frac{1}{4} & \frac{1}{4} & \frac{1}{4} & \frac{1}{4} \\ 0 & 0 & \frac{1}{2} & \frac{1}{2} \end{matrix} \right\|
\end{array}.
$$

4.2 Examples

Markov chains arising in meteorology, reliability, statistical quality control, and management science are presented next, and the long run behavior of each Markov chain is developed and interpreted in terms of the phenomenon under study.

4.2.1 Including History in the State Description

Often a phenomenon that is not naturally a Markov process can be modeled as a Markov process by including part of the past history in the state description. To illustrate this technique, we suppose that the weather on any day depends on the weather conditions for the previous two days. To be exact, we suppose that if it was sunny today and yesterday, then it will be sunny tomorrow with probability .8; if it was sunny today but cloudy yesterday, then it will be sunny tomorrow with probability .6; if it was cloudy today but sunny yesterday, then it will be sunny tomorrow with probability .4; if it was cloudy for the last two days, then it will be sunny tomorrow with probability .1.

Such a model can be transformed into a Markov chain provided we say that the state at any time is determined by the weather conditions during both that day and the previous day. We say the process is in

State (S, S) if it was sunny both today and yesterday,
State (S, C) if it was sunny yesterday but cloudy today,
State (C, S) if it was cloudy yesterday but sunny today,
State (C, C) if it was cloudy both today and yesterday.

Then the transition probability matrix is

Today's State

		(S, S)	(S, C)	(C, S)	(C, C)
	(S, S)	.8	.2		
Yesterday's	(S, C)			.4	.6
State	(C, S)	.6	.4		
	(C, C)			.1	.9

The equations determining the limiting distribution are

$$
\begin{aligned}
.8\pi_0 && + .6\pi_2 && &= \pi_0 \\
.2\pi_0 && + .4\pi_2 && &= \pi_1 \\
& .4\pi_1 && + .1\pi_3 &= \pi_2 \\
& .6\pi_1 && + .9\pi_3 &= \pi_3 \\
\pi_0 + & \pi_1 + & \pi_2 + & \pi_3 &= 1.
\end{aligned}
$$

Again, one of the top four equations is redundant. Striking out the first equation and solving the remaining four equations gives $\pi_0 = \frac{3}{11}$, $\pi_1 = \frac{1}{11}$, $\pi_2 = \frac{1}{11}$, and $\pi_3 = \frac{6}{11}$.

We recover the fraction of days, in the long run, on which it is sunny by summing the appropriate terms in the limiting distribution. It can be sunny today in conjunction with either being sunny or cloudy tomorrow. Therefore, the long run fraction of days on which it is sunny is $\pi_0 + \pi_1 = \pi(S, S) + \pi(S, C) = \frac{4}{11}$. Formally, $\lim_{n \to \infty} \Pr\{X_n = S\} = \lim_{n \to \infty} [\Pr\{X_n = S, X_{n+1} = S\} + \Pr\{X_n = S, X_{n+1} = C\}] = \pi_0 + \pi_1$.

4.2.2 Reliability and Redundancy

An airline reservation system has two computers only one of which is in operation at any given time. A computer may break down on any given day with probability p. There is a single repair facility which takes 2 days to restore a computer to normal. The facilities are such that only one computer at a time can be dealt with. Form a Markov chain by taking as states the pairs (x, y) where x is the number of machines in operating condition at the end of a day and y is 1 if a day's labor has been expended on a machine not yet repaired and 0 otherwise. The transition matrix is

To State
————— → (2, 0)   (1, 0)   (1, 1)   (0, 1)

From State
↓

$$\mathbf{P} = \begin{array}{c} (2,0) \\ (1,0) \\ (1,1) \\ (0,1) \end{array} \left\| \begin{array}{cccc} q & p & 0 & 0 \\ 0 & 0 & q & p \\ q & p & 0 & 0 \\ 0 & 1 & 0 & 0 \end{array} \right\|$$

where $p + q = 1$.

We are interested in the long run probability that both machines are inoperative. Let $(\pi_0, \pi_1, \pi_2, \pi_3)$ be the limiting distribution of the Markov chain. Then the long run probability that neither computer is operating is π_3, and the reliability, the probability that at least one computer is operating is $1 - \pi_3 = \pi_0 + \pi_1 + \pi_2$.

The equations for the limiting distributions are

$$\begin{aligned}
q\pi_0 \quad + q\pi_2 \quad\quad &= \pi_0 \\
p\pi_0 \quad + p\pi_2 + \pi_3 &= \pi_1 \\
q\pi_1 \quad\quad &= \pi_2 \\
p\pi_1 \quad\quad &= \pi_3
\end{aligned}$$

and

$$\pi_0 + \pi_1 + \pi_2 + \pi_3 = 1.$$

The solution is

$$\pi_0 = \frac{q^2}{1 + p^2} \qquad \pi_2 = \frac{qp}{1 + p^2}$$

$$\pi_1 = \frac{p}{1 + p^2} \qquad \pi_3 = \frac{p^2}{1 + p^2}.$$

The reliability is $R_1 = 1 - \pi_3 = 1/(1 + p^2)$.

In order to increase the system reliability it is proposed to add a duplicate repair facility so that both computers can be repaired simultaneously. The corresponding transition matrix is now

To State
————— → (2, 0)   (1, 0)   (1, 1)   (0, 1)

From State
↓

$$\mathbf{P} = \begin{array}{c} (2,0) \\ (1,0) \\ (1,1) \\ (0,1) \end{array} \left\| \begin{array}{cccc} q & p & 0 & 0 \\ 0 & 0 & q & p \\ q & p & 0 & 0 \\ 0 & 0 & 1 & 0 \end{array} \right\| ,$$

and the limiting distribution is

$$\pi_0 = \frac{q}{1 + p + p^2} \qquad \pi_2 = \frac{p}{1 + p + p^2}$$

$$\pi_1 = \frac{p}{1 + p + p^2} \qquad \pi_3 = \frac{p^2}{1 + p + p^2}.$$

The reliability has increased to $R_2 = 1 - \pi_3 = (1 + p)/(1 + p + p^2)$.

4.2.3 A Continuous Sampling Plan

Consider a production line where each item has probability p of being defective. Assume that the condition of a particular item (defective or nondefective) does not depend on the conditions of other items. The following sampling plan is used.

Initially every item is sampled as it is produced; this procedure continues until i consecutive nondefective items are found. Then the sampling plan calls for sampling only one out of every r items at random until a defective one is found. When this happens the plan calls for reverting to 100 percent sampling until i consecutive nondefective items are found. The process continues in the same way.

State E_k ($k = 0, 1, \ldots, i - 1$) denotes that k consecutive nondefective items have been found in the 100 percent sampling portion of the plan, while state E_i denotes that the plan is in the second stage (sampling one out of r). Time m is considered to follow the mth item, whether sampled or not. Then the sequence of states is a Markov chain with

$$P_{jk} = \Pr\{\text{in state } E_k \text{ after } m + 1 \text{ items} | \text{in state } E_j \text{ after } m \text{ items}\}$$

$$= \begin{cases} p & \text{for} \quad k = 0, 0 \le j < i, \\ 1 - p & \text{for} \quad k = j + 1 \le i, \\ \dfrac{p}{r} & \text{for} \quad k = 0, j = i, \\ 1 - \dfrac{p}{r} & \text{for} \quad k = j = i, \\ 0 & \text{otherwise.} \end{cases}$$

Let π_k be the limiting probability that the system is in state E_k for $k = 0, 1, \ldots, i$. The equations determining these limiting probabilities are

(0) $\qquad p\pi_0 + \qquad p\pi_1 + \cdots + \qquad p\pi_{i-1} + \qquad (p/r)\pi_i = \pi_0$

(1) $(1 - p)\pi_0 \qquad\qquad\qquad\qquad\qquad\qquad\qquad = \pi_1$

(2) $\qquad\qquad\qquad (1 - p)\pi_1 \qquad\qquad\qquad\qquad\qquad = \pi_2$

.

.

.

(i) $\qquad\qquad\qquad\qquad\qquad (1 - p)\pi_{i-1} + (1 - p/r)\pi_i = \pi_i$

together with

(★) $$\pi_0 + \pi_1 + \cdots + \pi_i = 1.$$

From Equations (1) through $(i - 1)$ we deduce that $\pi_k = (1 - p)\pi_{k-1}$ so that $\pi_k = (1 - p)^k\pi_0$ for $k = 0, \ldots, i - 1$, while Equation (i) yields $\pi_i = (r/p)(1 - p)\pi_{i-1}$ or $\pi_i = (r/p)(1 - p)^i\pi_0$. Having determined π_k in terms of π_0 for $k = 0, \ldots, i$, we place these values in (★) to obtain

$$\{[1 + \cdots + (1 - p)^{i-1}] + \frac{r}{p}(1 - p)^i\}\pi_0 = 1.$$

The geometric series simplifies, and after elementary algebra the solution is

$$\pi_0 = \frac{p}{1 + (r - 1)(1 - p)^i},$$

whence

$$\pi_k = \frac{p(1 - p)^k}{1 + (r - 1)(1 - p)^i} \qquad \text{for} \quad k = 0, \ldots, i - 1,$$

while

$$\pi_i = \frac{r(1 - p)^i}{1 + (r - 1)(1 - p)^i}.$$

Let AFI (Average Fraction Inspected) denote the long run fraction of items that are inspected. Since each item is inspected while in states $E_0, \ldots, E_{i-1}$ but only one out of r are inspected in state E_i we have

$$\begin{aligned}
\text{AFI} &= (\pi_0 + \cdots + \pi_{i-1}) + (1/r)\pi_i \\
&= (1 - \pi_i) + (1/r)\pi_i \\
&= \frac{1}{1 + (r - 1)(1 - p)^i}.
\end{aligned}$$

Let us assume that each item found to be defective is replaced by an item known to be good. The average outgoing quality (AOQ) is defined to be the fraction of defectives in the output of such an inspection scheme. The average fraction not inspected is

$$1 - \text{AFI} = \frac{(r - 1)(1 - p)^i}{1 + (r - 1)(1 - p)^i},$$

and of these on the average p are defective. Hence

$$\text{AOQ} = \frac{(r - 1)(1 - p)^i p}{1 + (r - 1)(1 - p)^i}.$$

This average outgoing quality is zero if $p = 0$ or $p = 1$, and rises to a maximum at some intermediate value, as shown in Figure 4.1. The maximum

AOQ is called the *average outgoing quality limit* (AOQL), and it has been determined numerically and tabulated as a function of i and r.

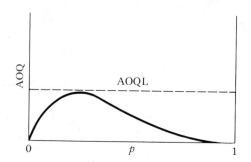

Figure 4.1   The average outgoing quality (AOQ) as a function of the input quality level p

This quality control scheme guarantees an output quality better than the AOQL regardless of the input fraction defective, as shown in Figure 4.2.

Figure 4.2   Blackbox picture of a continuous inspection scheme as a method of guaranteeing outgoing quality

4.2.4 Age Replacement Policies

A component of a computer has an active life, measured in discrete units, that is a random variable T where $\Pr[T = k] = a_k$ for $k = 1, 2, \ldots$. Suppose one starts with a fresh component and each component is replaced by a new component upon failure. Let X_n be the age of the component in service at time n. Then (X_n) is a success runs Markov chain. (See Section 3.5.4.)

In an attempt to replace components before they fail in service, an *age replacement* policy is instituted. This policy calls for replacing the component upon its failure, or upon its reaching age N, whichever occurs first. Under this age replacement policy, the Markov chain $\{X_n\}$ has the transition probability matrix

$$\mathbf{P} = \begin{array}{c} \\ 0 \\ 1 \\ 2 \\ \\ \\ \\ N-1 \end{array} \begin{array}{|ccccccc|} \multicolumn{1}{c}{0} & \multicolumn{1}{c}{1} & \multicolumn{1}{c}{2} & \multicolumn{1}{c}{3} & & \multicolumn{1}{c}{N-1} \\ \hline p_0 & 1-p_0 & 0 & 0 & \cdots & 0 \\ p_1 & 0 & 1-p_1 & 0 & & 0 \\ p_2 & 0 & 0 & 1-p_2 & \cdots & 0 \\ \cdot & & & & & \cdot \\ \cdot & & & & & \cdot \\ \cdot & & & & & \cdot \\ 1 & 0 & 0 & 0 & & 0 \end{array}$$

where

$$p_k = \frac{a_{k+1}}{a_{k+1} + a_{k+2} + \cdots} \qquad \text{for} \quad k = 0, 1, \ldots, N-2.$$

State 0 corresponds to a new component and therefore the limiting probability π_0 corresponds to the long run probability of replacement during any single time unit, or the long run replacement per unit time. Some of these replacements are planned or age replacements, and some correspond to failures in service. A planned replacement occurs in each period for which $X_n = N - 1$, and therefore the long run planned replacements per unit time is the limiting probability π_{N-1}. The difference $\pi_0 - \pi_{N-1}$ is the long run rate of failures in service. The equations for the limiting distribution $\pi = (\pi_0, \pi_1, \ldots, \pi_{N-1})$ are

$$
\begin{aligned}
p_0\pi_0 + \quad p_1\pi_1 + \cdots + \quad p_{N-2}\pi_{N-2} + \pi_{N-1} &= \pi_0 \\
(1 - p_0)\pi_0 \qquad\qquad\qquad\qquad\qquad &= \pi_1 \\
(1 - p_1)\pi_1 \qquad\qquad\qquad &= \pi_2 \\
&\vdots \\
(1 - p_{N-2})\pi_{N-2} \quad &= \pi_{N-1} \\
\pi_0 + \quad \pi_1 + \cdots \qquad\qquad + \pi_{N-1} &= 1.
\end{aligned}
$$

Solving in terms of π_0 we obtain

$$
\begin{aligned}
\pi_0 &= \pi_0 \\
\pi_1 &= (1 - p_0)\pi_0 \\
\pi_2 &= (1 - p_1)\pi_1 = (1 - p_1)(1 - p_0)\pi_0 \\
&\vdots \\
\pi_k &= (1 - p_{k-1})\pi_{k-1} = (1 - p_{k-1})(1 - p_{k-2}) \cdots (1 - p_0)\pi_0 \\
&\vdots \\
\pi_{N-1} &= (1 - p_{N-2})\pi_{N-2} = (1 - p_{N-2})(1 - p_{N-3}) \cdots (1 - p_0)\pi_0
\end{aligned}
$$

and since $\pi_0 + \pi_1 + \ldots + \pi_{N-1} = 1$, then

$$1 = [1 + (1 - p_0) + (1 - p_0)(1 - p_1) + \cdots \\ + (1 - p_0)(1 - p_1) \cdots (1 - p_{N-2})]\pi_0$$

or

$$\pi_0 = \\ \frac{1}{1 + (1 - p_0) + (1 - p_0)(1 - p_1) + \cdots + (1 - p_0)(1 - p_1) \cdots (1 - p_{N-2})}.$$

If $A_j = a_j + a_{j+1} + \ldots$ for $j = 1, 2, \ldots$ where $A_1 = 1$, then $p_k = a_{k+1}/A_{k+1}$ and $1 - p_k = A_{k+2}/A_{k+1}$ which simplifies the expression for π_0 to

$$\pi_0 = \frac{1}{A_1 + A_2 + \cdots + A_N},$$

and then

$$\pi_{N-1} = A_N \pi_0 = \frac{A_N}{A_1 + A_2 + \cdots + A_N}.$$

In practice, one determines the cost C of a replacement and the additional cost K that is incurred when a failure in service occurs. Then the long run total cost per unit time is $C\pi_0 + K(\pi_0 - \pi_{N-1})$ and the replacement age N is chosen so as to minimize this total cost per unit time.

Observe that

$$\frac{1}{\pi_0} = A_1 + A_2 + \cdots + A_N = \sum_{j=1}^{N} \Pr\{T \geq j\} = \sum_{k=0}^{N-1} \Pr\{T > k\}$$

$$= \sum_{k=0}^{\infty} \Pr\{\min\{T, N\} > k\} = E[\min\{T, N\}].$$

In words, the reciprocal of the mean time between replacements $E[\min\{T, N\}]$ yields the long run replacements per unit time π_0. This relation will be further explored in the chapter on renewal processes.

4.2.5 Optimal Replacement Rules

A common industrial activity is the periodic inspection of some system as part of a procedure for keeping it operative. After each inspection, a decision must be made whether or not to alter the system at that time. If the inspection procedure and the ways of modifying the system are fixed, an important problem is that of determining, according to some cost criterion, the optimal rule for making the appropriate decision. Here we consider the case in which the only possible act is to replace the system with a new one.

Suppose that the system is inspected at equally spaced points in time and that after each inspection it is classified into one of the $L + 1$ possible states $0, 1, \ldots, L$. A system is in state 0 if and only if it is new and is in state L if and only if it is inoperative. Let the inspection times be $n = 0$, $1, \ldots$, and let X_n denote the observed state of the system at time n. In the absence of replacement we assume that $\{X_n\}$ is a Markov chain with transition probabilities $p_{ij} = \Pr\{X_{n+1} = j | X_n = i\}$ for all i, j and n.

It is possible to replace the system at any time before failure. The motivation for doing so may be to avoid the possible consequences of further deterioration or of failure of the system. A replacement rule, denoted by R, is a specification of those states at which the system will be replaced. Replacement takes place at the next inspection time. A replacement rule R modifies the behavior of the system and results in a modified Markov chain $\{X_n(R); n = 0, 1, \ldots\}$. The corresponding modified transition probabilities $p_{ij}(R)$ are given by

$$\begin{aligned} p_{ij}(R) &= p_{ij} \\ p_{i0}(R) &= 1 \end{aligned} \qquad \text{if the system is not replaced at state } i;$$

and

$$p_{ij}(R) = 0, \quad j \neq 0 \qquad \text{if the system is replaced at state } i.$$

It is assumed that each time the equipment is replaced, a replacement cost of K units is incurred. Further it is assumed that each unit of time the system is in state j incurs an operating cost of a_j. Note that a_L may be interpreted as failure (inoperative) cost. This interpretation leads to the one period cost function $c_i(R)$ given for $i = 0, \ldots, L$ by

$$c_i(R) = \begin{cases} a_i & \text{if } p_{i0}(R) = 0; \\ \\ K + a_i & \text{if } p_{i0}(R) = 1. \end{cases}$$

We are interested in replacement rules that minimize the expected long run time average cost. This cost is given by the expected cost under the limiting distribution for the Markov chain $\{X_n(R)\}$. Denoting this average cost by $\phi(R)$, we have

$$\phi(R) = \sum_{i=0}^{L} \pi_i(R) c_i(R)$$

where $\pi_i(R) = \lim_{n \to \infty} \Pr\{X_n(R) = i\}$. The limiting distribution $\pi_i(R)$ is determined by the equations

$$\pi_i(R) = \sum_{k=0}^{L} \pi_k(R) p_{ki}(R), \qquad i = 0, \ldots, L,$$

and

$$\pi_0(R) + \pi_1(R) + \cdots + \pi_L(R) = 1.$$

We define a control limit rule to be a replacement rule of the form

"Replace the system if and only if $X_n \geq k$"

where k, called the control limit, is some fixed state between 0 and L. We let R_k denote the control limit rule with control limit equal to k. Then R_0 is the rule "Replace the system at every step" and R_L is the rule: "Replace only upon failure (State L)."

Control limit rules seem reasonable provided that the states are labeled monotonically from best (0) to worst (L) in some sense. Indeed, it can be shown that a control limit rule is optimal whenever the following two conditions hold:

(1) $a_0 \leq a_1 \leq \cdot \cdot \cdot \leq a_L$.

(2) If $i \leq j$, then $\displaystyle\sum_{m=k}^{L} p_{im} \leq \sum_{m=k}^{L} p_{jm}$ for every $k = 0, \ldots, L$.

Condition (1) asserts that the one stage costs are higher in the "worse" states. Condition (2) asserts that further deterioration is more likely in the "worse" states.

Let us suppose that conditions (1) and (2) prevail. Then we need only check the $L + 1$ control limit rules $R_0, \ldots, R_L$ in order to find an optimal rule. Futhermore, it can be shown that a control limit $k^\star$ satisfying $\phi(R_{k^\star-1}) \geq \phi(R_{k^\star}) \leq \phi(R_{k^\star+1})$ is optimal, so that not always do all $L + 1$ control limit rules need to be checked.

Under control limit k we have the cost vector

$$c(R_k) = (a_0, \ldots, a_{k-1}, K + a_k, \ldots, K + a_L),$$

and the transition probabilities

		0	1		$k-1$	k		L
	0	0	p_{01}	$\cdots$	$p_{0,k-1}$	p_{0k}	$\cdots$	p_{0L}
	1	0	p_{11}	$\cdots$	$p_{1,k-1}$	p_{1k}	$\cdots$	p_{1L}
$\mathbf{P}(R_k) =$	$k-1$	0	$p_{k-1,1}$	$\cdots$	$p_{k-1,k-1}$	$p_{k-1,k}$	$\cdots$	$p_{k-1,L}$
	k	1	0	$\cdots$	0	0	$\cdots$	0
	L	1	0	$\cdots$	0	0	$\cdots$	0

To look at a numerical example we will find the optimal control limit $k^\star$ for the following data: $L = 5$ and

$$\mathbf{P} = \begin{array}{c} \\ 0 \\ 1 \\ 2 \\ 3 \\ 4 \\ 5 \end{array} \begin{array}{cccccc} 0 & 1 & 2 & 3 & 4 & 5 \\ \left\|\begin{array}{c} 0 \\ 0 \\ 0 \\ 0 \\ 0 \\ 0 \end{array}\right. & \begin{array}{c} .2 \\ .1 \\ 0 \\ 0 \\ 0 \\ 0 \end{array} & \begin{array}{c} .2 \\ .2 \\ .1 \\ 0 \\ 0 \\ 0 \end{array} & \begin{array}{c} .2 \\ .2 \\ .2 \\ .1 \\ 0 \\ 0 \end{array} & \begin{array}{c} .2 \\ .2 \\ .3 \\ .4 \\ .4 \\ 0 \end{array} & \left.\begin{array}{c} .2 \\ .3 \\ .4 \\ .5 \\ .6 \\ 1 \end{array}\right\| \end{array},$$

$a_0 = \ldots = a_{L-1} = 0$, $a_L = 5$, and $K = 3$. When $k = 1$, the transition matrix is

$$\mathbf{P}(R_1) = \begin{array}{c} \\ 0 \\ 1 \\ 2 \\ 3 \\ 4 \\ 5 \end{array} \begin{array}{cccccc} 0 & 1 & 2 & 3 & 4 & 5 \\ \left\|\begin{array}{c} 0 \\ 1 \\ 1 \\ 1 \\ 1 \\ 1 \end{array}\right. & \begin{array}{c} .2 \\ 0 \\ 0 \\ 0 \\ 0 \\ 0 \end{array} & \begin{array}{c} .2 \\ 0 \\ 0 \\ 0 \\ 0 \\ 0 \end{array} & \begin{array}{c} .2 \\ 0 \\ 0 \\ 0 \\ 0 \\ 0 \end{array} & \begin{array}{c} .2 \\ 0 \\ 0 \\ 0 \\ 0 \\ 0 \end{array} & \left.\begin{array}{c} .2 \\ 0 \\ 0 \\ 0 \\ 0 \\ 0 \end{array}\right\| \end{array}$$

which implies the following equations for the stationary distribution:

$$\pi_1 + \pi_2 + \pi_3 + \pi_4 + \pi_5 = \pi_0$$
$$.2\pi_0 = \pi_1$$
$$.2\pi_0 = \pi_2$$
$$.2\pi_0 = \pi_3$$
$$.2\pi_0 = \pi_4$$
$$.2\pi_0 = \pi_5$$

and
$$\pi_0 + \pi_1 + \pi_2 + \pi_3 + \pi_4 + \pi_5 = 1.$$

The solution is $\pi_0 = .5$ and $\pi_1 = \pi_2 = \pi_3 = \pi_4 = \pi_5 = .1$. The average cost associated with $k = 1$ is

$$\phi_1 = .5(0) + .1(3) + .1(3) + .1(3) + .1(3) + .1(3 + 5)$$
$$= 2.0.$$

When $k = 2$, the transition matrix is

$$\mathbf{P}(R_2) = \begin{array}{c} \\ 0 \\ 1 \\ 2 \\ 3 \\ 4 \\ 5 \end{array} \begin{array}{cccccc} 0 & 1 & 2 & 3 & 4 & 5 \\ \left\|\begin{array}{c} 0 \\ 0 \\ 1 \\ 1 \\ 1 \\ 1 \end{array}\right. & \begin{array}{c} .2 \\ .1 \\ 0 \\ 0 \\ 0 \\ 0 \end{array} & \begin{array}{c} .2 \\ .2 \\ 0 \\ 0 \\ 0 \\ 0 \end{array} & \begin{array}{c} .2 \\ .2 \\ 0 \\ 0 \\ 0 \\ 0 \end{array} & \begin{array}{c} .2 \\ .2 \\ 0 \\ 0 \\ 0 \\ 0 \end{array} & \left.\begin{array}{c} .2 \\ .3 \\ 0 \\ 0 \\ 0 \\ 0 \end{array}\right\| \end{array}$$

and the associated stationary distribution is $\pi_0 = .450$, $\pi_1 = .100$, $\pi_2 = \pi_3 = \pi_4 = .110$, $\pi_5 = .120$. We evaluate the average cost to be

$$\phi_2 = .45(0) + .10(0) + .11(3) + .11(3) + .11(3) + .12(8)$$
$$= 1.95.$$

Continuing in this manner, we obtain the following table:

Control Limit	Stationary Distribution						Average Cost
k	π_0	π_1	π_2	π_3	π_4	π_5	ϕ_k
1	.5000	.1000	.1000	.1000	.1000	.1000	2.0000
2	.4500	.1000	.1100	.1100	.1100	.1200	1.9500
3	.4010	.0891	.1089	.1198	.1307	.1505	1.9555
4	.3539	.9786	.0961	.1175	.1623	.1916	2.0197
5	.2785	.06189	.0756	.0925	.2139	.2785	2.2280

The optimal control limit is $k^\star = 2$, and the corresponding minimum average cost per unit time is $\phi_2 = 1.95$.

Problems 4.2

1. Suppose that the weather on any day depends on the weather conditions during the previous two days. We form a Markov chain with the states:

State (S, S) if it was sunny both today and yesterday,
State (S, C) if it was sunny yesterday but cloudy today,
State (C, S) if it was cloudy yesterday but sunny today,
State (C, C) if it was cloudy both today and yesterday,

and transition probability matrix

Today's State

$$\mathbf{P} = \begin{array}{c} \\ (S, S) \\ (S, C) \\ (C, S) \\ (C, C) \end{array} \begin{array}{|cccc|} (S, S) & (S, C) & (C, S) & (C, C) \\ \hline .7 & .3 & 0 & 0 \\ 0 & 0 & .4 & .6 \\ .5 & .5 & 0 & 0 \\ 0 & 0 & .2 & .8 \end{array}.$$

(a) Given that it is sunny on days 0 and 1, what is the probability it is sunny on day 5?

(b) In the long run, what fraction of days are sunny?

2. An airline reservation system has a single computer which breaks down on any given day with probability p. It takes two days to restore a failed computer to normal service. Form a Markov chain by taking as states the pairs (x, y) where x is the number of machines in operating condition at the end of a day and y is 1 if a day's labor has been expended on a machine, and 0 otherwise. The transition probability matrix is

To State
$$\longrightarrow (1, 0) \qquad (0, 0) \qquad (0, 1)$$

From State
$$\downarrow$$

$$\mathbf{P} = \begin{matrix} (1, 0) \\ (0, 0) \\ (0, 1) \end{matrix} \left\| \begin{matrix} q & p & 0 \\ 0 & 0 & 1 \\ 1 & 0 & 0 \end{matrix} \right\|$$

Compute the system reliability π_0 for $p = .01, .02, .05,$ and $.10$.

3. Section 4.2.2 determined the reliability R of a certain computer system to be

$$R_1 = \frac{1}{1 + p^2} \qquad \text{for one repair facility}$$

$$R_2 = \frac{1 + p}{1 + p + p^2} \qquad \text{for two repair facilities,}$$

where p is the computer failure probability on a single day. Compute and compare R_1 and R_2 for $p = .01, .02, .05,$ and $.10$.

4. Consider a computer system that fails on a given day with probability p and remains "up" with probability $q = 1 - p$. Suppose the repair time is a random variable N having the probability mass function $p(k) = \beta(1 - \beta)^{k-1}$ for $k = 1, 2, \ldots$, where $0 < \beta < 1$. Let $X_n = 1$ if the computer is operating on day n and $X_n = 0$ if not. Show that $\{X_n\}$ is a Markov chain with transition matrix

$$\begin{matrix} & 0 & 1 \end{matrix}$$
$$\begin{matrix} 0 \\ 1 \end{matrix} \left\| \begin{matrix} \alpha & \beta \\ p & q \end{matrix} \right\|$$

and $\alpha = 1 - \beta$. Determine the long run probability that the computer is operating in terms of $\alpha, \beta, p,$ and q.

5. Determine the average fraction inspected, AFI, and the average outgoing quality, AOQ, of Section 4.2.3 for $p = 0, .05, .10, .15, \ldots, .50$ when
   (a) $r = 10$ and $i = 5$
   (b) $r = 5$ and $i = 10$

6. A component of a computer has an active life, measured in discrete units, that is a random variable T where

$$\Pr\{T = 1\} = .1 \qquad \Pr\{T = 3\} = .3$$
$$\Pr\{T = 2\} = .2 \qquad \Pr\{T = 4\} = .4.$$

Suppose one starts with a fresh component and each component is replaced by a new component upon failure. Determine the long run probability that a failure occurs in a given period.

7. Consider a machine whose condition at any time can be observed and classified as being in one of the following three states:

State 1: Good operating order
State 2: Deteriorated operating order
State 3: In repair

We observe the condition of the machine at the end of each period in a sequence of periods. Let X_n denote the condition of the machine at the end of period n for $n = 1, 2, \ldots$. Let X_0 be the condition of the machine at the start. We assume that the sequence of machine conditions is a Markov chain with transition probabilities:

$$
\begin{array}{ccc}
P_{11} = .9 & P_{12} = .1 & P_{13} = 0 \\
P_{21} = 0 & P_{22} = .9 & P_{23} = .1 \\
P_{31} = 1 & P_{32} = 0 & P_{33} = 0
\end{array}
$$

and that the process starts in state $X_0 = 1$.
(a) Find $\Pr\{X_4 = 1\}$.
(b) Calculate the limiting distribution.
(c) What is the long run repairs per unit time?

8. At the end of a month, a large retail store classifies each receivable account according to

0: Current
1: 30–60 days overdue
2: 60–90 days overdue
3: Over 90 days.

Each such account moves from state to state according to a Markov chain with transition probability matrix

	0	1	2	3
0	.95	.05	0	0
1	.50	0	.50	0
2	.20	0	0	.80
3	.10	0	0	.90

In the long run, what fraction of accounts are over 90 days overdue?

9. Customers arrive for service and take their place in a waiting line. There is a single service facility, and a customer undergoing service at the beginning of a period will complete service and depart at the end of the period with probability β and will continue service into the next period with probability $\alpha = 1 - \beta$, and then the process repeats. This description implies that the service time η of an individual is a random variable with the geometric distribution,

$$\Pr\{\eta = k\} = \beta\alpha^{k-1} \qquad \text{for} \quad k = 1, 2, \ldots,$$

and the service times of distinct customers are independent random variables.

At most a single customer can arrive during a period. We suppose that the actual number of arrivals during the nth period is a random variable ξ_n taking on the values 0 or 1 according to

$$\Pr\{\xi_n = 0\} = p$$

and

$$\Pr\{\xi_n = 1\} = q = 1 - p \qquad \text{for} \quad n = 0, 1, \ldots.$$

The state X_n of the system at the start of period n is defined to be the number of customers in the system, either waiting or being served. Then $\{X_n\}$ is a Markov chain. Specify the following transition probabilities in terms of α, β, p, and q: P_{00}, P_{01}, P_{02}, P_{10}, P_{11}, and P_{12}.

State any additional assumptions that you make.

10. From purchase to purchase, a particular customer switches brands among products A, B, and C according to a Markov chain whose transition probability matrix is

$$\mathbf{P} = \begin{array}{c} A \\ B \\ C \end{array} \begin{array}{ccc} A & B & C \\ \left\| \begin{array}{ccc} .6 & .2 & .2 \\ .1 & .7 & .2 \\ .1 & .1 & .8 \end{array} \right\| \end{array}$$

In the long run, what fraction of time does this customer purchase brand A?

4.3 The Classification of States

Not all Markov chains are regular. We consider some examples.

The Markov chain whose transition probability matrix is the identity matrix

$$\mathbf{P} = \begin{array}{c} \\ 0 \\ 1 \end{array} \begin{array}{cc} 0 & 1 \\ \left\| \begin{array}{cc} 1 & 0 \\ 0 & 1 \end{array} \right\| \end{array}$$

remains always in the state in which it starts. Since trivially $\mathbf{P}^n = \mathbf{P}$ for all n, the Markov chain X_n has a limiting distribution, but it obviously depends on the initial state.

The Markov chain whose transition probability matrix is

$$\mathbf{P} = \begin{array}{c} \\ 0 \\ 1 \end{array} \begin{array}{cc} 0 & 1 \\ \left\| \begin{array}{cc} 0 & 1 \\ 1 & 0 \end{array} \right\| \end{array}$$

oscillates deterministically between the two states. The Markov chain is *periodic* and no limiting distribution exists. When n is an odd number, then $\mathbf{P}^n = \mathbf{P}$, but when n is even, then $\mathbf{P}^n$ is the 2×2 identity matrix.

When $\mathbf{P}$ is the matrix

$$\mathbf{P} = \begin{array}{c} \\ 0 \\ 1 \end{array} \begin{array}{cc} 0 & 1 \\ \left\| \begin{array}{cc} \frac{1}{2} & \frac{1}{2} \\ 0 & 1 \end{array} \right\| \end{array}$$

then $\mathbf{P}^n$ is given by

$$\mathbf{P}^n = \begin{array}{c} \\ 0 \\ 1 \end{array} \begin{array}{cc} 0 & 1 \\ \left\| \begin{array}{cc} (\frac{1}{2})^n & 1 - (\frac{1}{2})^n \\ 0 & 1 \end{array} \right\| \end{array}$$

and the limit is

$$\lim_{n \to \infty} \mathbf{P}^n = \begin{array}{c} \\ 0 \\ 1 \end{array} \begin{array}{cc} 0 & 1 \\ \left\| \begin{array}{cc} 0 & 1 \\ 0 & 1 \end{array} \right\| \end{array}.$$

Here state 0 is *transient;* after the process starts from state 0 there is a positive probability that it will never return to that state.

The three matrices just presented illustrated three distinct types of behavior in addition to the convergence exemplified by a regular Markov chain. Various and more elaborate combinations of these behaviors are also possible. Some definitions and classifications of states and matrices are needed in order to sort out the variety of possibilities.

4.3.1 Irreducible Markov Chains

State j is said to be accessible from state i if $P_{ij}^{(n)} > 0$ for some integer $n \geq 0$, i.e., state j is accessible from state i if there is positive probability that state j can be reached starting from state i in some finite number of transitions.

Two states i and j, each accessible to the other, are said to *communicate* and we write $i \leftrightarrow j$. If two states i and j do not communicate, then either

$$P_{ij}^{(n)} = 0 \qquad \text{for all} \quad n \geq 0$$

or

$$P_{ji}^{(n)} = 0 \qquad \text{for all} \quad n \geq 0$$

or both relations are true. The concept of communication is an equivalence relation:

(i) $i \leftrightarrow i$ (reflexivity), a consequence of the definition of

$$P_{ij}^{(0)} = \delta_{ij} = \begin{cases} 1, & i = j \\ \\ 0, & i \neq j \end{cases}.$$

(ii) If $i \leftrightarrow j$, then $j \leftrightarrow i$ (symmetry), from the definition of communication.

(iii) If $i \leftrightarrow j$ and $j \leftrightarrow k$, then $i \leftrightarrow k$ (transitivity).

The proof of transitivity proceeds as follows: $i \leftrightarrow j$ and $j \leftrightarrow k$ imply that there exist integers n and m such that $P_{ij}^{(n)} > 0$ and $P_{jk}^{(m)} > 0$. Consequently by the nonnegativity of each $P_{rs}^{(t)}$, we conclude that

$$P_{ik}^{(n+m)} = \sum_{r=0}^{\infty} P_{ir}^{(n)} P_{rk}^{(m)} \geq P_{ij}^{(n)} P_{jk}^{(m)} > 0.$$

A similar argument shows the existence of an integer v such that $P_{ki}^{(v)} > 0$, as desired.

We can now partition the totality of states into equivalence classes. The states in an equivalence class are those that communicate with each other. It may be possible, starting in one class, to enter some other class with positive probability; if so, however, it is clearly not possible to return to the initial class, or else the two classes would together form a single class. We say that the Markov chain is irreducible if the equivalence relation induces only one class. In other words, a process is irreducible if all states communicate with each other.

To illustrate this concept, we consider the transition probability matrix

$$\mathbf{P} = \begin{Vmatrix} \frac{1}{2} & \frac{1}{2} & \vdots & 0 & 0 & 0 \\ \frac{1}{4} & \frac{3}{4} & \vdots & 0 & 0 & 0 \\ \cdots & \cdots & \cdots & \cdots & \cdots & \cdots \\ 0 & 0 & \vdots & 0 & 1 & 0 \\ 0 & 0 & \vdots & \frac{1}{2} & 0 & \frac{1}{2} \\ 0 & 0 & \vdots & 0 & 1 & 0 \end{Vmatrix} = \begin{Vmatrix} \mathbf{P}_1 & 0 \\ 0 & \mathbf{P}_2 \end{Vmatrix},$$

where $\mathbf{P}_1$ is an abbreviation for the matrix formed from the initial two rows and columns of **P**, and similarly for $\mathbf{P}_2$. This Markov chain clearly divides into the two classes composed of states {1, 2} and states {3, 4, 5}.

If the state of X_0 lies in the first class, then the state of the system thereafter remains in this class and for all purposes the relevant transition matrix is $\mathbf{P}_1$. Similarly, if the initial state belongs to the second class, then the relevant transition matrix is $\mathbf{P}_2$. This is a situation where we have two completely unrelated processes labeled together.

In the random walk model with transition matrix

$$
\mathbf{P} = \begin{Vmatrix}
1 & 0 & 0 & 0 & \cdots\cdots & 0 & 0 & 0 \\
q & 0 & p & 0 & \cdots\cdots & 0 & 0 & 0 \\
0 & q & 0 & p & \cdots\cdots & 0 & 0 & 0 \\
\cdot & & & & & \cdot & \cdot & \cdot \\
0 & \cdots\cdots\cdots\cdots & & & & q & 0 & p \\
0 & \cdots\cdots\cdots\cdots & & & & 0 & 0 & 1
\end{Vmatrix}
\begin{matrix}
\text{states} \\
0 \\ 1 \\ 2 \\ \cdot \\ a-1 \\ a
\end{matrix}
\tag{4.15}
$$

we have the three classes {0}, {1, 2, . . . , $a - 1$}, and {a}. In this example it is possible to reach the first class or third class from the second class, but it is not possible to return to the second class from either the first or the third class.

4.3.2 Periodicity of a Markov Chain

We define the period of state i, written $d(i)$, to be the greatest common divisor (g.c.d.) of all integers $n \geq 1$ for which $P_{ii}^{(n)} > 0$. [If $P_{ii}^{(n)} = 0$ for all $n \geq 1$ define $d(i) = 0$.] In the random walk (4.15), every transient state 1, 2, . . . , $N - 1$ has period 2. If $P_{ii} > 0$ for some single state i, then that state now has period 1, since the system can remain in this state any length of time.

In a finite Markov chain of n states with transition matrix

$$
\mathbf{P} = \overbrace{\begin{Vmatrix}
0 & 1 & 0 & 0 & \cdots\cdots & 0 \\
0 & 0 & 1 & 0 & \cdots\cdots & 0 \\
\cdot & & & & & \cdot \\
\cdot & & & & & \cdot \\
0 & 0 & \cdots\cdots\cdots\cdots & & & 1 \\
1 & 0 & 0 & \cdots\cdots\cdots & & 0
\end{Vmatrix}}^{n}
$$

each state has period n.

Consider the Markov chain whose transition probability matrix is

$$\begin{array}{c} \begin{array}{cccc} 0 & 1 & 2 & 3 \end{array} \\ \mathbf{P} = \begin{array}{c} 0 \\ 1 \\ 2 \\ 3 \end{array} \left\| \begin{array}{cccc} 0 & 1 & 0 & 0 \\ 0 & 0 & 1 & 0 \\ 0 & 0 & 0 & 1 \\ \frac{1}{2} & 0 & \frac{1}{2} & 0 \end{array} \right\| . \end{array}$$

We evaluate $P_{00} = 0$, $P_{00}^{(2)} = 0$, $P_{00}^{(3)} = 0$, $P_{00}^{(4)} = \frac{1}{2}$, $P_{00}^{(5)} = 0$, $P_{00}^{(6)} = \frac{1}{4}$. The set of integers $n \geq 1$ for which $P_{00}^{(n)} > 0$ is $\{4, 6, 8, \ldots\}$. The period of state 0 is $d(0) = 2$, the greatest common divisor of this set.

Example Suppose that the precipitation in a certain locale depends on the season (WET or DRY) as well as on the precipitation level (HIGH, LOW) during the preceding season. We model the process as a Markov chain whose states are of the form (x, y) where x denotes the season ($W = $ Wet, $D = $ Dry) and y denotes the precipitation level ($H = $ High, $L = $ Low). Suppose the transition probability matrix is

$$\mathbf{P} = \begin{array}{c} (W, H) \\ (W, L) \\ (D, H) \\ (D, L) \end{array} \begin{array}{c} \begin{array}{cccc} (W, H) & (W, L) & (D, H) & (D, L) \end{array} \\ \left\| \begin{array}{cccc} 0 & 0 & .8 & .2 \\ 0 & 0 & .4 & .6 \\ .7 & .3 & 0 & 0 \\ .2 & .8 & 0 & 0 \end{array} \right\| . \end{array}$$

All states are periodic with period $d = 2$.

A situation in which the demand for an inventory item depends on the month of the year as well as on the demand during the previous month would lead to a Markov chain whose states had period $d = 12$.

The random walk on the states $0, \pm 1, \pm 2, \ldots$ with probabilities $P_{i,i+1} = p$, $P_{i,i-1} = q = 1 - p$ is periodic with period $d = 2$.

We state, without proof, three basic properties of the period of a state:

(1) If $i \leftrightarrow j$ then $d(i) = d(j)$.

This assertion shows that the period is a constant in each class of communicating states.

(2) If state i has period $d(i)$ then there exists an integer N depending on i such that for all integers $n \geq N$

$$P_{ii}^{(nd(i))} > 0.$$

This asserts that a return to state i can occur at all sufficiently large multiples of the period $d(i)$.

(3) If $P_{ji}^{(m)} > 0$, then $P_{ji}^{(m+nd(i))} > 0$ for all n (a positive integer) sufficiently large.

A Markov chain in which each state has period 1 is called *aperiodic*. The vast majority of Markov chain processes we deal with are aperiodic. Results will be developed for the aperiodic case and the modified conclusions for the general case will be stated, usually without proof.

4.3.3 Recurrent and Transient States

Consider an arbitrary, but fixed, state i. We define for each integer $n \geq 1$.

$$f_{ii}^{(n)} = \Pr\{X_n = i, X_v \neq i, v = 1, 2, \ldots, n - 1 | X_0 = i\}.$$

In other words, $f_{ii}^{(n)}$ is the probability that, starting from state i, the first return to state i occurs at the nth transition. Clearly $f_{ii}^{(1)} = P_{ii}$, and $f_{ii}^{(n)}$ may be calculated recursively according to

$$P_{ii}^{(n)} = \sum_{k=0}^{n} f_{ii}^{(k)} P_{ii}^{(n-k)}, \qquad n \geq 1, \tag{4.16}$$

where we define $f_{ii}^{(0)} = 0$ for all i. Equation (4.16) is derived by decomposing the event from which $P_{ii}^{(n)}$ is computed according to the time of the first return to state i. Indeed, consider all the possible realizations of the process for which $X_0 = i$, $X_n = i$, and the first return to state i occurs at the kth transition. Call this event E_k. The events E_k $(k = 1, 2, \ldots, n)$ are clearly mutually exclusive. The probability of the event that the first return is at the kth transition is by definition $f_{ii}^{(k)}$. In the remaining $n - k$ transitions, we are dealing only with those realizations for which $X_n = i$. Using the Markov property, we have

$$\Pr\{E_k\} = \Pr\{\text{first return is at } k\text{th transition} | X_0 = i\}\Pr\{X_n = i | X_k = i\}$$

$$= f_{ii}^{(k)} P_{ii}^{(n-k)}, \qquad 1 \leq k \leq n$$

(recall that $P_{ii}^0 = 1$). Hence

$$\Pr\{X_n = i | X_0 = i\} = \sum_{k=1}^{n} \Pr\{E_k\} = \sum_{k=1}^{n} f_{ii}^{(k)} P_{ii}^{(n-k)} = \sum_{k=0}^{n} f_{ii}^{(k)} P_{ii}^{(n-k)},$$

since by definition $f_{ii}^{(0)} = 0$.

When the process starts from state i, the probability that it returns to state i at some finite time is

$$f_{ii} = \sum_{n=0}^{\infty} f_{ii}^{(n)} = \lim_{N \to \infty} \sum_{n=0}^{N} f_{ii}^{(n)}. \tag{4.17}$$

We say that a state i is *recurrent* if $f_{ii} = 1$. This statement says that a state i is recurrent if and only if, after the process starts from state i, the probability of its returning to state i after some finite length of time is one. A nonrecurrent state is said to be *transient*.

Consider a transient state i. Then the probability that a process starting from state i returns to state i at least once is $f_{ii} < 1$. Because of the Markov property, the probability that the process returns to state i at least twice is $(f_{ii})^2$, and, repeating the argument, we see that the probability that the process returns to i at least k times is $(f_{ii})^k$ for $k = 1, 2, \ldots$. Let M be the random variable that counts the number of times that the process returns to i. Then we have shown that M has the geometric distribution in which

$$\Pr\{M \geq k | X_0 = i\} = (f_{ii})^k \qquad \text{for} \quad k = 1, 2, \ldots \tag{4.18}$$

and

$$E[M|X_0 = i] = \frac{f_{ii}}{1 - f_{ii}}. \qquad (4.19)$$

Theorem 4.2 establishes a criterion for the recurrence of a state i in terms of the transition probabilities $P_{ii}^{(n)}$.

Theorem 4.2   *A state i is recurrent if and only if*

$$\sum_{n=1}^{\infty} P_{ii}^{(n)} = \infty.$$

Equivalently, state i is transient if and only if $\sum_{n=1}^{\infty} P_{ii}^{(n)} < \infty$.

Proof   Suppose first that state i is transient so that, by definition, $f_{ii} < 1$, and let M count the total number of returns to state i. We write M in terms of indicator random variables as

$$M = \sum_{n=1}^{\infty} \mathbf{1}\{X_n = i\},$$

where

$$\mathbf{1}\{X_n = i\} = \begin{cases} 1 & \text{if} \quad X_n = i, \\ 0 & \text{if} \quad X_n \neq i. \end{cases}$$

Now equation (4.19) shows that $E[M|X_0 = i] < \infty$ when i is transient. But then

$$\infty > E[M|X_0 = i] = \sum_{n=1}^{\infty} E[\mathbf{1}\{X_n = i\}|X_0 = i]$$
$$= \sum_{n=1}^{\infty} P_{ii}^{(n)}$$

as claimed.

Conversely, suppose $\sum_{n=1}^{\infty} P_{ii}^{(n)} < \infty$. Then M is a random variable whose mean is finite, and thus M must be finite. That is, starting from state i, the process returns to state i only a finite number of times. Then there must be a positive probability that, starting from state i, the process never returns to that state. In other words $1 - f_{ii} > 0$ or $f_{ii} < 1$, as claimed. $\square$

Corollary 4.1   If $i \leftrightarrow j$ and if i is recurrent then j is recurrent.

Proof   Since $i \leftrightarrow j$ there exists $m, n \geq 1$ such that

$$P_{ij}^{(n)} > 0 \quad \text{and} \quad P_{ji}^{(m)} > 0.$$

Let $v > 0$. We obtain, by the usual argument (see page 145), $P_{jj}^{(m+n+v)} \geq P_{ji}^{(m)} P_{ii}^{(v)} P_{ij}^{(n)}$ and, on summing,

$$\sum_{v=0}^{\infty} P_{jj}^{(m+n+v)} \geq \sum_{v=0}^{\infty} P_{ji}^{(m)} P_{ii}^{(v)} P_{ij}^{(n)} = P_{ji}^{(m)} P_{ij}^{(n)} \sum_{v=0}^{\infty} P_{ii}^{(v)}.$$

Hence if $\sum_{v=0}^{\infty} P_{ii}^{(v)}$ diverges, then $\sum_{v=0}^{\infty} P_{jj}^{(v)}$ also diverges. $\square$

This corollary proves that recurrence, like periodicity, is a class property: that is, all states in an equivalence class are either recurrent or nonrecurrent.

Example Consider the one-dimensional random walk on the positive and negative integers, where at each transition the particle moves with probability p one unit to the right and with probability q one unit to the left ($p + q = 1$). Hence

$$P_{00}^{(2n+1)} = 0, \qquad n = 0, 1, 2, \ldots$$

and

$$P_{00}^{(2n)} = \binom{2n}{n} p^n q^n = \frac{(2n)!}{n!n!} p^n q^n. \tag{4.20}$$

We appeal now to Stirling's formula (see page 42),

$$n! \sim n^{n+1/2} e^{-n} \sqrt{2\pi}. \tag{4.21}$$

Applying (4.21) to (4.20) we obtain

$$P_{00}^{(2n)} \sim \frac{(pq)^n 2^{2n}}{\sqrt{\pi n}} = \frac{(4pq)^n}{\sqrt{\pi n}}.$$

It is readily verified that $p(1 - p) = pq \leq \frac{1}{4}$ with equality holding if and only if $p = q = \frac{1}{2}$. Hence $\sum_{n=0}^{\infty} P_{00}^{(n)} = \infty$ if and only if $p = \frac{1}{2}$. Therefore, from Theorem 4.2, the one-dimensional random walk is recurrent if and only if $p = q = \frac{1}{2}$. Remember that recurrence is a class property. Intuitively, if $p \neq q$ there is positive probability that a particle initially at the origin will drift to $+\infty$ if $p > q$ (to $-\infty$ if $p < q$) without ever returning to the origin.

Problems 4.3

1. Determine the communicating classes and period for each state of the Markov chain whose transition probability matrix is

	0	1	2	3	4	5
0	$\frac{1}{2}$	0	0	0	$\frac{1}{2}$	0
1	0	0	1	0	0	0
2	0	0	0	1	0	0
3	0	0	0	0	1	0
4	0	0	0	0	0	1
5	0	0	$\frac{1}{3}$	$\frac{1}{3}$	0	$\frac{1}{3}$

2. Which states are transient and which are recurrent in the Markov chain whose transition probability matrix is

$$
\begin{array}{c|cccccc}
 & 0 & 1 & 2 & 3 & 4 & 5 \\
\hline
0 & \frac{1}{3} & 0 & \frac{1}{3} & 0 & 0 & \frac{1}{3} \\
1 & \frac{1}{2} & \frac{1}{4} & \frac{1}{4} & 0 & 0 & 0 \\
2 & 0 & 0 & 0 & 0 & 1 & 0 \\
3 & \frac{1}{4} & \frac{1}{4} & \frac{1}{4} & 0 & 0 & \frac{1}{4} \\
4 & 0 & 0 & 1 & 0 & 0 & 0 \\
5 & 0 & 0 & 0 & 0 & 0 & 1
\end{array}.
$$

3. A Markov chain on states $\{0, 1, 2, 3, 4, 5\}$ has transition probability matrix

$$
\text{(a)} \quad
\begin{Vmatrix}
\frac{1}{3} & 0 & \frac{2}{3} & 0 & 0 & 0 \\
0 & \frac{1}{4} & 0 & \frac{3}{4} & 0 & 0 \\
\frac{2}{3} & 0 & \frac{1}{3} & 0 & 0 & 0 \\
0 & \frac{1}{5} & 0 & \frac{4}{5} & 0 & 0 \\
\frac{1}{4} & \frac{1}{4} & 0 & 0 & \frac{1}{4} & \frac{1}{4} \\
\frac{1}{6} & \frac{1}{6} & \frac{1}{6} & \frac{1}{6} & \frac{1}{6} & \frac{1}{6}
\end{Vmatrix}
$$

$$
\text{(b)} \quad
\begin{Vmatrix}
1 & 0 & 0 & 0 & 0 & 0 \\
0 & \frac{3}{4} & \frac{1}{4} & 0 & 0 & 0 \\
0 & \frac{1}{8} & \frac{7}{8} & 0 & 0 & 0 \\
\frac{1}{4} & \frac{1}{4} & 0 & \frac{1}{8} & \frac{3}{8} & 0 \\
\frac{1}{3} & 0 & \frac{1}{6} & \frac{1}{4} & \frac{1}{4} & 0 \\
0 & 0 & 0 & 0 & 0 & 1
\end{Vmatrix}
$$

Find all communicating classes.

4. Show that a finite state aperiodic irreducible Markov chain is regular and recurrent.

5. Recall the first return distribution (Section 4.3.3)

$$
f_{ii}^{(n)} = \Pr\{X_1 \neq i, X_2 \neq i, \ldots, X_{n-1} \neq i, X_n = i \mid X_0 = i\}
$$

$$
\text{for} \quad n = 1, 2, \ldots,
$$

with $f_{ii}^{(0)} = 0$ by convention. Using Equation (4.16), determine $f_{00}^{(n)}$, $n = 1, 2, 3, 4, 5$ for the Markov chain whose transition probability matrix is

$$
\begin{array}{c|cccc}
 & 0 & 1 & 2 & 3 \\
\hline
0 & 0 & \frac{1}{2} & 0 & \frac{1}{2} \\
1 & 0 & 0 & 1 & 0 \\
2 & 0 & 0 & 0 & 1 \\
3 & \frac{1}{2} & 0 & 0 & \frac{1}{2}
\end{array}.
$$

4.4 The Basic Limit Theorem of Markov Chains

Consider a recurrent state i. Then

$$f_{ii}^{(n)} = \Pr\{X_n = i, X_v \neq i \text{ for } v = 1, \ldots, n - 1 | X_0 = i\} \quad (4.22)$$

is the probability distribution of the *first return time*

$$R_i = \min\{n \geq 1; X_n = i\}. \quad (4.23)$$

This is

$$f_{ii}^{(n)} = \Pr\{R_i = n | X_0 = i\} \quad \text{for} \quad n = 1, 2, \ldots \quad (4.24)$$

Since state i is recurrent by assumption, then $f_{ii} = \sum_{n=1}^{\infty} f_{ii}^{(n)} = 1$ and R_i is a finite valued random variable. The mean duration between visits to state i is

$$m_i = E[R_i | X_0 = i] = \sum_{n=1}^{\infty} n f_{ii}^{(n)}. \quad (4.25)$$

After starting in i, then, on the average, the process is in state i once every $m_i = E[R_i | X_0 = i]$ units of time. The basic limit theorem of Markov chains states this result in a sharpened form.

Theorem 4.3   The basic limit theorem of Markov chains
    (a) *Consider a recurrent irreducible aperiodic Markov chain. Let $P_{ii}^{(n)}$ be the probability of entering state i at the nth transition, $n = 0, 1, 2, \ldots$, given that $X_0 = i$ (the initial state is i). By our earlier convention $P_{ii}^{(0)} = 1$. Let $f_{ii}^{(n)}$ be the probability of first returning to state i at the nth transition, $n = 0, 1, 2, \ldots$, where $f_{ii}^{(0)} = 0$. Then,*

$$\lim_{n \to \infty} P_{ii}^{(n)} = \frac{1}{\sum_{n=0}^{\infty} n f_{ii}^{(n)}} = \frac{1}{m_i}. \quad (4.26)$$

    (b) *Under the same conditions as in (a), $\lim_{n \to \infty} P_{ji}^{(n)} = \lim_{n \to \infty} P_{ii}^{(n)}$ for all states j.*

Remark   Let C be a recurrent class. Then $P_{ij}^{(n)} = 0$ for $i \in C, j \notin C$, and every n. Hence, once in C it is not possible to leave C. It follows that the sub-matrix $\|P_{ij}\|$, $i, j \in C$, is a transition probability matrix and the associated Markov chain is irreducible and recurrent. The limit theorem, therefore, applies verbatim to any aperiodic recurrent class.
    If $\lim_{n \to \infty} P_{ii}^{(n)} > 0$ for one i in an aperiodic recurrent class, then $\pi_j > 0$ for all j in the class of i. In this case, we call the class *positive recurrent* or strongly ergodic. If each $\pi_i = 0$ and the class is recurrent we speak of the class as *null recurrent* or weakly ergodic. In terms of the first return time $R_i = \min\{n \geq 1; X_n = i\}$, then state i is positive recurrent if $m_i = E[R_i | X_0 = i] < \infty$ and null recurrent if $m_i = \infty$. This statement is immediate from the equality $\lim_{n \to \infty} P_{ii}^{(n)} = \pi_i = 1/m_i$. An alternative method for determining the limiting distribution π_i for a positive recurrent aperiodic class is given in Theorem 4.4.

Theorem 4.4 *In a positive recurrent aperiodic class with states* $j =$ 0, 1, 2, . . . ,

$$\lim_{n \to \infty} P_{jj}^{(n)} = \pi_j = \sum_{j=0}^{\infty} \pi_i P_{ij}, \qquad \sum_{i=0}^{\infty} \pi_i = 1$$

and the π's are uniquely determined by the set of equations

$$\pi_i \geq 0, \quad \sum_{i=0}^{\infty} \pi_i = 1, \quad \text{and} \quad \pi_j = \sum_{i=0}^{\infty} \pi_i P_{ij} \quad \text{for} \quad j = 0, 1, \ldots \quad (4.27)$$

Any set $(\pi_i)_{i=0}^{\infty}$ satisfying (4.27) is called a *stationary probability distribution* of the Markov chain. The term "stationary" derives from the property that a Markov chain started according to a stationary distribution will follow this distribution at all points of time. Formally, if $\Pr\{X_0 = i\} = \pi_i$ then $\Pr\{X_n = i\} = \pi_i$ for all $n = 1, 2, \ldots$. We check this for the case $n = 1$; the general case follows by induction. We write

$$\Pr\{X_1 = i\} = \sum_{k=0}^{\infty} \Pr\{X_0 = k\}\Pr\{X_1 = i | X_0 = k\}$$

$$= \sum_{k=0}^{\infty} \pi_k P_{ki} = \pi_i$$

where the last equality follows because $\boldsymbol{\pi} = (\pi_0, \pi_1, \ldots)$ is a stationary distribution. When the initial state X_0 is selected according to the stationary distribution, then the joint probability distribution of (X_n, X_{n+1}) is given by

$$\Pr\{X_n = i, X_{n+1} = j\} = \pi_i P_{ij}.$$

The reader should supply the proof.

A limiting distribution, when it exists, is always a stationary distribution, but the converse is not true. There may exist a stationary distribution but no limiting distribution. For example, there is no limiting distribution for the periodic Markov chain whose transition probability matrix is

$$\mathbf{P} = \begin{Vmatrix} 0 & 1 \\ 1 & 0 \end{Vmatrix}$$

but $\boldsymbol{\pi} = (\tfrac{1}{2}, \tfrac{1}{2})$ is a stationary distribution since

$$(\tfrac{1}{2}, \tfrac{1}{2}) \begin{Vmatrix} 0 & 1 \\ 1 & 0 \end{Vmatrix} = (\tfrac{1}{2}, \tfrac{1}{2}).$$

Example Consider the class of random walks whose transition matrices are given by

$$\mathbf{P} = \|P_{ij}\| = \begin{Vmatrix} 0 & 1 & 0 & \cdots \cdots \\ q_1 & 0 & p_1 & \cdots \cdots \\ 0 & q_2 & 0 & p_2 & \cdots \\ \vdots & & & & \\ \vdots & & & & \end{Vmatrix}.$$

This Markov chain has period 2. Nevertheless we investigate the existence of a stationary probability distribution; i.e., we wish to determine the positive solutions of

$$x_i = \sum_{j=0}^{\infty} x_j P_{ji} = p_{i-1} x_{i-1} + q_{i+1} x_{i+1}, \qquad i = 0, 1, \ldots, \qquad (4.28)$$

under the normalization

$$\sum_{i=0}^{\infty} x_i = 1,$$

where $p_{-1} = 0$ and $p_0 = 1$, and thus $x_0 = q_1 x_1$. Using Equation (4.28) for $i = 1$, we could determine x_2 in terms of x_0. Equation (4.28) for $i = 2$ determines x_3 in terms of x_0, and so forth. It is immediately verified that

$$x_i = \frac{p_{i-1} p_{i-2} \cdots p_1}{q_i q_{i-1} \cdots q_1} x_0 = x_0 \prod_{k=0}^{i-1} \frac{p_k}{q_{k+1}}, \qquad i \geq 1,$$

is a solution of (4.28), with x_0 still to be determined. Now since

$$1 = x_0 + \sum_{i=1}^{\infty} x_0 \prod_{k=0}^{i-1} \frac{p_k}{q_{k+1}},$$

we have

$$x_0 = \frac{1}{1 + \sum\limits_{i=1}^{\infty} \prod\limits_{k=0}^{i-1} \frac{p_k}{q_{k+1}}}$$

and so

$$x_0 > 0 \text{ if and only if } \sum_{i=1}^{\infty} \prod_{k=0}^{i-1} \frac{p_k}{q_{k+1}} < \infty.$$

In particular, if $p_k = p$ and $q_k = q = 1 - p$ for $k \geq 1$, the series

$$\sum_{i=1}^{\infty} \prod_{k=0}^{i-1} \frac{p_k}{q_{k+1}} = \frac{1}{p} \sum_{i=1}^{\infty} \left(\frac{p}{q} \right)^i$$

converges only when $p < q$, and then

$$\frac{1}{p} \sum_{i=1}^{\infty} \left(\frac{p}{q} \right)^i = \frac{1}{p} \frac{p/q}{1 - p/q} = \frac{1}{q - p},$$

and

$$x_0 = \frac{1}{1 + 1/(q - p)} = \frac{q - p}{1 + q - p} = \frac{1}{2} \left(1 - \frac{p}{q} \right),$$

$$x_k = \frac{1}{p} \left(\frac{p}{q} \right)^k x_0 = \frac{1}{2p} \left(1 - \frac{p}{q} \right) \left(\frac{p}{q} \right)^k \qquad \text{for } k = 1, 2, \ldots.$$

Example Consider now the Markov chain that represents the success runs of binomial trials. The transition probability matrix is

$$\left\| \begin{matrix} p_0 & 1-p_0 & 0 & 0 & \cdots \\ p_1 & 0 & 1-p_1 & 0 & \cdots \\ p_2 & 0 & 0 & 1-p_2 & \cdots \\ \cdot & & & & \\ \cdot & \cdot & \cdot & \cdot & \\ \cdot & \cdot & \cdot & \cdot & \end{matrix} \right\| \qquad (0 < p_k < 1).$$

The states of this Markov chain all belong to the same equivalence class (any state can be reached from any other state). Since recurrence is a class property (see Corollary 4.1), we will investigate recurrence for the zeroth state.

Let $R_0 = \min\{n \geq 1; X_n = 0\}$ be the time of first return to state 0. It is easy to evaluate

$$\Pr\{R_0 > 1 | X_0 = 0\} = (1 - p_0)$$
$$\Pr\{R_0 > 2 | X_0 = 0\} = (1 - p_0)(1 - p_1)$$
$$\Pr\{R_0 > 3 | X_0 = 0\} = (1 - p_0)(1 - p_1)(1 - p_2)$$

$$\cdot$$
$$\cdot$$
$$\cdot$$

$$\Pr\{R_0 > k | X_0 = 0\} = (1 - p_0)(1 - p_1) \cdots (1 - p_{k-1}) = \prod_{i=0}^{k-1} (1 - p_i).$$

In terms of the first return distribution

$$f_{00}^{(n)} = \Pr\{R_0 = n | X_0 = 0\}$$

we have

$$\Pr\{R_0 > k | X_0 = 0\} = 1 - \sum_{n=1}^{k} f_{00}^{(n)}$$

or

$$\sum_{n=1}^{k} f_{00}^{(n)} = 1 - \Pr\{R_0 > k | X_0 = 0\} = 1 - \prod_{i=0}^{k-1} (1 - p_i).$$

By definition, state 0 is recurrent provided $\sum_{n=1}^{\infty} f_{00}^{(n)} = 1$. In terms of $p_0, p_1,$ . . . then, state 0 is recurrent whenever $\lim_{k \to \infty} \prod_{i=0}^{k-1} (1 - p_i) = \prod_{i=0}^{\infty} (1 - p_i) = 0$. Lemma 4.1 shows that $\prod_{i=0}^{\infty} (1 - p_i) = 0$ is equivalent, in this case, to the condition $\sum_{i=0}^{\infty} p_i = \infty$.

Lemma 4.1 If $0 < p_i < 1, i = 0, 1, 2, \ldots,$ then $u_m = \prod_{i=0}^{m} (1 - p_i) \to 0$ as $m \to \infty$ if and only if $\sum_{i=0}^{\infty} p_i = \infty$.

Proof Assume $\sum_{i=0}^{\infty} p_i = \infty$. Since the series expansion for $\exp(-p_i)$ is an alternating series with terms decreasing in absolute value, we can write

$$1 - p_i < 1 - p_i + \frac{p_i^2}{2!} - \frac{p_i^3}{3!} + \cdots = \exp(-p_i),$$

$$i = 0, 1, 2, \ldots \quad (4.29)$$

Since (4.29) holds for all i, we obtain $\Pi_{i=0}^{m} (1 - p_i) < \exp(-\Sigma_{i=0}^{m} p_i)$. But, by assumption,

$$\lim_{m \to \infty} \sum_{i=0}^{m} p_i = \infty; \qquad \text{hence } \lim_{m \to \infty} \prod_{i=0}^{m} (1 - p_i) = 0.$$

To prove necessity, observe that from a straightforward induction

$$\prod_{i=j}^{m} (1 - p_i) > (1 - p_j - p_{j+1} - \cdots - p_m)$$

for any j and all $m = j + 1, j + 2, \ldots$ Assume now that $\Sigma_{i=1}^{\infty} p_i < \infty$; then $0 < \Sigma_{i=j}^{\infty} p_i < 1$ for some $j > 1$. Thus

$$\lim_{m \to \infty} \prod_{i=j}^{m} (1 - p_i) > \lim_{m \to \infty} \left(1 - \sum_{i=j}^{m} p_i\right) > 0,$$

which contradicts $u_m \to 0$. $\square$

State 0 is recurrent when $\Pi_{i=0}^{\infty} (1 - p_i) = 0$, or equivalently, when $\Sigma_{i=0}^{\infty} p_i = \infty$. The state is positive recurrent when $m_0 = E[R_0 | X_0 = 0] < \infty$. But

$$m_0 = \sum_{k=0}^{\infty} \Pr\{R_0 > k | X_0 = 0\}$$

$$= 1 + \sum_{k=1}^{\infty} \prod_{i=0}^{k-1} (1 - p_i).$$

Thus positive recurrence requires the stronger condition that $\Sigma_{k=1}^{\infty} \Pi_{i=0}^{k-1} (1 - p_i) < \infty$ and in this case, the stationary probability π_0 is given by

$$\pi_0 = \frac{1}{m_0} = \frac{1}{1 + \displaystyle\sum_{k=1}^{\infty} \prod_{i=0}^{k-1} (1 - p_i)}.$$

From the equations for the stationary distribution we have

$$(1 - p_0)\pi_0 = \pi_1$$
$$(1 - p_1)\pi_1 = \pi_2$$
$$(1 - p_2)\pi_2 = \pi_3$$

$$\cdot$$
$$\cdot$$
$$\cdot$$

or

$$\pi_1 = (1 - p_0)\pi_0$$
$$\pi_2 = (1 - p_1)\pi_1 = (1 - p_1)(1 - p_0)\pi_0$$
$$\pi_3 = (1 - p_2)\pi_2 = (1 - p_2)(1 - p_1)(1 - p_0)\pi_0$$

and, in general

$$\pi_k = \pi_0 \prod_{i=0}^{k-1} (1 - p_i) \qquad \text{for} \quad k \geq 1.$$

In the special case where $p_i = p = 1 - q$ for $i = 0, 1, \ldots$, then $\Pi_{i=0}^{k-1} (1 - p_i) = q^k$,

$$m_0 = 1 + \sum_{k=1}^{\infty} q^k = \frac{1}{p}$$

so that $\pi_k = pq^k$ for $k = 0, 1, \ldots$.

Remark Suppose $a_0, a_1, a_2, \ldots$ is a convergent sequence of real numbers where $a_n \to a$ as $n \to \infty$. Then it can be proved by elementary methods that the partial averages of the sequence also converge in the form

$$\lim_{n \to \infty} \frac{1}{n} \sum_{k=0}^{n-1} a_k = a. \tag{4.30}$$

Applying (4.30) with $a_n = P_{ii}^{(n)}$, where i is a member of a positive recurrent aperiodic class, we obtain

$$\lim_{n \to \infty} \frac{1}{n} \sum_{m=0}^{n-1} P_{ii}^{(m)} = \pi_i = \frac{1}{m_i} > 0, \tag{4.31}$$

where $\pi = (\pi_0, \pi_1, \ldots)$ is the stationary distribution and where m_i is the mean return time for state i. Let $M_i^{(n)}$ be the random variable that counts the total number of visits to state i during time periods $0, 1, \ldots, n - 1$. We may write

$$M_i^{(n)} = \sum_{k=0}^{n-1} \mathbf{1}\{X_k = i\} \tag{4.32}$$

where

$$\mathbf{1}\{X_k = i\} = \begin{cases} 1 & \text{if } X_k = i, \\ 0 & \text{if } X_k \neq i, \end{cases} \tag{4.33}$$

and then see that

$$E[M_i^{(n)}|X_0 = i] = \sum_{k=0}^{n-1} E[\mathbf{1}\{X_k = i\}|X_0 = i] = \sum_{k=0}^{n-1} P_{ii}^{(k)}. \tag{4.34}$$

Then, referring to (4.31) we have

$$\lim_{n \to \infty} \frac{1}{n} E[M_i^{(n)}|X_0 = i] = \frac{1}{m_i}. \tag{4.35}$$

In words, the long run ($n \to \infty$) mean visits to state i per unit time equals π_i, the probability of state i under the stationary distribution.

Next, let $r(i)$ define a cost or rate to be accumulated upon each visit to state i. The total cost accumulated during the first n-stages is

$$R^{(n-1)} = \sum_{k=0}^{n-1} r(X_k) = \sum_{k=0}^{n-1} \sum_i \mathbf{1}\{X_k = i\}r(i)$$

$$= \sum_{i=0}^{\infty} M_i^{(n)} r(i). \tag{4.36}$$

This leads to the following derivation showing that the long run mean cost per unit time equals the mean cost evaluated over the stationary distribution:

$$\lim_{n\to\infty} \frac{1}{n} E[R^{(n-1)}|X_0 = i] = \lim_{n\to\infty} \sum_{i=0}^{\infty} \frac{1}{n} E[M_i^{(n)}|X_0 = i]r(i)$$

$$= \sum_{i=0}^{\infty} \pi_i r(i). \qquad (4.37)$$

(When the Markov chain has an infinite number of states, then the derivation requires that a limit and infinite sum be interchanged. A sufficient condition to justify this interchange is that $r(i)$ be a bounded function of i.)

Remark   *The periodic case*   If i is a member of a recurrent periodic irreducible Markov chain with period d, one can show that $P_{ii}^m = 0$ if m is not a multiple of d (i.e., if $m \neq nd$ for any n), and that

$$\lim_{n\to\infty} P_{ii}^{nd} = \frac{d}{m_i}.$$

These last two results are easily combined with (4.30) to show that (4.31) also holds in the periodic case. If $m_i < \infty$, then the chain is positive recurrent and

$$\lim_{n\to\infty} \frac{1}{n} \sum_{m=0}^{n-1} P_{ii}^{(m)} = \pi_i = \frac{1}{m_i} \qquad (4.38)$$

where $\pi = (\pi_0, \pi_1, \ldots)$ is given as the unique nonnegative solution to

$$\pi_j = \sum_{k=0}^{\infty} \pi_k P_{kj}, \qquad j = 0, 1, \ldots$$

and

$$\sum_{j=0}^{\infty} \pi_j = 1.$$

That is, a unique stationary distribution $\pi = (\pi_0, \pi_1, \ldots)$ exists for a positive recurrent periodic irreducible Markov chain and the mean fraction of time in state i converges to π_i as the number of stages n grows to infinity.

   The convergence of (4.38) does not require the chain to start in state i. Under the same conditions

$$\lim_{n\to\infty} \frac{1}{n} \sum_{m=0}^{n-1} P_{ki}^{(m)} = \pi_i = \frac{1}{m_i}$$

holds for all states $k = 0, 1, \ldots$ as well.

Problems 4.4

1. Determine the stationary distribution for the periodic Markov chain whose transition probability matrix is

$$
\begin{array}{c c c c c}
 & 0 & 1 & 2 & 3 \\
0 & 0 & \frac{1}{2} & 0 & \frac{1}{2} \\
1 & \frac{1}{4} & 0 & \frac{3}{4} & 0 \\
2 & 0 & \frac{1}{3} & 0 & \frac{2}{3} \\
3 & \frac{1}{2} & 0 & \frac{1}{2} & 0
\end{array}.
$$

2. Consider the Markov chain on $\{0, 1\}$ whose transition probability matrix is

$$
\begin{array}{c c c}
 & 0 & 1 \\
0 & 1 - \alpha & \alpha \\
1 & \beta & 1 - \beta
\end{array}, \qquad 0 < \alpha, \beta < 1.
$$

(a) Verify that $(\pi_0, \pi_1) = (\beta/(\alpha + \beta), \alpha/(\alpha + \beta))$ is a stationary distribution.

(b) Show that the first return distribution to state 0 is given by $f_{00}^{(1)} = (1 - \alpha)$ and $f_{00}^{(n)} = \alpha\beta(1 - \beta)^{n-2}$ for $n = 2, 3, \ldots$.

(c) Calculate the mean return time $m_0 = \sum_{n=1}^{\infty} n f_{00}^{(n)}$ and verify that $\pi_0 = 1/m_0$.

3. Determine the stationary distribution for the Markov chain whose transition probability matrix is

$$
\mathbf{P} =
\begin{array}{c c c c c}
 & 0 & 1 & 2 & 3 \\
0 & 0 & 0 & \frac{1}{2} & \frac{1}{2} \\
1 & 0 & 0 & \frac{1}{3} & \frac{2}{3} \\
2 & \frac{1}{4} & \frac{3}{4} & 0 & 0 \\
3 & \frac{1}{3} & \frac{2}{3} & 0 & 0
\end{array}.
$$

4. Determine the period of state 0 in the Markov chain whose transition probability matrix is

$$
\mathbf{P} =
\begin{array}{c c c c c c c c c}
 & 3 & 2 & 1 & 0 & -1 & -2 & -3 & -4 \\
3 & 0 & 0 & 0 & 1 & 0 & 0 & 0 & 0 \\
2 & 1 & 0 & 0 & 0 & 0 & 0 & 0 & 0 \\
1 & 0 & 1 & 0 & 0 & 0 & 0 & 0 & 0 \\
0 & 0 & 0 & \frac{1}{2} & 0 & \frac{1}{2} & 0 & 0 & 0 \\
-1 & 0 & 0 & 0 & 0 & 0 & 1 & 0 & 0 \\
-2 & 0 & 0 & 0 & 0 & 0 & 0 & 1 & 0 \\
-3 & 0 & 0 & 0 & 0 & 0 & 0 & 0 & 1 \\
-4 & 0 & 0 & 0 & 1 & 0 & 0 & 0 & 0
\end{array}.
$$

4.5 Reducible Markov Chains

Recall that states i and j *communicate* if it is possible to reach state j starting from state i, and vice versa, and a Markov chain is *irreducible* if all pairs of states communicate. In this section we show, mostly by example, how to analyze more general Markov chains.

Consider first the Markov chain whose transition probability matrix is

$$
\mathbf{P} = \left\|
\begin{array}{cccc}
\frac{1}{2} & \frac{1}{2} & 0 & 0 \\
\frac{1}{4} & \frac{3}{4} & 0 & 0 \\
0 & 0 & \frac{1}{3} & \frac{2}{3} \\
0 & 0 & \frac{2}{3} & \frac{1}{3}
\end{array}
\right\|
$$

which we write in the form

$$
\mathbf{P} = \left\|
\begin{array}{cc}
\mathbf{P}_1 & \mathbf{0} \\
\mathbf{0} & \mathbf{P}_2
\end{array}
\right\|
$$

where

$$
\mathbf{P}_1 = \left\|
\begin{array}{cc}
\frac{1}{2} & \frac{1}{2} \\
\frac{1}{4} & \frac{3}{4}
\end{array}
\right\|
\quad \text{and} \quad
\mathbf{P}_2 = \left\|
\begin{array}{cc}
\frac{1}{3} & \frac{2}{3} \\
\frac{2}{3} & \frac{1}{3}
\end{array}
\right\|.
$$

The chain has two communicating classes, the first two states forming one class, and the last two states forming the other. Then

$$
\mathbf{P}^2 = \left\|
\begin{array}{cccc}
\frac{1}{2} & \frac{1}{2} & 0 & 0 \\
\frac{1}{4} & \frac{3}{4} & 0 & 0 \\
0 & 0 & \frac{1}{3} & \frac{2}{3} \\
0 & 0 & \frac{2}{3} & \frac{1}{3}
\end{array}
\right\|
\times
\left\|
\begin{array}{cccc}
\frac{1}{2} & \frac{1}{2} & 0 & 0 \\
\frac{1}{4} & \frac{3}{4} & 0 & 0 \\
0 & 0 & \frac{1}{3} & \frac{2}{3} \\
0 & 0 & \frac{2}{3} & \frac{1}{3}
\end{array}
\right\|
$$

$$
= \left\|
\begin{array}{cccc}
\frac{3}{8} & \frac{5}{8} & 0 & 0 \\
\frac{5}{16} & \frac{11}{16} & 0 & 0 \\
0 & 0 & \frac{5}{9} & \frac{4}{9} \\
0 & 0 & \frac{4}{9} & \frac{5}{9}
\end{array}
\right\|
= \left\|
\begin{array}{cc}
\mathbf{P}_1^2 & \mathbf{0} \\
\mathbf{0} & \mathbf{P}_2^2
\end{array}
\right\|,
$$

and, in general,

$$
\mathbf{P}^n = \left\|
\begin{array}{cc}
\mathbf{P}_1^n & \mathbf{0} \\
\mathbf{0} & \mathbf{P}_2^n
\end{array}
\right\|, \quad n \geq 1. \tag{4.39}
$$

Equation (4.39) is the mathematical expression of the property that it is not possible to communicate back and forth between distinct communicating classes; once in the first class the process remains there thereafter, and, similarly, once in the second class, the process remains there. In effect, two completely unrelated processes have been labeled together. The transition probability matrix $\mathbf{P}$ is reducible to the irreducible matrices $\mathbf{P}_1$ and $\mathbf{P}_2$. It follows from (4.39) that

$$
\lim_{n \to \infty} \mathbf{P}^n = \left\|
\begin{array}{cccc}
\pi_0^{(1)} & \pi_1^{(1)} & 0 & 0 \\
\pi_0^{(1)} & \pi_1^{(1)} & 0 & 0 \\
0 & 0 & \pi_0^{(2)} & \pi_1^{(2)} \\
0 & 0 & \pi_0^{(2)} & \pi_1^{(2)}
\end{array}
\right\|,
$$

where

$$\lim_{n\to\infty} \mathbf{P}_1^n = \begin{Vmatrix} \pi_0^{(1)} & \pi_1^{(1)} \\ \pi_0^{(1)} & \pi_1^{(1)} \end{Vmatrix} \quad \text{and} \quad \lim_{n\to\infty} \mathbf{P}_2^n = \begin{Vmatrix} \pi_0^{(2)} & \pi_1^{(2)} \\ \pi_0^{(2)} & \pi_1^{(2)} \end{Vmatrix}.$$

We solve for $\boldsymbol{\pi}^{(1)} = (\pi_0^{(1)}, \pi_1^{(1)})$ and $\boldsymbol{\pi}^{(2)} = (\pi_0^{(2)}, \pi_1^{(2)})$ in the usual way:

$$\tfrac{1}{2}\pi_0^{(1)} + \tfrac{1}{4}\pi_1^{(1)} = \pi_0^{(1)}$$

$$\tfrac{1}{2}\pi_0^{(1)} + \tfrac{3}{4}\pi_1^{(1)} = \pi_1^{(1)}$$

$$\pi_0^{(1)} + \pi_1^{(1)} = 1$$

or

$$\pi_0^{(1)} = \tfrac{1}{3}, \ \pi_1^{(1)} = \tfrac{2}{3}; \tag{4.40}$$

and because $\mathbf{P}_2$ is doubly stochastic (see p. 125), then $\pi_0^{(2)} = \tfrac{1}{2}$, $\pi_1^{(2)} = \tfrac{1}{2}$.

The basic limit theorem of Markov chains, Theorem 4.3, referred to an irreducible Markov chain. The limit theorem applies verbatim to any aperiodic recurrent class in a reducible Markov chain. If i, j are in the same aperiodic recurrent class, then $P_{ij}^{(n)} \to 1/m_j \geq 0$ as $n \to \infty$. If i, j are in the same periodic recurrent class, then $n^{-1} \sum_{m=0}^{n-1} P_{ij}^{(m)} \to 1/m_j \geq 0$ as $n \to \infty$.

If j is a transient state, then $P_{ij}^{(n)} \to 0$ as $n \to \infty$, and, more generally, $P_{ij}^{(n)} \to 0$ as $n \to \infty$ for all initial states i.

In order to complete the discussion of the limiting behavior of $P_{ij}^{(n)}$, we still must consider the case where i is transient and j is recurrent. Consider the transition probability matrix

$$\mathbf{P} = \begin{array}{c} \\ 0 \\ 1 \\ 2 \\ 3 \end{array} \begin{Vmatrix} \begin{array}{cccc} 0 & 1 & 2 & 3 \end{array} \\ \begin{array}{cccc} \tfrac{1}{2} & \tfrac{1}{2} & 0 & 0 \\ \tfrac{1}{4} & \tfrac{3}{4} & 0 & 0 \\ \tfrac{1}{4} & \tfrac{1}{4} & \tfrac{1}{4} & \tfrac{1}{4} \\ 0 & 0 & 0 & 1 \end{array} \end{Vmatrix}.$$

There are three classes: $\{0, 1\}$, $\{2\}$, and $\{3\}$ and of these $\{0, 1\}$ and $\{3\}$ are recurrent while $\{2\}$ is transient. Starting from state 2, the process ultimately gets absorbed in one of the other classes. The question is, which one, or more precisely, what are the probabilities of absorption in the two recurrent classes starting from state 2?

A first step analysis answers the question. Let u denote the probability of absorption in class $\{0, 1\}$ starting from state 2. Then $1 - u$ is the probability of absorption in class $\{3\}$. Conditioning on the first step, we have

$$u = (\tfrac{1}{4} + \tfrac{1}{4})1 + \tfrac{1}{4}u + \tfrac{1}{4}(0) = \tfrac{1}{2} + \tfrac{1}{4}u$$

or $u = \tfrac{2}{3}$. With probability $\tfrac{2}{3}$ the process enters $\{0, 1\}$ and remains there ever after. The stationary distribution for the recurrent class $\{0, 1\}$, computed in

(4.40), is $\pi_0 = \frac{1}{3}$, $\pi_1 = \frac{2}{3}$. Therefore $\lim_{n\to\infty} P_{20}^{(n)} = \frac{2}{3} \times \frac{1}{3} = \frac{2}{9}$, $\lim_{n\to\infty}$ $P_{21}^{(n)} = \frac{2}{3} \times \frac{2}{3} = \frac{4}{9}$. That is, we multiply the probability of entering the class $\{0, 1\}$ by the appropriate probabilities under the stationary distribution for the various states in the class. In matrix form, the limiting behavior of $\mathbf{P}^n$ is given by

$$\lim_{n\to\infty} \mathbf{P}^n = \begin{Vmatrix} \frac{1}{3} & \frac{2}{3} & 0 & 0 \\ \frac{1}{3} & \frac{2}{3} & 0 & 0 \\ \frac{2}{9} & \frac{4}{9} & 0 & \frac{1}{3} \\ 0 & 0 & 0 & 1 \end{Vmatrix}.$$

To firm up the principles, consider one last example:

$$\mathbf{P} = \begin{array}{c} \\ 0 \\ 1 \\ 2 \\ 3 \\ 4 \\ 5 \end{array} \begin{Vmatrix} \frac{1}{2} & \frac{1}{2} & 0 & 0 & 0 & 0 \\ \frac{1}{3} & \frac{2}{3} & 0 & 0 & 0 & 0 \\ \frac{1}{3} & 0 & 0 & \frac{1}{3} & \frac{1}{6} & \frac{1}{6} \\ \frac{1}{6} & \frac{1}{6} & \frac{1}{6} & 0 & \frac{1}{3} & \frac{1}{6} \\ 0 & 0 & 0 & 0 & 0 & 1 \\ 0 & 0 & 0 & 0 & 1 & 0 \end{Vmatrix}.$$

There are three classes: $C_1 = \{0, 1\}$, $C_2 = \{2, 3\}$, and $C_3 = \{4, 5\}$. The stationary distribution in C_1 is (π_0, π_1) where

$$\tfrac{1}{2}\pi_0 + \tfrac{1}{3}\pi_1 = \pi_0$$

$$\tfrac{1}{2}\pi_0 + \tfrac{2}{3}\pi_1 = \pi_1$$

$$\pi_0 + \pi_1 = 1.$$

Then $\pi_0 = \frac{2}{5}$ and $\pi_1 = \frac{3}{5}$.

Class C_3 is periodic, and $P_{ij}^{(n)}$ does not converge for i, j in $C_3 = \{4, 5\}$. The time averages do converge, however; and $\lim_{n\to\infty} n^{-1} \sum_{m=0}^{n-1} P_{ij}^{(m)} = \frac{1}{2}$ for $i = 3, 4$ and $j = 3, 4$.

For the transient class $C_2 = \{2, 3\}$, let u_i be the probability of ultimate absorption in class $C_1 = \{0, 1\}$ starting from state i for $i = 2, 3$. From a first step analysis, then

$$u_2 = \tfrac{1}{3}(1) + 0(1) + 0\,u_2 + \tfrac{1}{3}\,u_3 + \tfrac{1}{6}(0) + \tfrac{1}{6}(0)$$

$$u_3 = \tfrac{1}{6}(1) + \tfrac{1}{6}(1) + \tfrac{1}{6}\,u_2 + 0\,u_3 + \tfrac{1}{3}(0) + \tfrac{1}{6}(0),$$

or

$$u_2 = \tfrac{1}{3} + \tfrac{1}{3}\,u_3; \qquad u_3 = \tfrac{1}{3} + \tfrac{1}{6}\,u_2.$$

The solution is $u_2 = \frac{8}{17}$ and $u_3 = \frac{7}{17}$. Combining these partial answers in matrix form, we have

$$
\lim_{n\to\infty} \mathbf{P}^n =
\begin{array}{c|cccccc}
 & 0 & 1 & 2 & 3 & 4 & 5 \\
\hline
0 & \frac{2}{5} & \frac{3}{5} & 0 & 0 & 0 & 0 \\
1 & \frac{2}{5} & \frac{3}{5} & 0 & 0 & 0 & 0 \\
2 & (\frac{8}{17})(\frac{2}{5}) & (\frac{8}{17})(\frac{3}{5}) & 0 & 0 & X & X \\
3 & (\frac{7}{17})(\frac{2}{5}) & (\frac{7}{17})(\frac{3}{5}) & 0 & 0 & X & X \\
4 & 0 & 0 & 0 & 0 & X & X \\
5 & 0 & 0 & 0 & 0 & X & X
\end{array}
$$

where X denotes that the limit does not exist. For the time averages, we have

$$
\lim_{n\to\infty} \frac{1}{n}\sum_{m=0}^{n-1} P^m =
\begin{array}{c|cccccc}
 & 0 & 1 & 2 & 3 & 4 & 5 \\
\hline
0 & \frac{2}{5} & \frac{3}{5} & 0 & 0 & 0 & 0 \\
1 & \frac{2}{5} & \frac{3}{5} & 0 & 0 & 0 & 0 \\
2 & (\frac{8}{17})(\frac{2}{5}) & (\frac{8}{17})(\frac{3}{5}) & 0 & 0 & (\frac{9}{17})(\frac{1}{2}) & (\frac{9}{17})(\frac{1}{2}) \\
3 & (\frac{7}{17})(\frac{2}{5}) & (\frac{7}{17})(\frac{3}{5}) & 0 & 0 & (\frac{10}{17})(\frac{1}{2}) & (\frac{10}{17})(\frac{1}{2}) \\
4 & 0 & 0 & 0 & 0 & \frac{1}{2} & \frac{1}{2} \\
5 & 0 & 0 & 0 & 0 & \frac{1}{2} & \frac{1}{2}
\end{array}.
$$

One possible behavior remains to be illustrated. It can occur only when there are an infinite number of states. In this case it is possible that all states are transient or null recurrent and $\lim_{n\to\infty} P_{ij}^{(n)} = 0$ for all states i, j. For example, consider the deterministic Markov chain described by $X_n = X_0 + n$. The transition probability matrix is

$$
\mathbf{P} =
\begin{array}{c|ccccc}
 & 0 & 1 & 2 & 3 & \\
\hline
0 & 0 & 1 & 0 & 0 & \cdots \\
1 & 0 & 0 & 1 & 0 & \cdots \\
2 & 0 & 0 & 0 & 1 & \cdots \\
3 & 0 & 0 & 0 & 0 & \cdots \\
 & \vdots & \vdots & \vdots & \vdots &
\end{array}.
$$

Then all states are transient and $\lim_{n\to\infty} P_{ij}^{(n)} = \lim_{n\to\infty} \Pr\{X_n = j | X_0 = i\} = 0$ for all states i, j.

If there are only a finite number M of states, then there are no null recurrent states and not all states can be transient. In fact, since $\sum_{j=0}^{M-1} P_{ij}^{(n)} = 1$ for all n, it cannot happen that $\lim_{n\to\infty} P_{ij}^{(n)} = 0$ for all j.

Problems 4.5

1. Given the transition matrix

$$
\mathbf{P} =
\begin{array}{c|ccccc}
 & 0 & 1 & 2 & 3 & 4 \\
\hline
0 & \frac{1}{4} & \frac{3}{4} & 0 & 0 & 0 \\
1 & \frac{1}{2} & \frac{1}{2} & 0 & 0 & 0 \\
2 & 0 & 0 & 1 & 0 & 0 \\
3 & 0 & 0 & \frac{1}{3} & \frac{2}{3} & 0 \\
4 & 1 & 0 & 0 & 0 & 0
\end{array},
$$

determine the limits, as $n \to \infty$, of $P_{i0}^{(n)}$ for $i = 0, 1, \ldots, 4$.

2. Given the transition matrix

$$
\mathbf{P} =
\begin{array}{c|ccccccc}
 & 1 & 2 & 3 & 4 & 5 & 6 & 7 \\
\hline
1 & \frac{1}{3} & \frac{2}{3} & 0 & 0 & 0 & 0 & 0 \\
2 & \frac{1}{4} & \frac{3}{4} & 0 & 0 & 0 & 0 & 0 \\
3 & 0 & 0 & 0 & \frac{2}{3} & \frac{1}{3} & 0 & 0 \\
4 & 0 & 0 & 1 & 0 & 0 & 0 & 0 \\
5 & 0 & 0 & 1 & 0 & 0 & 0 & 0 \\
6 & \frac{1}{6} & 0 & \frac{1}{6} & \frac{1}{6} & 0 & \frac{1}{4} & \frac{1}{4} \\
7 & 0 & 0 & 0 & 0 & 0 & 0 & 1
\end{array},
$$

derive the following limits, where they exist

(a) $\lim_{n\to\infty} P_{11}^{(n)}$

(b) $\lim_{n\to\infty} P_{31}^{(n)}$

(c) $\lim_{n\to\infty} P_{61}^{(n)}$

(d) $\lim_{n\to\infty} P_{63}^{(n)}$

(e) $\lim_{n\to\infty} P_{21}^{(n)}$

(f) $\lim_{n\to\infty} P_{33}^{(n)}$

(g) $\lim_{n\to\infty} P_{67}^{(n)}$

(h) $\lim_{n\to\infty} P_{64}^{(n)}$

4.6 Sequential Decisions and Markov Chains*

This section is of primary interest to students of engineering and management science, although recent applications in the biological and social sciences have appeared. The problems treated here go under a variety of names, including dynamic programming, controlled Markov chains, and Markov decision models.

Consider a system with a finite number S of states, labeled by the integers $1, 2, \ldots, S$. Periodically, say once a day, we observe the current state

---

*This section contains material at a more difficult level. It is not prerequisite to what follows.

of the system, and then choose an action from a set containing a finite number A of possible actions, labeled by 1, 2, . . . , A. As a joint result of the current state s and the chosen action a, two things happen: (i) we receive an immediate income $i(s, a)$, and (ii) the system moves to a new state, where the probability of a particular state s' being chosen is given by a function $q = q(s'|s, a)$. Our problem is to choose a policy that maximizes the long run time average expected income.

In many applications it is helpful to restrict the allowable actions according to the current state. That is, in state s, the possible actions are labeled 1, 2, . . . , A_s where A_s may vary with s. This is a straightforward generalization that the reader can readily supply in any particular instance. To keep an already cumbersome notation from becoming even more complex, we treat only the simpler case in our exposition.

Example *A Water Resources System* Consider a water reservoir system with the dual purpose of generating electricity and supplying water for agricultural irrigation. Let S_n be the level (quantity) of water in the system at the start of period n, and let A_n be the amount released during the period.

Let I_n be the input (rainfall) to the storage system during period n, and suppose that I_n is a random variable, independent from period to period, and following the probability distribution

$$\Pr\{I_n = k\} = p(k) \qquad \text{for} \quad k = 0, 1, ...$$

The mass balance equation OLD LEVEL + INFLOW − OUTFLOW = NEW LEVEL is written $S_n + I_n - A_n = S_{n+1}$. If we assume a reservoir maximum capacity of M, the balance equation becomes

$$S_{n+1} = \min\{M, S_n + I_n - A_n\}, \tag{4.41}$$

where the excess over capacity is assumed lost. The relation expressed in (4.41) leads to the transition law

$$q(s'|s, a) = \Pr\{I_n = s' - s + a\} = p(s' - s + a) \qquad \text{if} \quad 0 \le s' < M$$

and

$$q(M|s, a) = \Pr\{I_n \ge M - s + a\} = \sum_{s' \ge M} p(s' - s + a).$$

We turn to describing the income function. Let us suppose there is a target level of e units per period for electrical generation, and that there is no value ascribed to exceeding this target level but a proportional penalty for not meeting it. With

$$E = \text{Unit income for electrical use}$$
$$K = \text{Unit penalty for electrical shortage}$$

we have the total income from electrical generation given by

$$i_E(s, a) = E \min\{s, e\} - K \max\{0, e - s\}.$$

Water released for irrigation has an assumed unit value of W, leading to the total income allocated to irrigation of

$$i_W(s, a) = Wa.$$

Combining the two sources of income yields the per period income function

$$i(s, a) = i_E(s, a) + i_W(s, a)$$

$$= E \min\{s, e\} - K \max\{0, e - s\} + Wa.$$

Specifying the law of motion $q(s'|s, a)$ and the income function $i(s, a)$ places the water resources problem in the framework of a Markov decision model.

Example   *An Inventory Model*   Let $D_1, D_2, \ldots$ be the demands in successive periods for a certain stocked commodity. We assume that D_1, $D_2, \ldots$ are independent and identically distributed random variables following the probability distribution

$$\Pr\{D_n = k\} = p(k) \qquad \text{for} \quad k = 0, 1, \ldots . \qquad (4.42)$$

Let S_0 be the number of items on hand initially, and for $n = 1, 2, \ldots,$ let S_n be the number of items on hand at the end of period n. We assume that delivery of replenishment items is instantaneous, and let A_n be the amount ordered and delivered immediately, available at the beginning of period $n + 1$. The inventory movement is shown in Figure 4.3.

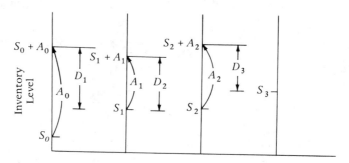

Figure 4.3   An inventory problem as a Markov decision model

The mass balance equation BEGINNING INVENTORY + AMOUNT ORDERED − DEMAND = ENDING INVENTORY translates into $S_n + A_n - D_{n+1} = S_{n+1}$. Suppose, however, that negative inventory levels are not allowed (no backlogging). Then the amount sold is the smaller of the amount demanded and the amount available, or what is the same thing,

$$S_{n+1} = \max\{0, S_n + A_n - D_{n+1}\}. \qquad (4.43)$$

Considering the two possibilities for the maximum in (4.43), we obtain the transition law

$$q(0|s, a) = \sum_{k \geq s+a} p(k)$$

and

$$q(s'|s, a) = p(s + a - s') \qquad \text{if} \quad 0 < s' \leq s + a.$$

We turn to describing the cost structure by first introducing a penalty cost p per unit of unsatisfied demand. Then the expected total penalty cost from a starting inventory of $s + a$ is

$$c_p(s, a) = p \sum_{k=s+a+1}^{\infty} (k - s - a)p(k).$$

If it costs K units to place an order and c per item ordered, then the total ordering cost is

$$c_0(s, a) = K\mathbf{1}\{a > 0\} + ca$$

where

$$\mathbf{1}\{a > 0\} = \begin{cases} 1 & \text{if} \quad a > 0, \\ 0 & \text{if} \quad a = 0. \end{cases}$$

Note that there is no cost if no order is placed ($a = 0$).

The one period income function is the negative of the total costs, or

$$i(s, a) = -\{K\mathbf{1}\{a > 0\} + ca + p \sum_{k=s+a+1}^{\infty} (k - s - a)p(k)\}.$$

Having specified the transition law and the one period income function, we have formulated the inventory model as a Markov decision problem.

Having exhibited some motivating examples, we return to the analysis of the general problem of sequential decisions and Markov chains.

A *policy* is a function f that specifies the probability $f(a|s)$ of selecting action a whenever the system is in state s. Every policy f satisfies

$$f(a|s) \geq 0 \text{ for all } a, s, \quad \text{and} \quad \sum_a f(a|s) = 1 \text{ for all } s. \qquad (4.44)$$

Let S_n be the state of the system at time n and let A_n be the action chosen. Then under any fixed policy f the pair $X_n = (S_n, A_n)$ forms a two-dimensional Markov chain with transition function

$$P[S_{n+1} = s', A_{n+1} = a'|S_n = s, A_n = a] = q(s'|s, a) f(a'|s'), \qquad (4.45)$$

and if this Markov chain is irreducible, then by the basic limit theorem of Markov chains [as stated in Equation (4.37)], the long run mean income per unit time $I(f)$ is given by the mean income under the stationary distribution and is independent of the starting state and action. That is

$$I(f) = \lim_{n \to \infty} \frac{1}{n} \sum_{m=0}^{n-1} E[i(S_m, A_m)] = \sum_{s=1}^{S} \sum_{a=1}^{A} \pi(s, a) i(s, a), \qquad (4.46)$$

where $\pi(s, a)$ is the stationary distribution associated with the transition probabilities in (4.45). We know that $\pi(s, a)$ is given as the unique solution to

$$\pi(s, a) \geq 0 \text{ for all } s, a, \quad \text{and} \quad \sum_{s=1}^{S} \sum_{a=1}^{A} \pi(s, a) = 1, \qquad (4.47)$$

and

$$\pi(s', a') = \sum_{s=1}^{S} \sum_{a=1}^{A} \pi(s, a) q(s'|s, a) f(a'|s')$$
$$\text{for} \quad s' = 1, \ldots, S \text{ and } a' = 1, \ldots, A. \quad (4.48)$$

The objective is to find a policy f that maximizes the long run expected income per unit time. Equivalently, the task is to find a policy f satisfying (4.44) that maximizes $I(f)$ in (4.46), where $\pi(s, a)$ is related to the policy f through (4.47) and (4.48). Stated thus, the problem is very difficult because of the nonlinear connection between the policy f and the income $I(f)$. Fortunately there is a way to reduce the problem to an equivalent simpler one in which a linear objective function is maximized subject to linear constraints. Standard computer programs are available for solving such *linear programming* problems. The linear programming problem is:

Problem 1 Find $\pi = \pi(s, a) \geq 0$ that will maximize

$$I = \sum_{s=1}^{S} \sum_{a=1}^{A} i(s, a) \pi(s, a) \qquad (4.49)$$

subject to

$$\sum_{s=1}^{S} \sum_{a=1}^{A} \pi(s, a) = 1 \qquad (4.50)$$

and

$$\sum_{a'=1}^{A} \pi(s', a') = \sum_{s=1}^{S} \sum_{a=1}^{A} \pi(s, a) q(s'|s, a) \qquad \text{for} \quad s' = 1, \ldots, S. \quad (4.51)$$

Equation (4.51) results from summing (4.48) over $a' = 1, \ldots, A$ and using (4.44). To show that the linear programming problem is equivalent to the original problem, we need only show how to recover the policy f from the linear programming solution. Consider a state s' for which there is some a' satisfying $\pi(s', a') > 0$. Then from (4.48) and (4.51) we obtain

$$f(a'|s') = \frac{\pi(s', a')}{\sum_{a=1}^{A} \pi(s', a)}. \qquad (4.52)$$

Equation (4.52) recovers the policy $f(a'|s')$ from the stationary distribution $\pi(s', a')$, at least for all states s' that are recurrent $[\pi(s', a') > 0$ for some $a']$.

One can take any action whatsoever in transient states, since income earned there will not affect the asymptotic long run income per unit time. That is, if for some state s we have $\pi(s, a) = 0$ for all actions a, then that state s cannot influence the income rate as given in (4.49).

Every linear programming problem has associated with it a dual problem that is equivalent to it in the sense that the solution of one problem provides the solution of the other, and vice versa. The dual to Problem 1 is

Problem 2 Find $v(1), \ldots, v(S)$ and a constant g for which

$$v(s) = \max_{a=1,\ldots,A} \left\{ i(s, a) - g + \sum_{s'=1}^{S} v(s')q(s'|s, a) \right\}$$
$$\text{for} \quad s = 1, \ldots, S. \quad (4.53)$$

If $v(1), \ldots, v(S)$ solves this problem then so does $v(1) + c, \ldots, v(S) + c$ for any constant c. Thus one may arbitrarily set $v(1) = 0$. The variable g is the maximum mean income per period or maximal gain rate $[g = \max_f I(f)]$.

Problem 2 is often called the dynamic programming formulation. In each state s, the optimal act is the one that maximizes the right side of (4.53).

Problem 1 can be solved by using standard computer routines for linear programming problems. Certain intelligent trial and error techniques (called Howard's algorithm) can be used to solve Problem 2. A numerical example illustrating this technique follows.

Example Consider the Markov sequential decision problem on states $s = 1, 2, 3$ and having acts $a = 1, 2$ for which the one period income function is

		Acts	
	States	1	2
$i(s, a) =$	1	1	2
	2	5	3
	3	3	4

and the transition law is

		Next State s'		
		1	2	3
Current	1, 1	.2	.6	.2
State-Act	1, 2	.4	.3	.3
(s, a)	2, 1	.7	.2	.1
$q(s'|s, a) =$	2, 2	.2	.5	.3
	3, 1	.1	.7	.2
	3, 2	.8	.1	.1

We attempt a trial and error solution of Problem 2. Consider the trial policy $f_0(s)$ given by

State s	1	2	3
Act $f_0(s)$	2	1	2.

Looking at Problem 2, we solve the equations

$$v_0(s) = i(s, f_0(s)) - g_0 + \sum_{s'} v_0(s')g(s'|s, f_0(s))$$

for $v_0(1) = 0$, $v_0(2)$, $v_0(3)$, and g_0. The equations in this instance are

$$0 = 2 - g_0 + 0 + .3v_0(2) + .3v_0(3)$$
$$v_0(2) = 5 - g_0 + 0 + .2v_0(2) + .1v_0(3)$$
$$v_0(3) = 4 - g_0 + 0 + .1v_0(2) + .1v_0(3),$$

which solves to $v_0(1) = 0$, $v_0(2) = 2.50$, $v_0(3) = 1.25$ and $g_0 = 3.125$.

We check to see if these values satisfy the maximal property expressed in Problem 2. For each state-action pair s, a, we evaluate $i(s, a) - g_0 + \sum_{s'} v_0(s')q(s'|s, a)$ and then for each state s choose the maximizing act a. This leads to

State-Act	$i(s, a) - g_0 + \sum_{s'} v_0(s')q(s'|s, a)$
(1, 1)	$1 - g_0 + .6v_0(2) + .2v_0(3) = 2.750 - g_0$
(1, 2)	$2 - g_0 + .3v_0(2) + .3v_0(3) = 3.125 - g_0$ ★
(2, 1)	$5 - g_0 + .2v_0(2) + .1v_0(3) = 5.625 - g_0$ ★
(2, 2)	$3 - g_0 + .5v_0(2) + .3v_0(3) = 4.625 - g_0$
(3, 1)	$3 - g_0 + .7v_0(2) + .2v_0(3) = 5.000 - g_0$ ★
(3, 2)	$4 - g_0 + .1v_0(2) + .1v_0(3) = 4.375 - g_0$

This suggests an improved policy, indicated by the asterisks and given by

State s	1	2	3
Act $f_1 (s)$	2	1	1.

We repeat, solving now

$$v_1(s) = i(s, f_1(s)) - g_1 + \sum_{s'} v_1(s')q(s'|s, f_1(s))$$

for $v_1(1) = 0$, $v_1(2)$, $v_1(3)$, and g_1. The equations are

$$0 = 2 - g_1 + .3v_1(2) + .3v_1(3)$$
$$v_1(2) = 5 - g_1 + .2v_1(2) + .1v_1(3)$$
$$v_1(3) = 3 - g_1 + .7v_1(2) + .2v_1(3)$$

which yields $g_1 = 3.2558$, $v_1(1) = 0$, $v_1(2) = 2.4031$, and $v_1(3) = 1.7829$. We again check the maximal property expressed in Problem 2, evaluating $i(s, a) - g_1 + \sum_{s'} v_1(s')q(s'|s, a)$ for each state-act pair (s, a). This results in

State-Act	$i(s, a) - g_1 + \sum_{s'} v_1(s')q(s'|s, a)$
(1, 1)	$1 - g_1 + .6v_1(2) + .2v_1(3) = 2.7984$
(1, 2)	$2 - g_1 + .3v_1(2) + .3v_1(3) = 3.2558$ ★
(2, 1)	$5 - g_1 + .2v_1(2) + .1v_1(3) = 5.6589$ ★
(2, 2)	$3 - g_1 + .5v_1(2) + .3v_1(3) = 4.7364$

$$(3, 1) \quad 3 - g_1 + .7v_1(2) + .2v_1(3) = 5.0308 \star$$
$$(3, 2) \quad 4 - g_1 + .1v_1(2) + .1v_1(3) = 4.4186$$

We see that we have solved completely the requirements expressed by Problem 2. The optimal policy is to choose Act 2 in State 1 and choose Act 1 in States 2 and 3. The maximal income per unit time is $g = 3.2558$.

Problems 4.6

1. A Markov decision problem on states $s = 1, 2, 3$ and having acts $a = 1, 2$ is specified by the data

		Acts	
	States	1	2
$i(s, a) =$	1	1	2
	2	6	3
	3	3	3

and

		New State s'		
		1	2	3
Current	1, 1	.7	.2	.1
State-Act	1, 2	.2	.5	.3
(s, a)	2, 1	.2	.6	.2
$q(s'|s, a) =$	2, 2	.4	.3	.3
	3, 1	.1	.7	.2
	3, 2	.8	.1	.1

Determine an optimal policy and the maximum long run income per unit time.

2. Solve the following Markov decision problem for the optimal policy that achieves a long run maximal income per unit time.

States: 1, 2, 3
Acts: 1, 2
Income $i(s, a) =$

		Acts	
	States	1	2
	1	1	3
	2	5	2
	3	3	4

Transition Law
$q(s'|s, a) =$

		Next State		
		1	2	3
Current	1, 1	.2	.6	.2
State–Act	1, 2	.4	.3	.3
	2, 1	.3	.2	.5
	2, 2	.2	.5	.3
	3, 1	.4	.4	.2
	3, 2	.6	.3	.1

Chapter 5 | Poisson Processes

5.1 The Poisson Distribution and the Poisson Process

Poisson behavior is so pervasive in natural phenomena and the Poisson distribution is so amenable to extensive and elaborate analysis as to make the Poisson process a cornerstone of stochastic modeling.

5.1.1 The Poisson Distribution

The Poisson distribution with parameter $\mu > 0$ is given by

$$p_k = \frac{e^{-\mu}\mu^k}{k!} \qquad \text{for} \quad k = 0, 1, \ldots \qquad (5.1)$$

Let X be a random variable having the Poisson distribution in (5.1). We evaluate the mean or first moment via

$$E[X] = \sum_{k=0}^{\infty} k p_k = \sum_{k=1}^{\infty} \frac{k e^{-\mu}\mu^k}{k!}$$

$$= \mu e^{-\mu} \sum_{k=1}^{\infty} \frac{\mu^{(k-1)}}{(k-1)!}$$

$$= \mu.$$

To evaluate the variance, it is easier first to determine

$$E[X(X-1)] = \sum_{k=2}^{\infty} k(k-1) p_k$$

$$= \mu^2 e^{-\mu} \sum_{k=2}^{\infty} \frac{\mu^{(k-2)}}{(k-2)!}$$

$$= \mu^2.$$

Then

$$E[X^2] = E[X(X-1)] + E[X]$$
$$= \mu^2 + \mu$$

while

$$\sigma_X^2 = \text{Var}[X] = E[X^2] - \{E[X]\}^2$$
$$= \mu^2 + \mu - \mu^2 = \mu.$$

Thus the Poisson distribution has the unusual characteristic that both the mean and the variance are given by the same value μ.

Two fundamental properties of the Poisson distribution, which will arise later in a variety of forms, concern the sum of independent Poisson random variables and certain random decompositions of Poisson phenomena. We state these properties formally as Theorems 5.1 and 5.2.

Theorem 5.1 *Let X and Y be independent random variables having Poisson distributions with parameters μ and v, respectively. Then the sum $X + Y$ has a Poisson distribution with parameter $\mu + v$.*

Proof By the law of total probability,

$$\Pr\{X + Y = n\} = \sum_{k=0}^{n} \Pr\{X = k, Y = n - k\}$$

$$= \sum_{k=0}^{n} \Pr\{X = k\} \Pr\{Y = n - k\}$$

$$(X \text{ and } Y \text{ are independent})$$

$$= \sum_{k=0}^{n} \left\{\frac{\mu^k e^{-\mu}}{k!}\right\} \left\{\frac{v^{n-k} e^{-v}}{(n-k)!}\right\}$$

$$= \frac{e^{-(\mu+v)}}{n!} \sum_{k=0}^{n} \frac{n!}{k!(n-k)!} \mu^k v^{n-k}. \tag{5.2}$$

The binomial expansion of $(\mu + v)^n$ is, of course,

$$(\mu + v)^n = \sum_{k=0}^{n} \frac{n!}{k!(n-k)!} \mu^k v^{n-k}$$

and so (5.2) simplifies to

$$\Pr\{X + Y = n\} = \frac{e^{-(\mu+\nu)} (\mu + \nu)^n}{n!}, \qquad n = 0, 1, \ldots,$$

the desired Poisson distribution. $\square$

To describe the second result, we consider first a Poisson random variable N where the parameter is $\mu > 0$. Write N as a sum of ones in the form

$$N = \underbrace{1 + 1 + \cdots + 1}_{N \text{ ones}}$$

and next, considering each one separately and independently, erase it with probability $1 - p$ and keep it with probability p. What is the distribution of the resulting sum M, of the form $M = 1 + 0 + 0 + 1 + \ldots + 1$?

The next theorem states and answers the question in a more precise wording.

Theorem 5.2 *Let N be a Poisson random variable with parameter μ, and conditional on N, let M have a binomial distribution with parameters N and p. Then the unconditional distribution of M is Poisson with parameter μp.*

Proof The verification proceeds via a direct application of the law of total probability. Then

$$\Pr\{M = k\} = \sum_{n=0}^{\infty} \Pr\{M = k | N = n\} \Pr\{N = n\}$$

$$= \sum_{n=k}^{\infty} \left\{ \frac{n!}{k!(n - k)!} \, p^k (1 - p)^{n-k} \right\} \left\{ \frac{\mu^n e^{-\mu}}{n!} \right\}$$

$$= \frac{e^{-\mu}(\mu p)^k}{k!} \sum_{n=k}^{\infty} \frac{[\mu(1 - p)]^{n-k}}{(n - k)!}$$

$$= \frac{e^{-\mu}(\mu p)^k}{k!} \, e^{\mu(1-p)}$$

$$= \frac{e^{-\mu p}(\mu p)^k}{k!} \qquad \text{for} \quad k = 0, 1, \ldots,$$

which is the claimed Poisson distribution. $\square$

5.1.2 The Poisson Process

The Poisson process entails notions of both independence and the Poisson distribution.

Definition *A Poisson process of intensity or rate $\lambda > 0$ is an integer-valued stochastic process $\{X(t); t \geq 0\}$ for which*

(i) *for any time points $t_0 = 0 < t_1 < t_2 < . . . < t_n$, the process increments*

$$X(t_1) - X(t_0), X(t_2) - X(t_1), . . . , X(t_n) - X(t_{n-1})$$

are independent random variables;
(ii) *for $s \geq 0$ and $t > 0$, the random variable $X(s + t) - X(s)$ has the Poisson distribution*

$$\Pr\{X(s + t) - X(s) = k\} = \frac{(\lambda t)^k e^{-\lambda t}}{k!} \qquad \text{for} \quad k = 0, 1, . . . ;$$

and
(iii) $X(0) = 0$.

In particular, observe that if $X(t)$ is a Poisson process of rate $\lambda > 0$, then the moments are

$$E[X(t)] = \lambda t \quad \text{and} \quad \text{Var}[X(t)] = \sigma^2_{X(t)} = \lambda t.$$

Example Defects occur along an undersea cable according to a Poisson process of rate $\lambda = .1$ per mile. (a) What is the probability that no defects appear in the first two miles of cable? (b) Given that there are no defects in the first two miles of cable, what is the conditional probability of no defects between mile points two and three? To answer (a) we observe that $X(2)$ has a Poisson distribution whose parameter is $(.1)(2) = .2$. Thus $\Pr\{X(2) = 0\} = e^{-.2} = .8187$. In part (b) we use the independence of $X(3) - X(2)$ and $X(2) - X(0) = X(2)$. Thus the conditional probability is the same as the unconditional probability and

$$\Pr\{X(3) - X(2) = 0\} = \Pr\{X(1) = 0\} = e^{-.1} = .9048.$$

Example Customers arrive in a certain store according to a Poisson process of rate $\lambda = 4$ per hour. Given that the store opens at 9:00 A.M., what is the probability that exactly one customer has arrived by 9:30 and a total of five has arrived by 11:30 A.M.

Measuring time t in hours from 9:00 A.M., we are asked to determine $\Pr\{X(\frac{1}{2}) = 1, X(\frac{5}{2}) = 5\}$. We use the independence of $X(\frac{5}{2}) - X(\frac{1}{2})$ and $X(\frac{1}{2})$ to reformulate the question thus:

$$\Pr\{X(\tfrac{1}{2}) = 1, X(\tfrac{5}{2}) = 5\} = \Pr\{X(\tfrac{1}{2}) = 1, X(\tfrac{5}{2}) - X(\tfrac{1}{2}) = 4\}$$

$$= \left\{\frac{e^{-4(1/2)}4(1/2)}{1!}\right\}\left\{\frac{e^{-4(2)}[4(2)]^4}{4!}\right\}$$

$$= (2e^{-2})(\tfrac{512}{3}e^{-8}) = .0154965.$$

5.1.3 Nonhomogeneous Processes

The rate λ in a Poisson process $X(t)$ is the proportionality constant in the probability of an event occurring during an arbitrarily small interval. To explain this more precisely,

$$\Pr\{X(t + h) - X(t) = 1\} = \frac{(\lambda h)e^{-\lambda h}}{1!}$$

$$= (\lambda h)(1 - \lambda h + \tfrac{1}{2}\lambda^2 h^2 - \cdots)$$

$$= \lambda h + o(h)$$

where $o(h)$ denotes a general and unspecified remainder term of smaller order than h $[o(h)/h \to 0$ as $h \to 0]$.

It is pertinent in many applications to consider rates $\lambda = \lambda(t)$ that vary with time. Such a process is termed a *nonhomogeneous* or *nonstationary* Poisson process to distinguish it from the stationary or homogeneous process that we primarily consider. If $X(t)$ is a nonhomogeneous Poisson process with rate $\lambda(t)$, then an increment $X(t) - X(s)$, giving the number of events in an interval $(s, t]$, has a Poisson distribution with parameter $\int_s^t \lambda(u)du$, and increments over disjoint intervals are independent random variables.

Example Demands on a first aid facility in a certain location occur according to a nonhomogeneous Poisson process having the rate function

$$\lambda(t) = \begin{cases} 2t & \text{for} \quad 0 \le t < 1 \\ 2 & \text{for} \quad 1 \le t < 2 \\ 4 - t & \text{for} \quad 2 \le t \le 4 \end{cases}$$

where t is measured in hours from the opening time of the facility. What is the probability that two demands occur in the first two hours of operation and two in the second two hours? Since demands during disjoint intervals are independent random variables we can answer the two questions separately. The mean for the first two hours is $\mu = \int_0^1 2t\,dt + \int_1^2 2\,dt = 3$ and thus

$$\Pr\{X(2) = 2\} = \frac{e^{-3}(3)^2}{2!} = .2240.$$

For the second two hours, $\mu = \int_2^4 (4 - t)dt = 2$ and

$$\Pr\{X(4) - X(2) = 2\} = \frac{e^{-2}(2)^2}{2!} = .2707.$$

Let $X(t)$ be a nonhomogeneous Poisson process of rate $\lambda(t) > 0$ and define $\Lambda(t) = \int_0^t \lambda(u)du$. Make a deterministic change in the time scale and define a new process $Y(s) = X(t)$ where $s = \Lambda(t)$. Observe that $\Delta s = \lambda(t)\Delta t + o(\Delta t)$. Then

$$\Pr\{Y(s + \Delta s) - Y(s) = 1\} = \Pr\{X(t + \Delta t) - X(t) = 1\}$$

$$= \lambda(t)\Delta t + o(\Delta t)$$

$$= \Delta s + o(\Delta s)$$

so that $Y(s)$ is a homogeneous Poisson process of unit rate. By this means, questions about nonhomogeneous Poisson processes can be transformed into corresponding questions about homogeneous processes. For this reason we concentrate our exposition on the latter.

Problems 5.1

1. Suppose that a random variable X is distributed according to a Poisson distribution with parameter λ. The parameter λ is itself a random variable, exponentially distributed with density $f(x) = \theta e^{-\theta x}$ for $x \geq 0$. Find the probability mass function for X.

2. Assume a device fails when a cumulative effect of k shocks occurs. If the shocks happen according to a Poisson process of parameter λ, what is the density function for the life T of the device?

3. Messages arrive at a telegraph office as a Poisson process with mean rate of 3 messages per hour.
   (a) What is the probability that no messages arrive during the morning hours 8:00 A.M. to noon?
   (b) What is the distribution of the time at which the first afternoon message arrives?

4. Let $X(t)$ be a homogeneous Poisson process with parameter λ. For $t, s \geq 0$, determine the product moment $E[X(t)X(t + s)]$.

5. Find the probability $\Pr\{X(t) = 1, 3, 5, \ldots\}$ that a Poisson process having rate λ is odd.

6. Arrivals of passengers at a bus stop form a Poisson process $X(t)$ with rate $\lambda = 2$ per unit time. Assume that a bus departed at time $t = 0$ leaving no customers behind. Let T denote the arrival time of the next bus. Then the number of passengers present when it arrives is $X(T)$. Suppose that the bus arrival time T is independent of the Poisson process and that T has the uniform probability density function

$$f_T(t) = \begin{cases} 1 & \text{for} \quad 1 \leq t \leq 2 \\ 0 & \text{elsewhere} \end{cases}$$

   (a) Determine the conditional moments $E[X(T)|T = t]$ and $E[\{X(T)\}^2|T = t]$.
   (b) Determine the mean $E[X(T)]$ and variance $\text{Var}[X(T)]$.

7. Suppose that customers arrive at a facility according to a Poisson process having rate $\lambda = 2$. Let $X(t)$ be the number of customers that have arrived up to time t. Determine the following probabilities and conditional probabilities:
   (a) $\Pr\{X(1) = 2\}$
   (b) $\Pr\{X(1) = 2 \text{ and } X(3) = 6\}$

(c) $\Pr\{X(1) = 2|X(3) = 6\}$
(d) $\Pr\{X(3) = 6|X(1) = 2\}$

8. Let $\{X(t); t \geq 0\}$ be a Poisson process having rate parameter $\lambda = 2$. Determine the numerical values to two decimal places for the following probabilities:
(a) $\Pr\{X(1) \leq 2\}$
(b) $\Pr\{X(1) = 1 \text{ and } X(2) = 3\}$
(c) $\Pr\{X(1) \geq 2|X(1) \geq 1\}$

9. Let $\{X(t); t \geq 0\}$ be a Poisson process having rate parameter $\lambda = 2$. Determine the following expectations:
(a) $E[X(2)]$
(b) $E[\{X(1)\}^2]$
(c) $E[X(1) X(2)]$

10. Customers arrive at a facility at random according to a Poisson process of rate λ. There is a waiting time cost of c per customer per unit time. The customers gather at the facility and are processed or dispatched in groups at fixed times $T, 2T, 3T, \ldots$. There is a dispatch cost of K. The process is depicted in the following graph.

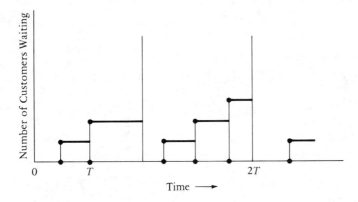

(a) What is the total dispatch cost during the first cycle from time 0 to time T?
(b) What is the mean total customer waiting cost during the first cycle?
(c) What is the mean total customer waiting + dispatch cost per unit time during the first cycle?
(d) What value of T minimizes this mean cost per unit time?

5.2 The Law of Rare Events

The common occurrence of the Poisson distribution in nature is explained by the Law of Rare Events. Informally this law asserts that where a certain event may occur in any of a large number of possibilities, but where the probability that the event does occur in any given possibility is small, then the total number of events that do happen should follow, approximately, the Poisson distribution.

A more formal statement in a particular instance follows. Consider a large number N of independent Bernoulli trials where the probability p of success on each trial is small and constant from trial to trial. Let $X_{N,p}$ denote the total number of successes in the N trials, where $X_{N,p}$ follows the binomial distribution

$$\Pr\{X_{N,p} = k\} = \frac{N!}{k!(N-k)!}\, p^k(1-p)^{N-k} \qquad \text{for} \quad k = 0, \ldots, N. \quad (5.3)$$

Now let us consider the limiting case in which $N \to \infty$ and $p \to 0$ in such a way that $Np = \lambda > 0$ where λ is constant. It is a familiar fact, one form of the Law of Rare Events, that the distribution for $X_{N,p}$ becomes, in the limit, the Poisson distribution

$$\Pr\{X_\lambda = k\} = \frac{e^{-\lambda}\lambda^k}{k!} \qquad \text{for} \quad k = 0, 1, \ldots \quad (5.4)$$

To derive (5.4), begin by writing (5.3) in the form

$$\Pr\{X_{N,p} = k\} = \frac{(Np)^k}{k!} \frac{N(N-1)\cdots(N-k+1)}{N^k} (1-p)^{N-k},$$

and first substitute $\lambda = Np$. Then note that

$$\frac{N(N-1)\cdots(N-k+1)}{N^k} = 1\left(1 - \frac{1}{N}\right)\cdots\left(1 - \frac{k-1}{N}\right) \to 1$$
$$\text{as} \quad N \to \infty$$

and recognize the familiar convergence $\left(1 - \dfrac{\lambda}{N}\right)^N \to e^{-\lambda}$ in

$$(1-p)^{N-k} = \left(1 - \frac{\lambda}{N}\right)^N \left(1 - \frac{\lambda}{N}\right)^{-k} \to e^{-\lambda} \times 1 \qquad \text{as} \quad N \to \infty.$$

Equation (5.4) follows instantly.

In words, where there are a large number N of independent trials and a small constant probability p of success on each trial, then one expects the total number of successes observed to follow approximately a Poisson distribution with parameter $\lambda = Np$.

The Law of Rare Events holds under more generality than indicated by the calculations leading to (5.4). In particular, consider events occurring along the positive axis $[0, \infty)$ in the manner shown in Figure 5.1. Concrete examples of such processes are the time points of the X-ray emissions of a substance undergoing radioactive decay, the instances of telephone calls originating in a given locality, the occurrence of accidents at a certain inter-

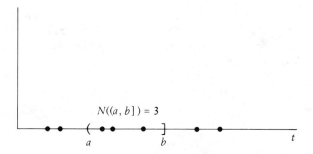

$$N((a, b]) = 3$$

Figure 5.1   A Poisson point process

section, the location of faults or defects along the length of a fiber of fila-ment, and the successive arrival times of customers for service.

Let $N((a, b])$ denote the number of events that occur during the interval $(a, b]$. That is, if $t_1 < t_2 < t_3 < \ldots$ denote the times (or locations, etc.) of successive events, then $N((a, b])$ is the number of values t_i for which $a < t_i \leq b$.

We make the following postulates:

(1) The numbers of events happening in disjoint intervals are indepen-dent random variables. That is, for every integer $m = 2, 3, \ldots$ and time points $t_0 = 0 < t_1 < t_2 < \ldots < t_m$, the random variables

$$N((t_0, t_1]), N((t_1, t_2]), \ldots, N((t_{m-1}, t_m])$$

are independent.

(2) For any time t and positive number h, the probability distribution of $N((t, t + h])$, the number of events occurring between time t and $t + h$, depends only on the interval length h and not on the time t.

(3) There is a positive constant λ for which the probability of at least one event happening in a time interval of length h is

$$\Pr\{N((t, t + h]) \geq 1\} = \lambda h + o(h) \qquad \text{as} \quad h \downarrow 0.$$

(Conforming to a common notation, here $o(h)$ as $h \downarrow 0$ stands for a general and unspecified remainder time for which $o(h)/h \to 0$ as $h \downarrow 0$. That is, a remainder term of smaller order than h as h vanishes.)
(4) The probability of two or more events occurring in an interval of length h is $o(h)$, or

$$\Pr\{N((t, t + h]) \geq 2\} = o(h), \qquad h \downarrow 0.$$

Postulate 3 is a specific formulation of the notion that events are rare. Postulate 4 is tantamount to excluding the possibility of the simultaneous occurrence of two or more events. In the concrete illustrations cited earlier, this requirement is usually satisfied.

   Disjoint intervals are independent by 1, and 2 asserts that the distribution of $N((s, t])$ is the same as that of $N((0, t - s])$. Therefore to describe the probability law of the system, it suffices to determine the probability distribution of $N((0, t])$ for an arbitrary value of t. Let

$$P_m(t) = \Pr\{N((0, t]) = m\}. \tag{5.5}$$

We will show that postulates 1 through 4 require that $P_m(t)$ be the Poisson distribution

$$P_m(t) = \frac{(\lambda t)^m e^{-\lambda t}}{m!} \quad \text{for} \quad m = 0, 1, \ldots \tag{5.6}$$

   To establish (5.6), let $p(h)$ be the probability of at least one event in an interval of length $h > 0$, or

$$p(h) = \Pr\{X(h) \geq 1\} = P_1(h) + P_2(h) + \cdots$$

Because of the assumption of independence,

$$P_0(t + h) = P_0(t)P_0(h) = P_0(t)[1 - p(h)],$$

and therefore

$$\frac{P_0(t + h) - P_0(t)}{h} = -P_0(t)\frac{p(h)}{h}.$$

We now let h decrease to zero whereupon the left side becomes the derivative $P_0'(t)$. For the right side, on the basis of Postulate 3 we know that $p(h)/h \to \lambda$. Therefore, the probability $P_0(t)$ that the event has not happened during $(0, t)$ satisfies the differential equation

$$P_0'(t) = -\lambda P_0(t),$$

whose well-known solution is $P_0(t) = ce^{-\lambda t}$. The constant c is determined by the initial condition $P_0(0) = 1$, which implies $c = 1$. Thus, $P_0(t) = e^{-\lambda t}$.

   Next we determine $P_1(t)$. A first step analysis at time t shows that

$$P_1(t + h) = P_1(t)P_0(h) + P_0(t)P_1(h). \tag{5.7}$$

By definition, $P_0(h) = 1 - p(h)$. Postulate 4 implies that

$$\begin{aligned} p(h) &= \Pr\{N((t, t + h]) \geq 1\} \\ &= P_1(h) + \Pr\{N((t, t + h]) \geq 2\} \\ &= P_1(h) + o(h), \end{aligned}$$

or

$$P_1(h) = p(h) + o(h).$$

Substitution for $P_1(h)$ and $P_0(h)$ in (5.7) gives

$$P_1(t + h) = P_1(t)[1 - p(h)] + P_0(t)[p(h) + o(h)]$$

which, after rearrangement, becomes

$$P_1(t + h) - P_1(t) = -p(h)P_1(t) + p(h)P_0(t) + o(h)P_0(t).$$

We divide both sides by h and let h vanish. The left side becomes the derivative $P_1'(t)$, and on the right side we use $p(h)/h \to \lambda$ and $o(h)/h \to 0$ to obtain the differential equation

$$P_1'(t) = -\lambda P_1(t) + \lambda P_0(t).$$

A similar approach works to determine $P_m(t)$ for a general $m > 1$. A first step analysis establishes that

$$P_m(t + h) = P_m(t)P_0(h) + P_{m-1}(t)P_1(h) + \sum_{i=2}^{m} P_{m-i}(t)P_i(h). \qquad (5.8)$$

Again we use the substitutions $P_0(h) = 1 - p(h)$ and $P_1(h) = p(h) + o(h)$, and this time we also use the inequality

$$\sum_{i=2}^{m} P_{m-i}(t)P_i(h) \le \sum_{i=2}^{\infty} P_i(h) = o(h)$$

which results from Postulate 4. After we subtract $P_m(t)$ from both sides, (5.8) becomes

$$P_m(t + h) - P_m(t) = P_m(t)[P_0(h) - 1] + P_{m-1}(t)P_1(h) + \sum_{i=2}^{m} P_{m-i}(t)P_i(h)$$
$$= -P_m(t)p(h) + P_{m-1}(t)p(h) + o(h).$$

Therefore

$$\frac{P_m(t + h) - P_m(t)}{h} = -P_m(t)\frac{p(h)}{h} + P_{m-1}(t)\frac{p(h)}{h} + \frac{o(h)}{h}.$$

Taking the limit as h decreases to zero, we obtain the differential equation

$$P_m'(t) = -\lambda P_m(t) + \lambda P_{m-1}(t), \qquad m = 1, 2, \ldots, \qquad (5.9)$$

subject to the initial conditions

$$P_m(0) = 0, \qquad m = 1, 2, \ldots$$

We turn to the solution of the differential equations (5.9) by introducing the functions

$$Q_m(t) = P_m(t)e^{\lambda t}, \qquad m = 0, 1, 2, \ldots$$

Substituting these functions in (5.9) gives

$$Q_m'(t) = \lambda Q_{m-1}(t), \qquad m = 1, 2, \ldots, \qquad (5.10)$$

where $Q_0(t) \equiv 1$ and the initial conditions are $Q_m(0) = 0$, $m = 1, 2, \ldots$. Solving (5.10) recursively we obtain

$$Q_1'(t) = \lambda \quad \text{or} \quad Q_1(t) = \lambda t + c \quad \text{so} \quad Q_1(t) = \lambda t$$

$$Q_2'(t) = \lambda^2 t \qquad\qquad \text{so} \quad Q_2(t) = \frac{\lambda^2 t^2}{2!}$$

$$\vdots \qquad\qquad\qquad\qquad \vdots$$

$$Q_m'(t) = \frac{\lambda^m t^{m-1}}{(m-1)!} \qquad\qquad Q_m(t) = \frac{\lambda^m t^m}{m!}.$$

Therefore

$$P_m(t) = Q_m(t)e^{-\lambda t} = \frac{\lambda^m t^m}{m!} e^{-\lambda t}.$$

In other words, for each t, the number of occurrences in time t follows a Poisson distribution with parameter λt. In particular, the mean number of occurrences in time t is λt.

Postulates 1 through 4 arise as physically plausible assumptions in many circumstances of stochastic modeling. The postulates seem rather weak. Surprisingly, they are sufficiently strong to force the Poisson behavior just derived. This motivates the following definition.

Definition   *Let $N((s, t])$ be a random variable counting the number of events occurring in an interval $(s, t]$. Then $N((s, t])$ is a Poisson point process of intensity $\lambda > 0$ if*

(i) *for every $m = 2, 3, \ldots$ and distinct time points $t_0 = 0 < t_1 < t_2 < \ldots < t_m$, the random variables*

$$N((t_0, t_1]), N((t_1, t_2]), \ldots, N((t_{m-1}, t_m])$$

are independent; and
(ii) *for any times $s < t$ the random variable $N((s, t])$ has the Poisson distribution*

$$\Pr\{N((s, t]) = k\} = \frac{[\lambda(t - s)]^k e^{-\lambda(t-s)}}{k!}, \qquad k = 0, 1, \ldots.$$

Poisson point processes often arise in a form where the time parameter is replaced by a suitable spatial parameter. The following formal example illustrates this vein of ideas. Consider an array of points distributed in a space E (E is a Euclidean space of dimension $d \geq 1$). Let $N(A)$ denote the number of points (finite or infinite) contained in the region A of E. We postulate that $N(A)$ is a random variable. The collection $\{N(A)\}$ of random variables, where A varies over all possible subsets of E, is said to be a homogeneous Poisson process if the following assumptions are fulfilled:

(i) The numbers of points in nonoverlapping regions are independent random variables.

(ii) For any region A of finite volume, $N(A)$ is Poisson distributed with mean $\lambda|A|$, where $|A|$ is the volume of A. The parameter λ is fixed and measures in a sense the intensity component of the distribution, which is independent of the size or shape. Spatial Poisson processes arise in considering such phenomena as the distribution of stars or galaxies in space, the spatial distribution of plants and animals, and the spatial distribution of bacteria on a slide. These ideas and concepts will be further studied in Section 5.5.

Problems 5.2

1. Certain computer coding systems use randomization to assign memory storage locations to account numbers. Suppose that $N = M\lambda$ different accounts are to be randomly located among M storage locations. Let X_i be the number of accounts assigned to the ith location. If the accounts are distributed independently and each location is equally likely to be chosen, show that $\Pr\{X_i = k\} \to e^{-\lambda}\lambda^k/k!$ as $N \to \infty$. Show that X_i and X_j are independent random variables in the limit, for distinct locations $i \neq j$. In the limit, what fraction of storage locations have two or more accounts assigned to them?

2. Suppose that a book of 600 pages contains a total of 240 typographical errors. Develop a Poisson approximation for the probability that three particular successive pages are error free.

3. N bacteria are spread independently with uniform distribution on a microscope slide of area A. An arbitrary region having area a is selected for observation. Determine the probability of k bacteria within the region of area a.

4. Show that as $N \to \infty$ and $a \to 0$ such that $(a/A)N \to c$ ($0 < c < \infty$) then $p(k) \to e^{-c}c^k/k!$. (See Problem 3.)

5.3 Distributions Associated with the Poisson Process

A *Poisson point process* $N((s, t])$ counts the number of events occurring in an interval $(s, t]$. A *Poisson counting process*, or more simply, a *Poisson process* $X(t)$ counts the number of events occurring up to time t. Formally, $X(t) = N((0, t])$.

Poisson events occurring in space can best be modeled as a point process. For Poisson events occurring on the positive time axis, whether we view them as a Poisson point process or Poisson counting process is largely a matter of convenience, and we will freely do both. The two descriptions are equivalent for Poisson events occurring along a line. The Poisson process is the more common and traditional description in this case because it allows a pictorial representation as an increasing integer-valued random function taking unit steps.

Figure 5.2 shows a typical sample path of a Poisson process where W_n is the time of occurrence of the nth event, the so-called *waiting time*. It is often convenient to set $W_0 = 0$. The differences $S_n = W_{n+1} - W_n$ are called *sojourn times*; S_n measures the duration that the Poisson process sojourns in state n.

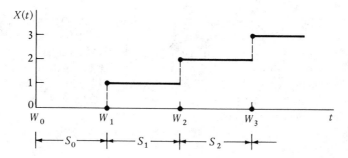

Figure 5.2   A typical sample path of a Poisson process showing the waiting times W_n and the sojourn times S_n

In this section we will determine a number of probability distributions associated with the Poisson process $X(t)$, the waiting times W_n, and the sojourn times S_n.

Theorem 5.3   *The waiting time W_n has the gamma distribution whose probability density function is*

$$f_{W_n}(t) = \frac{\lambda^n t^{n-1}}{(n-1)!} e^{-\lambda t}, \qquad n = 1, 2, \ldots, \quad t \geq 0. \qquad (5.11)$$

In particular W_1, the time to the first event, is exponentially distributed:

$$f_{W_1}(t) = \lambda e^{-\lambda t}, \qquad t \geq 0. \qquad (5.12)$$

Proof   The event $W_n \leq t$ occurs if and only if there are at least n events in the interval $(0, t]$, and since the number of events in $(0, t]$ has a Poisson distribution with mean λt we obtain the cumulative distribution function of W_n via

$$F_{W_n}(t) = \Pr\{W_n \le t\} = \Pr\{X(t) \ge n\}$$

$$= \sum_{k=n}^{\infty} \frac{(\lambda t)^k e^{-\lambda t}}{k!}$$

$$= 1 - \sum_{k=0}^{n-1} \frac{(\lambda t)^k e^{-\lambda t}}{k!}, \qquad n = 1, 2, \ldots, \quad t \ge 0.$$

We obtain the probability density function $f_{W_n}(t)$ by differentiating the cumulative distribution function. Then

$$f_{W_n}(t) = \frac{d}{dt} F_{W_n}(t)$$

$$= \frac{d}{dt}\left\{1 - e^{-\lambda t}\left[1 + \frac{\lambda t}{1!} + \frac{(\lambda t)^2}{2!} + \cdots + \frac{(\lambda t)^{n-1}}{(n-1)!}\right]\right\}$$

$$= -e^{-\lambda t}\left[\lambda + \lambda\frac{(\lambda t)}{1!} + \lambda\frac{(\lambda t)^2}{2!} + \cdots + \lambda\frac{(\lambda t)^{n-2}}{(n-2)!}\right]$$

$$\qquad + \lambda e^{-\lambda t}\left[1 + \frac{\lambda t}{1!} + \frac{(\lambda t)^2}{2!} + \cdots + \frac{(\lambda t)^{n-1}}{(n-1)!}\right]$$

$$= \frac{\lambda^n t^{n-1}}{(n-1)!}\, e^{-\lambda t}, \qquad n = 1, 2, \ldots, \quad t \ge 0.$$

There is an alternative derivation of the density in (5.11) that uses the Poisson point process $N((s, t])$ and proceeds directly without differentiation. The event $t < W_n \le t + \Delta t$ corresponds exactly to $n - 1$ occurrences in $(0, t]$ and one in $(t, t + \Delta t]$, as depicted in Figure 5.3.

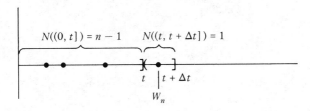

Figure 5.3

Then

$$f_{W_n}(t)\Delta t \doteq \Pr\{t < W_n \le t + \Delta t\} + o(\Delta t) \qquad [\text{see } (1.5), \text{ p. 7}]$$

$$= \Pr\{N((0, t]) = n - 1\}\Pr\{N((t, t + \Delta t]) = 1\} + o(\Delta t)$$

$$= \frac{(\lambda t)^{n-1} e^{-\lambda t}}{(n-1)!}\, \lambda(\Delta t) + o(\Delta t).$$

Dividing by Δt and passing to the limit as $\Delta t \to 0$ we obtain (5.11).

Observe that $\Pr\{N((t, t + \Delta t]) \geq 1\} = \Pr\{N((t, t + \Delta t]) = 1\} + o(\Delta t) = \lambda(\Delta t) + o(\Delta t)$. $\square$

Theorem 5.4   *The sojourn times S_0, S_1, . . . , S_{n-1} are independent random variables, each having the exponential probability density function*

$$f_{S_k}(s) = \lambda e^{-\lambda s}, \qquad s \geq 0. \tag{5.13}$$

Proof   We are being asked to show that the joint probability density function of S_0, S_1, . . . , S_{n-1} is the product of the exponential densities given by

$$f_{S_0, S_1, \ldots, S_{n-1}}(s_0, s_1, \ldots, s_{n-1}) = (\lambda e^{-\lambda s_0})(\lambda e^{-\lambda s_1}) \cdots (\lambda e^{-\lambda s_{n-1}}). \tag{5.14}$$

We give the proof only in the case $n = 2$, the general case being entirely similar. Referring to Figure 5.4 we see that the joint occurrence of

$$s_1 < S_1 < s_1 + \Delta s_1 \quad \text{and} \quad s_2 < S_2 < s_2 + \Delta s_2$$

corresponds to no events in the intervals $(0, s_1]$ and $(s_1 + \Delta s_1, s_1 + \Delta s_1 + s_2]$ and exactly one event in each of the intervals $(s_1, s_1 + \Delta s_1]$ and $(s_1 + \Delta s_1 + s_2, s_1 + \Delta s_1 + s_2 + \Delta s_2]$. Thus

$$
\begin{aligned}
f_{S_1, S_2}(s_1, s_2)\Delta s_1 \Delta s_2 &= \Pr\{s_1 < S_1 < s_1 + \Delta s_1, s_2 < S_2 < s_2 + \Delta s_2\} \\
&\quad + o(\Delta s_1 \Delta s_2) \\
&= \Pr\{N((0, s_1]) = 0\} \\
&\quad \times \Pr\{N((s_1 + \Delta s_1, s_1 + \Delta s_1 + s_2]) = 0\} \\
&\quad \times \Pr\{N((s_1, s_1 + \Delta s_1]) = 1\} \\
&\quad \times \Pr\{N((s_1 + \Delta s_1 + s_2, s_1 + \Delta s_1 + s_2 + \Delta s_2]) = 1\} \\
&\quad + o(\Delta s_1 \Delta s_2) \\
&= e^{-\lambda s_1} e^{-\lambda s_2} e^{-\lambda \Delta s_1} e^{-\lambda \Delta s_2} \lambda(\Delta s_1)\lambda(\Delta s_2) + o(\Delta s_1 \Delta s_2) \\
&= (\lambda e^{-\lambda s_1})(\lambda e^{-\lambda s_2})(\Delta s_1)(\Delta s_2) + o(\Delta s_1 \Delta s_2)
\end{aligned}
$$

Upon dividing both sides by $(\Delta s_1)(\Delta s_2)$ and passing to the limit as $\Delta s_1 \to 0$ and $\Delta s_2 \to 0$, we obtain (5.13) in the case $n = 2$. $\square$

The binomial distribution also arises in the context of Poisson processes.

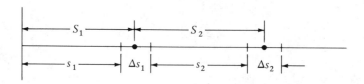

Figure 5.4

Theorem 5.5   *Let* $\{X(t)\}$ *be a Poisson process of rate* $\lambda > 0$. *Then for* $0 < u < t$
and $0 \le k \le n$,

$$\Pr\{X(u) = k | X(t) = n\} = \frac{n!}{k!(n-k)!} \left(\frac{u}{t}\right)^k \left(1 - \frac{u}{t}\right)^{n-k}. \quad (5.15)$$

Proof   Straightforward computations give

$$\Pr\{X(u) = k | X(t) = n\} = \frac{\Pr\{X(u) = k \text{ and } X(t) = n\}}{\Pr\{X(t) = n\}}$$

$$= \frac{\Pr\{X(u) = k \text{ and } X(t) - X(u) = n - k\}}{\Pr\{X(t) = n\}}$$

$$= \frac{\{e^{-\lambda u}(\lambda u)^k/k!\}\{e^{-\lambda(t-u)}[\lambda(t-u)]^{n-k}/(n-k)!\}}{e^{-\lambda t}(\lambda t)^n/n!}$$

$$= \frac{n!}{k!(n-k)!} \frac{u^k(t-u)^{n-k}}{t^n}$$

which establishes (5.15). $\square$

Problems 5.3

1. Consider a Poisson process with parameter λ. Given that $X(t) = n$ events occur in time t, find the density function for W_r, the time of occurrence of the rth event. Assume that $r \le n$.

2. The following calculations arise in certain highly simplified models of learning processes. Let $X_1(t)$ and $X_2(t)$ be independent Poisson processes having parameters λ_1 and λ_2, respectively.
   (a) What is the probability that $X_1(t) = 1$ before $X_2(t) = 1$?
   (b) What is the probability that $X_1(t) = 2$ before $X_2(t) = 2$?

3. Let $\{X_i(t), t \ge 0\}_{i=1}^n$ be independent Poisson processes with the same parameter λ. Find the distribution of the time until at least one event has occurred in every process.

4. Suppose a device is exposed to one of k possible environments $E_1, E_2, \ldots, E_k$, which can occur with respective probabilities $c_1, c_2, \ldots, c_k$ ($\sum_{j=1}^k c_j = 1$). In each environment dangerous peaks occur according to a Poisson process with parameter $\lambda_j, j = 1, 2, \ldots, k$. Within the environment E_j the conditional probability that the device fails, given that a peak occurs, is p_j. Find the probability that the device fails within a given length of time t.

5. Customers arrive at a service facility according to a Poisson process of rate $\lambda = 5$ per hour. Given that 12 customers arrived during the first two hours of service, what is the conditional probability that 5 customers arrived during the first hour?

6. A critical component on a submarine has an operating lifetime that is exponentially distributed with mean 0.50 years. As soon as a component fails, it is replaced by a new one having statistically identical properties. What is the smallest number of *spare* components that the submarine should stock if it is leaving for a one year tour and wishes the probability of having an inoperable unit caused by failures exceeding the spare inventory to be less than .02?

7. Customers arrive at a holding facility at random according to a Poisson process having rate λ. The facility processes in batches of size Q. That is, the first $Q - 1$ customers wait until the arrival of the Qth customer. Then all are passed simultaneously, and the process repeats. Service times are instantaneous. Let $N(t)$ be the number of customers in the holding facility at time t. Assume that $N(0) = 0$ and let $T = \min\{t \geq 0: N(t) = Q\}$ be the first dispatch time. Show that $E[T] = Q/\lambda$ and $E[\int_0^T N(t)dt] = [1 + 2 + \ldots + (Q - 1)]/\lambda = Q(Q - 1)/2\lambda$.

5.4 The Uniform Distribution and Poisson Processes

The major result of this section, Theorem 5.6, provides an important tool for computing certain functionals on a Poisson process. It asserts that, conditioned on a fixed total number of events in an interval, the locations of those events are uniformly distributed in a certain way.

After a complete discussion of the theorem and its proof, its application in a wide range of problems will be given.

In order to completely understand the theorem, consider first the following experiment. We begin with a line segment t units long and a fixed number n of darts and throw darts at the line segment in such a way that each dart's position upon landing is uniformly distributed along the segment, independent of the location of the other darts. Let U_1 be the position of the first dart thrown, U_2 the position of the second, and so on up to U_n. The probability density function is the uniform density

$$f_U(u) = \begin{cases} \dfrac{1}{t} & \text{for } 0 \leq u \leq t, \\ 0 & \text{elsewhere.} \end{cases}$$

Now let $W_1 \leq W_2 \leq \ldots \leq W_n$ denote these same positions, not in the order in which the darts were thrown, but instead in the order in which they appear along the line. Figure 5.5 depicts a typical relation between U_1, U_2, . . . , U_n and W_1, W_2, . . . , W_n.

Figure 5.5 W_1, W_2, . . . , W_n are the values U_1, U_2, . . . , U_n arranged in increasing order.

The joint probability density function for W_1, W_2, . . . , W_n is

$$f_{W_1,...,W_n}(w_1, . . . , w_n) = n!t^{-n}$$

$$\text{for} \quad 0 < w_1 < w_2 < \cdots < w_n \leq t. \quad (5.16)$$

For example, to establish (5.16) in the case $n = 2$ we have

$$f_{W_1,W_2}(w_1, w_2)\Delta w_1 \Delta w_2$$
$$\doteq \Pr\{w_1 < W_1 \leq w_1 + \Delta w_1, w_2 < W_2 \leq w_2 + \Delta w_2\}$$
$$= \Pr\{w_1 < U_1 \leq w_1 + \Delta w_1, w_2 < U_2 < w_2 + \Delta w_2\}$$
$$+ \Pr\{w_1 < U_2 \leq w_1 + \Delta w_1, w_2 < U_1 \leq w_2 + \Delta w_2\}$$
$$= 2\left(\frac{\Delta w_1}{t}\right)\left(\frac{\Delta w_2}{t}\right) = 2t^{-2}\Delta w_1 \Delta w_2.$$

Dividing by $\Delta w_1 \Delta w_2$ and passing to the limit gives (5.16). When $n = 2$, there are two ways that U_1 and U_2 can be ordered; either U_1 is less than U_2 or U_2 is less than U_1. In general there are $n!$ arrangements of U_1, . . . , U_n which lead to the same ordered values $W_1 \leq . . . \leq W_n$, thus giving (5.16).

Theorem 5.6 *Let W_1, W_2, . . . be the occurrence times in a Poisson process of rate $\lambda > 0$. Conditioned on $N(t) = n$, the random variables W_1, W_2, . . . , W_n have the joint probability density function*

$$f_{W_1,...,W_n|X(t)=n}(w_1, . . . , w_n) = n!t^{-n}$$

$$\text{for} \quad 0 < w_1 < \cdots < w_n \leq t. \quad (5.17)$$

Proof The event $w_i < W_i \leq w_i + \Delta w_i$ for $i = 1, . . . , n$ and $N(t) = n$ corresponds to no events occurring in any of the intervals $(0, w_1]$, $(w_1 + \Delta w_1, w_2]$, . . . , $(w_{n-1} + \Delta w_{n-1}, w_n]$, $(w_n + \Delta w_n, t]$, and exactly one event in each of the intervals $(w_1, w_1 + \Delta w_1]$, $(w_2, w_2 + \Delta w_2]$, . . . , $(w_n, w_n + \Delta w_n]$. These intervals are independent and

$$\Pr\{N((0, w_1]) = 0, \ldots, N((w_n + \Delta w_n, t]) = 0\}$$

$$= e^{-\lambda w_1} e^{-\lambda(w_2 - w_1 - \Delta w_1)} \cdots e^{-\lambda(w_n - w_{n-1} - \Delta w_{n-1})} e^{-\lambda(t - w_n - \Delta w_n)}$$

$$= e^{-\lambda t}[e^{\lambda(\Delta w_1 + \cdots + \Delta w_n)}]$$

$$= e^{-\lambda t} + O(\max\{\Delta w_i\}),$$

while

$$\Pr\{N((w_1, w_1 + \Delta w_1]) = 1, \ldots, N((w_n, w_n + \Delta w_n]) = 1\}$$

$$= \lambda(\Delta w_1) \cdots \lambda(\Delta w_n) + o(\Delta w_1 \Delta w_2 \ldots \Delta w_n).$$

Thus

$$f_{W_1, \ldots, W_n \mid X(t) = n}(w_1, \ldots, w_n) \Delta w_1 \ldots \Delta w_n$$

$$\doteq \Pr\{w_1 < W_1 \leq w_1 + \Delta w_1, \ldots, w_n < W_n \leq w_n + \Delta w_n \mid N(t) = n\}$$
$$\quad + o(\Delta w_1 \ldots \Delta w_n)$$

$$= \frac{\Pr\{w_i < W_i \leq w_i + \Delta w_i, i = 1, \ldots, n, N(t) = n\}}{\Pr\{N(t) = n\}}$$
$$\quad + o(\Delta w_1 \ldots \Delta w_n)$$

$$= \frac{e^{-\lambda t}\lambda(\Delta w_1) \cdots \lambda(\Delta w_n)}{e^{-\lambda t}(\lambda t)^n/n!} + o(\Delta w_1 \ldots \Delta w_n)$$

$$= n! t^{-n}(\Delta w_1) \cdots (\Delta w_n) + o(\Delta w_1 \ldots \Delta w_n).$$

Dividing both sides by $(\Delta w_1) \cdots (\Delta w_n)$ and letting $\Delta w_1 \to 0, \ldots,$ $\Delta w_n \to 0$ establishes (5.17). $\square$

Theorem 5.6 has important applications in evaluating certain symmetric functionals on Poisson processes. Some sample instances follow.

Example   Customers arrive at a facility according to a Poisson process of rate λ. Each customer pays \$1 on arrival, and it is desired to evaluate the expected value of the total sum collected during the interval $(0, t]$ discounted back to time 0. This quantity is given by

$$M = E\left[\sum_{k=1}^{X(t)} e^{-\beta W_k}\right]$$

where β is the discount rate, $W_1, W_2, \ldots$ are the arrival times, and $X(t)$ is the total number of arrivals in $(0, t]$. The process is shown in Figure 5.6.

We evaluate the mean total discounted sum M by conditioning on $X(t) = n$. Then

$$M = \sum_{n=1}^{\infty} E\left[\sum_{k=1}^{n} e^{-\beta W_k} \mid X(t) = n\right] \Pr\{X(t) = n\}. \tag{5.18}$$

Let $U_1, \ldots, U_n$ denote independent random variables that are uniformly distributed in $(0, t]$. Because of the symmetry of the functional $\sum_{k=1}^{n} \exp\{-\beta W_k\}$ and Theorem 5.6, we have

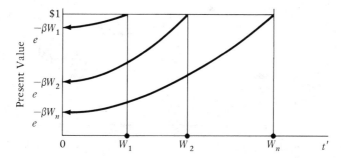

Figure 5.6 A dollar received at time W_k is discounted to a present value at time 0 of $\exp\{-\beta W_k\}$.

$$E\left[\sum_{k=1}^{n} e^{-\beta W_k}\Big|X(t) = n\right] = E\left[\sum_{k=1}^{n} e^{-\beta U_k}\right]$$

$$= nE[e^{-\beta U_1}]$$

$$= nt^{-1}\int_0^t e^{-\beta u}du$$

$$= \frac{n}{\beta t}[1 - e^{-\beta t}].$$

Substitution into (5.18) then gives

$$M = \frac{1}{\beta t}[1 - e^{-\beta t}]\sum_{n=1}^{\infty} n\Pr\{X(t) = n\}$$

$$= \frac{1}{\beta t}[1 - e^{-\beta t}]E[X(t)]$$

$$= \frac{\lambda}{\beta}[1 - e^{-\beta t}].$$

Example Viewing a fixed mass of a certain radioactive material, suppose that *alpha* particles appear in time according to a Poisson process of intensity λ. Each particle exists for a random duration and is then annihilated. Suppose that the successive lifetimes Y_1, Y_2, . . . of distinct particles are independent random variables having the common distribution function $G(y) = \Pr\{Y_k \le y\}$. Let $M(t)$ count the number of alpha particles existing at time t. The process is depicted in Figure 5.7.

We will use Theorem 5.6 to evaluate the probability distribution of $M(t)$ under the condition that $M(0) = 0$.

Let $X(t)$ be the number of particles created up to time t, by assumption, a Poisson process of intensity λ. Observe that $M(t) \le X(t)$; the number of existing particles cannot exceed the number of particles created. Condition

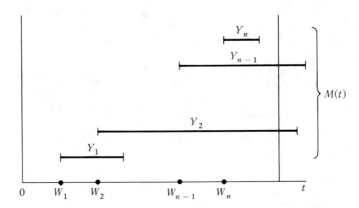

Figure 5.7   A particle created at time $W_k \leq t$ still exists at time t if $W_k + Y_k \geq t$.

on $X(t) = n$ and let $W_1, \ldots, W_n \leq t$ be the times of particle creation. Then particle k exists at time t if and only if $W_k + Y_k \geq t$. Let

$$\mathbf{1}\{W_k + Y_k \geq t\} = \begin{cases} 1 & \text{if} \quad W_k + Y_k \geq t \\ 0 & \text{if} \quad W_k + Y_k < t. \end{cases}$$

Then $\mathbf{1}\{W_k + Y_k \geq t\} = 1$ if and only if the kth particle is alive at time t. Thus

$$\Pr\{M(t) = m | X(t) = n\} = \Pr\left\{\sum_{k=1}^{n} \mathbf{1}\{W_k + Y_k \geq t\} = m | X(t) = n\right\}.$$

Invoking Theorem 5.6 and the symmetry among particles, we have

$$\Pr\left\{\sum_{k=1}^{n} \mathbf{1}\{W_k + Y_k \geq t\} = m | X(t) = n\right\}$$

$$= \Pr\left\{\sum_{k=1}^{n} \mathbf{1}\{U_k + Y_k \geq t\} = m\right\}, \tag{5.19}$$

where $U_1, U_2, \ldots, U_m$ are independent and uniformly distributed on $(0, t]$. The right-hand side of (5.19) is readily recognized as the binomial distribution in which

$$p = \Pr\{U_k + Y_k \geq t\} = \frac{1}{t}\int_0^t \Pr\{Y_k \geq t - u\}du$$

$$= \frac{1}{t}\int_0^t [1 - G(t - u)]du$$

$$= \frac{1}{t}\int_0^t [1 - G(z)]dz. \tag{5.20}$$

Thus, explicitly writing the binomial distribution, we have

$$\Pr\{M(t) = m | X(t) = n\} = \frac{n!}{m!(n - m)!} p^m(1 - p)^{n-m}$$

with p given by (5.20). Finally,

$$\Pr\{M(t) = m\} = \sum_{n=m}^{\infty} \Pr\{M(t) = m | X(t) = n\} \Pr\{X(t) = n\}$$

$$= \sum_{n=m}^{\infty} \frac{n!}{m!(n-m)!} p^m (1-p)^{n-m} \frac{(\lambda t)^n e^{-\lambda t}}{n!}$$

$$= e^{-\lambda t} \frac{(\lambda p t)^m}{m!} \sum_{n=m}^{\infty} \frac{(1-p)^{n-m}(\lambda t)^{n-m}}{(n-m)!}. \tag{5.21}$$

The infinite sum is an exponential series and reduces according to

$$\sum_{n=m}^{\infty} \frac{(1-p)^{n-m}(\lambda t)^{n-m}}{(n-m)!} = \sum_{j=0}^{\infty} \frac{[\lambda t(1-p)]^j}{j!} = e^{\lambda t(1-p)},$$

and this simplifies (5.21) to

$$\Pr\{M(t) = m\} = \frac{e^{-\lambda p t}(\lambda p t)^m}{m!} \qquad \text{for} \quad m = 0, 1, \ldots .$$

In words, the number of particles existing at time t has a Poisson distribution with mean

$$\lambda p t = \lambda \int_0^t [1 - G(y)] dy. \tag{5.22}$$

It is often relevant to let $t \to \infty$ in (5.22) and determine the corresponding long run distribution. Let $\mu = E[Y_k] = \int_0^\infty [1 - G(y)] dy$ be the mean lifetime of an alpha particle. It is immediate from (5.22) that as $t \to \infty$, the distribution of $M(t)$ converges to the Poisson distribution with parameter $\lambda\mu$. A great simplification has taken place. In the long run, the probability distribution for existing particles depends only on the mean lifetime μ, and not otherwise on the lifetime distribution $G(y)$. In practical terms this statement implies that in order to apply this model, only the mean lifetime μ need be known.

Problems 5.4

1. Electrical pulses with independent and identically distributed random amplitudes $\xi_1, \xi_2, \ldots$ arrive at a detector at random times $W_1, W_2,$ . . . according to a Poisson process of rate λ. The detector output $\Theta_k(t)$ for the kth pulse at time t is

$$\Theta_k(t) = \begin{cases} 0 & \text{for} \quad t < W_k, \\ \xi_k \exp\{-\alpha(t - W_k)\} & \text{for} \quad t \ge W_k. \end{cases}$$

That is, the amplitude impressed on the detector when the pulse arrives is ξ_k, and its effect thereafter decays exponentially at rate α.

Assume that the detector is additive so that if $N(t)$ pulses arrive during the time interval $[0, t]$, then the output at time t is

$$Z(t) = \sum_{k=1}^{N(t)} \Theta_k(t).$$

Determine the mean output $E[Z(t)]$ assuming $N(0) = 0$. Assume that the amplitudes $\xi_1, \xi_2, \ldots$ are independent of the arrival times $W_1, W_2, \ldots$.

2. Customers arrive at a service facility according to a Poisson process of intensity λ. The service times $Y_1, Y_2, \ldots$ of the arriving customers are independent random variables having the common probability distribution function $G(\gamma) = \Pr\{Y_k \leq \gamma\}$. Assume that there is no limit to the number of customers that can be serviced simultaneously; i.e., there are an infinite number of servers available. Let $M(t)$ count the number of customers in the system at time t. Argue that $M(t)$ has a Poisson distribution with mean λpt where

$$p = t^{-1} \int_0^t [1 - G(\gamma)] d\gamma.$$

5.5 Spatial Poisson Processes

In this section we define some versions of multidimensional Poisson processes and describe some examples and applications.

Let S be a set in n-dimensional space and let $\mathcal{A}$ be a family of subsets of S. A *point process* in S is a stochastic process $N(A)$ indexed by the sets A in $\mathcal{A}$ and having the set of nonnegative integers $\{0, 1, 2, \ldots\}$ as its possible values. We think of "points" being scattered over S in some random manner and of $N(A)$ as counting the number of points in the set A. Because $N(A)$ is a counting function, there are certain obvious requirements that it must satisfy. For example, if A and B are disjoint sets in $\mathcal{A}$ whose union $A \cup B$ is also in $\mathcal{A}$, then it must be that $N(A \cup B) = N(A) + N(B)$. In words, the number of points in A or B equals the number of points in A plus the number of points in B when A and B are disjoint.

The one-dimensional case, in which S is the positive half line and $\mathcal{A}$ comprises all intervals of the form $A = (s, t]$, for $0 \leq s < t$, was introduced in Section 5.3. The straightforward generalization to the plane and three-dimensional space that is now being discussed has relevance when we consider the spatial distribution of stars or galaxies in astronomy, of plants or animals in ecology, of bacteria on a slide in medicine, and of defects on a surface or in a volume in reliability engineering.

Let S be a subset of the real line, two-dimensional plane or three-dimensional space; let $\mathcal{A}$ be the family of subsets of S and for any set A in $\mathcal{A}$, let $|A|$ denote the size (length, area, or volume, respectively) of A. Then $\{N(A); A \text{ in } \mathcal{A}\}$ is a *homogeneous Poisson point process* of intensity $\lambda > 0$ if:

(i) for each A in $\mathcal{A}$, the random variable $N(A)$ has a Poisson distribution with parameter $\lambda|A|$;

(ii) for every finite collection $\{A_1, \ldots, A_n\}$ of disjoint subsets of S, the random variables $N(A_1), \ldots, N(A_n)$ are independent.

In Section 5.2, the Law of Rare Events was invoked to derive the Poisson process as a consequence of certain physically plausible postulates. This implication serves to justify the Poisson process as a model in those situations where the postulates may be expected to hold. An analogous result is available in the multidimensional case at hand. Given an arbitrary point process $\{N(A); A \text{ in } \mathcal{A}\}$, the required postulates are as follows:

(1) The possible values for $N(A)$ are the nonnegative integers $\{0, 1, 2, \ldots\}$ and $0 < \Pr\{N(A) = 0\} < 1$ if $0 < |A| < \infty$.

(2) The probability distribution of $N(A)$ depends on the set A only through its size (length, area, or volume) $|A|$ with the further property that $\Pr\{N(A) \geq 1\} = \lambda|A| + o(|A|)$ as $|A| \downarrow 0$.

(3) For $m = 2, 3, \ldots$, if $A_1, A_2, \ldots, A_m$ are disjoint regions, then $N(A_1), N(A_2), \ldots, N(A_m)$ are independent random variables and $N(A_1 \cup A_2 \cup \ldots \cup A_m) = N(A_1) + N(A_2) + \ldots + N(A_m)$.

(4)
$$\lim_{|A| \to 0} \frac{\Pr\{N(A) \geq 1\}}{\Pr\{N(A) = 1\}} = 1.$$

The motivation and interpretation of these postulates is quite evident. Axiom 2 asserts that the probability distribution of $N(A)$ does not depend on the shape or location of A, but only on its size. Postulate 3 requires that the outcome in one region not influence or be influenced by the outcome in a second region that does not overlap the first. Requirement 4 precludes the possibility of two points occupying the same location.

If a random point process $N(A)$ defined with respect to subsets A of Euclidean n space satisfies Postulates 1 through 4, then $N(A)$ is a homogeneous Poisson point process of intensity $\lambda > 0$ and

$$\Pr\{N(A) = k\} = \frac{e^{-\lambda|A|}(\lambda|A|)^k}{k!} \qquad \text{for} \quad k = 0, 1, \ldots . \qquad (5.23)$$

As in the one-dimensional case, homogeneous Poisson point processes in n-dimensions are highly amenable to analysis and many results are known for them. We elaborate a few of these consequences next, beginning with the uniform distribution of a single point. Consider a region A of positive size $|A| > 0$, and suppose it is known that A contains exactly one point, i.e., $N(A) = 1$. Where in A is this point located? We claim that the point is uniformly distributed in the sense that

$$\Pr\{N(B) = 1 | N(A) = 1\} = \frac{|B|}{|A|} \qquad \text{for any set} \quad B \subset A. \qquad (5.24)$$

In words, the probability of the point being in any subset B of A is proportional to the size of B; that is, the point is uniformly distributed in A. The uniform distribution expressed in (5.24) is an immediate consequence of elementary conditional probability manipulations. We write $A = B \cup C$ where B is an arbitrary subset of A and C is the portion of A not included in B. Then B and C are disjoint so that $N(B)$ and $N(C)$ are independent Poisson random variables with respective means $\lambda|B|$ and $\lambda|C|$. Then

$$\Pr\{N(B) = 1 | N(A) = 1\} = \frac{\Pr\{N(B) = 1, N(C) = 0\}}{\Pr\{N(A) = 1\}}$$

$$= \frac{\lambda|B|e^{-\lambda|B|}e^{-\lambda|C|}}{\lambda|A|e^{-\lambda|A|}}$$

$$= \frac{|B|}{|A|} \qquad \text{because } |B| + |C| = |A|,$$

and the proof is complete.

The generalization to n points in a region A is stated as follows. Consider a set A of positive size $|A| > 0$ and containing $N(A) = n \geq 1$ points. Then these n points are independent and uniformly distributed in A in the sense that, for any disjoint partition $A_1, \ldots, A_m$ of A where $A_1 \cup \ldots \cup A_m = A$, and any positive integers $k_1, \ldots, k_m$, where $k_1 + \ldots + k_m = n$, then

$$\Pr\{N(A_1) = k_1, \ldots, N(A_m) = k_m | N(A) = n\}$$

$$= \frac{n!}{k_1! \cdots k_m!} \left(\frac{|A_1|}{|A|}\right)^{k_1} \cdots \left(\frac{|A_m|}{|A|}\right)^{k_m}. \tag{5.25}$$

Equation (5.25) expresses the multinomial distribution for the conditional distribution of $N(A_1), \ldots, N(A_m)$ given that $N(A) = n$.

Example *An Application in Astronomy* Consider stars distributed in space in accordance with a three-dimensional Poisson point process of intensity $\lambda > 0$. Let $\mathbf{x}$ and $\mathbf{y}$ designate general three-dimensional vectors, and assume that the light intensity exerted at $\mathbf{x}$ by a star located at $\mathbf{y}$ is $f(\mathbf{x}, \mathbf{y}, \alpha) = \alpha/\|\mathbf{x} - \mathbf{y}\|^2 = \alpha/[(x_1 - y_1)^2 + (x_2 - y_2)^2 + (x_3 - y_3)^2]$, where α is a random parameter depending on the intensity of the star at $\mathbf{y}$. We assume that the intensities α associated with different stars are independent, identically distributed random variables possessing a common mean μ_α and variance σ_α^2. We also assume that the combined intensity exerted at the point $\mathbf{x}$ due to light created by different stars accumulates additively. Let $Z(\mathbf{x}, A)$ denote the total light intensity at the point $\mathbf{x}$ due to signals emanating from all sources located in region A. Then

$$Z(\mathbf{x}, A) = \sum_{r=1}^{N(A)} f(\mathbf{x}, \mathbf{y}_r, \alpha_r),$$

$$= \sum_{r=1}^{N(A)} \frac{\alpha_r}{\|\mathbf{x} - \mathbf{y}_r\|^2} \tag{5.26}$$

where $\mathbf{y}_r$ is the location of the rth star in A. We recognize (5.26) as a random sum, as discussed in Section 2.3.2. Accordingly, we have the mean intensity at $\mathbf{x}$ given by

$$E[Z(\mathbf{x}, A)] = (E[N(A)])(E[f(\mathbf{x}, \mathbf{y}, \alpha)]). \tag{5.27}$$

Note that $E[N(A)] = \lambda|A|$ while, because we have assumed α and $\mathbf{y}$ to be independent,

$$E[f(\mathbf{x}, \mathbf{y}, \alpha)] = E[\alpha]E[\|\mathbf{x} - \mathbf{y}\|^{-2}].$$

But as a consequence of the Poisson distribution of stars in space, we may take $\mathbf{y}$ to be uniformly distributed in A. Thus

$$E[\|\mathbf{x} - \mathbf{y}\|^{-2}] = \frac{1}{|A|} \int_A \frac{d\mathbf{y}}{\|\mathbf{x} - \mathbf{y}\|^2}.$$

With $\mu_\alpha = E[\alpha]$, then (5.27) reduces to

$$E[Z(\mathbf{x}, A)] = \lambda\mu_\alpha \int_A \frac{d\mathbf{y}}{\|\mathbf{x} - \mathbf{y}\|^2}.$$

Problems 5.5

1. Suppose that stars are distributed in space following a Poisson point process of intensity λ. Fix a star *alpha* and let R be the distance from alpha to its nearest neighbor. Show that R has the probability density function

$$f_R(x) = (4\lambda\pi x^2)\exp\left\{\frac{-4\lambda\pi x^3}{3}\right\}, \qquad x > 0.$$

2. Consider a collection of circles in the plane whose centers are distributed according to a spatial Poisson process with parameter $\lambda|A|$, where $|A|$ denotes the area of the set A. (In particular, the number of centers $\xi(A)$ in the set A follows the distribution law $\Pr\{\xi(A) = k\} = e^{-\lambda|A|}[(\lambda|A|)^k/k!]$.) The radius of each circle is assumed to be a random variable independent of the location of the center of the circle with density function $f(r)$ and finite second moment. Show that $C(r)$, defined to be the number of circles that cover the origin and have centers at a distance less than r from the origin, determines a variable time

Poisson process where the time variable is now taken to be the distance r.

Hint: Prove that an event occurring between r and $r + dr$ (i.e., there is a circle that covers the origin and whose center is in the ring of radius r to $r + dr$) has probability $\lambda 2\pi r \, dr \int_r^\infty f(\rho) d\rho + o(dr)$ and events occurring over disjoint intervals constitute independent r.v.'s. Show that $C(r)$ is a variable time (nonhomogeneous) Poisson process with parameter

$$\lambda(r) = 2\pi\lambda r \int_r^\infty f(\rho) d\rho.$$

3. Show that the number of circles that cover the origin is a Poisson random variable with parameter $\lambda \int_0^\infty \pi r^2 f(r) dr$.

4. Consider spheres in three-dimensional space with centers distributed according to a Poisson distribution with parameter $\lambda|A|$ where $|A|$ now represents the volume of the set A. If the radii of all spheres are distributed according to $F(r)$ with density $f(r)$ and finite third moment, show that the number of spheres that cover a point t is a Poisson random variable with parameter $\frac{4}{3}\lambda\pi \int_0^\infty r^3 f(r) dr$.

5. Consider a two-dimensional Poisson process of particles in the plane with intensity parameter ν. Determine the distribution $F_D(x)$ of the distance between a particle and its nearest neighbor. Compute the mean distance.

5.6 Compound and Marked Poisson Processes

Given a Poisson process $X(t)$ of rate $\lambda > 0$, suppose that each event has associated with it a random variable, possibly representing a value or a cost. Examples will appear shortly. The successive values $Y_1, Y_2, \ldots$ are assumed to be independent random variables sharing the common distribution function

$$G(y) = \Pr\{Y_k \le y\}.$$

A *compound Poisson process* is the cumulative value process defined by

$$Z(t) = \sum_{k=1}^{X(t)} Y_k \quad \text{for} \quad t \ge 0. \tag{5.28}$$

A *marked Poisson process* is the sequence of pairs (W_1, Y_1), (W_2, Y_2), $\ldots$, where $W_1, W_2, \ldots$ are the waiting times or event times in the Poisson process $X(t)$.

Both compound Poisson and marked Poisson processes appear often as models of physical phenomena.

5.6.1 Compound Poisson Processes

Consider the compound Poisson process $Z(t) = \sum_{k=1}^{X(t)} Y_k$. If $\lambda > 0$ is the rate for the process $X(t)$ and $\mu = E[Y_1]$ and $\nu^2 = \mathrm{Var}[Y_1]$ are the common mean and variance for $Y_1, Y_2, \ldots$, then the moments of $Z(t)$ can be determined from the random sums formulas of Section 2.3.2 and are

$$E[Z(t)] = \lambda\mu t; \qquad \mathrm{Var}[Z(t)] = \lambda(\nu^2 + \mu^2)t. \qquad (5.29)$$

Examples

   (a) *Risk Theory*   Suppose claims arrive at an insurance company in accordance with a Poisson process having rate λ. Let Y_k be the magnitude of the kth claim. Then $Z(t) = \sum_{k=1}^{X(t)} Y_k$ represents the cumulative amount claimed up to time t.

   (b) *Stock Prices*   Suppose that transactions in a certain stock take place according to a Poisson process of rate λ. Let Y_k denote the change in market price of the stock between the kth and $(k - 1)$st transaction. The *random walk hypothesis* asserts that $Y_1, Y_2, \ldots$ are independent random variables. The random walk hypothesis, which has a history dating back to 1900, can be deduced formally from certain assumptions describing a "perfect market."

   Then $Z(t) = \sum_{k=1}^{X(t)} Y_k$ represents the total price change up to time t.

   This stock price model has been proposed as an explanation why stock price changes do not follow a Gaussian (normal) distribution.

   The distribution function for the compound Poisson process $Z(t) = \sum_{k=1}^{X(t)} Y_k$ can be represented explicitly after conditioning on the values of $X(t)$. Recall the convolution notation

$$G^{(n)}(y) = \Pr\{Y_1 + \cdots + Y_n \le y\} \qquad (5.30)$$

$$= \int_{-\infty}^{+\infty} G^{(n-1)}(y - z)dG(z)$$

with

$$G^{(0)}(y) = \begin{cases} 1 & \text{for } y \ge 0 \\ 0 & \text{for } y < 0. \end{cases}$$

Then

$$\Pr\{Z(t) \le z\} = \Pr\left\{ \sum_{k=1}^{X(t)} Y_k \le z \right\}$$

$$= \sum_{n=0}^{\infty} \Pr\left\{ \sum_{k=1}^{X(t)} Y_k \le z \,\middle|\, X(t) = n \right\} \frac{(\lambda t)^n e^{-\lambda t}}{n!} \qquad (5.31)$$

$$= \sum_{n=0}^{\infty} \frac{(\lambda t)^n e^{-\lambda t}}{n!} G^{(n)}(z). \quad \text{(since } X(t) \text{ is independent of } Y_1, Y_2, \ldots)$$

Example   *A Shock Model*   Let $X(t)$ be the number of shocks to a system up to time t and let Y_k be the damage or wear incurred by the kth shock. We assume that damage is positive, i.e., that $\Pr\{Y_k \geq 0\} = 1$, and that the damage accumulates additively so that $Z(t) = \sum_{k=1}^{X(t)} Y_k$ represents the total damage sustained up to time t. Suppose that the system continues to operate as long as this total damage is less than some critical value a, and fails in the contrary circumstance. Let T be the time of system failure. Then,

$$\{T > t\} \quad \text{if and only if} \quad \{Z(t) < a\}. \text{ (Why?)} \qquad (5.32)$$

In view of (5.31) and (5.32), we have

$$\Pr\{T > t\} = \sum_{n=0}^{\infty} \frac{(\lambda t)^n e^{-\lambda t}}{n!} \, G^{(n)}(a).$$

All summands are nonnegative, so we may interchange integration and summation to get the mean system failure time

$$E[T] = \int_0^{\infty} \Pr\{T > t\} dt$$

$$= \sum_{n=0}^{\infty} \left(\int_0^{\infty} \frac{(\lambda t)^n e^{-\lambda t}}{n!} \, dt \right) G^{(n)}(a)$$

$$= \lambda^{-1} \sum_{n=0}^{\infty} G^{(n)}(a).$$

This expression simplifies greatly in the special case in which Y_1, Y_2, $\ldots$ are each exponentially distributed according to the density $g_Y(y) = \mu e^{-\lambda y}$ for $y \geq 0$. Then the sum $Y_1 + \ldots + Y_n$ has the gamma distribution

$$G^{(n)}(z) = 1 - \sum_{k=0}^{n-1} \frac{(\mu z)^k e^{-\mu z}}{k!} = \sum_{k=n}^{\infty} \frac{(\mu z)^k e^{-\mu z}}{k!}$$

and

$$\sum_{n=0}^{\infty} G^{(n)}(a) = \sum_{n=0}^{\infty} \sum_{k=n}^{\infty} \frac{(\mu a)^k e^{-\mu a}}{k!}$$

$$= \sum_{k=0}^{\infty} \sum_{n=0}^{k} \frac{(\mu a)^k e^{-\mu a}}{k!}$$

$$= \sum_{k=0}^{\infty} (1 + k) \frac{(\mu a)^k e^{-\mu a}}{k!}$$

$$= 1 + \mu a.$$

When Y_1, Y_2, $\ldots$ are exponentially distributed, then

$$E[T] = \frac{1 + \mu a}{\lambda}.$$

5.6.2 Marked Poisson Processes

Again suppose that a random variable Y_k is associated with the kth event in a Poisson process of rate λ. We stipulate that Y_1, Y_2, . . . are independent and share the common distribution function

$$G(y) = \Pr\{Y_k \le y\}.$$

The sequence of pairs (W_1, Y_1), (W_2, Y_2), . . . is called a *marked Poisson process*.

We begin the analysis of marked Poisson processes with one of the simplest cases. For a fixed value p $(0 < p < 1)$ suppose

$$\Pr\{Y_k = 1\} = p, \qquad \Pr\{Y_k = 0\} = q = 1 - p.$$

Now consider separately the processes of points marked with ones and of points marked with zeros. In this case we can define the relevant Poisson processes explicitly by

$$X_1(t) = \sum_{k=1}^{X(t)} Y_k \quad \text{and} \quad X_0(t) = X(t) - X_1(t).$$

Then nonoverlapping increments in $X_1(t)$ are independent random variables, $X_1(0) = 0$, and finally Theorem 5.2 applies to assert that $X_1(t)$ has a Poisson distribution with mean $\lambda p t$. In summary, $X_1(t)$ is a Poisson process with rate λp, and the parallel argument shows that $X_0(t)$ is a Poisson process with rate $\lambda(1 - p)$. *What is even more interesting and surprising is that $X_0(t)$ and $X_1(t)$ are independent processes!* The relevant property to check is that $\Pr\{X_0(t) = j \text{ and } X_1(t) = k\} = \Pr\{X_0(t) = j\} \times \Pr\{X_1(t) = k\}$ for j, $k = 0, 1, ...$ We establish this independence by writing

$$\Pr\{X_0(t) = j, X_1(t) = k\} = \Pr\{X(t) = j + k, X_1(t) = k\}$$
$$= \Pr\{X_1(t) = k|X(t) = j + k\}\Pr\{X(t) = j + k\}$$
$$= \frac{(j + k)!}{j!k!} p^k(1 - p)^j \frac{(\lambda t)^{j+k} e^{-\lambda t}}{(j + k)!}$$
$$= \left[\frac{e^{-\lambda p t}(\lambda p t)^k}{k!}\right]\left[\frac{e^{-\lambda(1-p)t}(\lambda(1 - p)t)^j}{j!}\right]$$
$$= \Pr\{X_1(t) = k\}\Pr\{X_0(t) = j\}$$

$$\text{for} \quad j, k = 0, 1, . . .$$

Example   Customers enter a store according to a Poisson process of rate $\lambda = 10$ per hour. Independently, each customer buys something with probability $p = .3$ and leaves without making a purchase with probability $q = 1 - p = .7$. What is the probability that during the first hour 9 people enter the store and that 3 of these people make a purchase and 6 do not?

Let $X_1 = X_1(1)$ be the number of customers who make a purchase during the first hour and $X_0 = X_0(1)$ be the number of people who do not. Then X_1 and X_0 are independent Poisson random variables having respective rates $.3(10) = 3$ and $.7(10) = 7$. According to the Poisson distribution, then

$$\Pr\{X_1 = 3\} = \frac{3^3 e^{-3}}{3!} = .2240,$$

$$\Pr\{X_0 = 6\} = \frac{7^6 e^{-7}}{6!} = .1490,$$

and

$$\Pr\{X_1 = 3, X_0 = 6\} = \Pr\{X_1 = 3\}\Pr\{X_0 = 6\} = (.2240)(.1490) = .0334.$$

In our study of marked Poisson processes, let us next consider the case where the value random variables $Y_1, Y_2, \ldots$ are discrete, with possible values $0, 1, 2, \ldots$ and

$$\Pr\{Y_n = k\} = a_k > 0 \qquad \text{for} \quad k = 0, 1, \ldots, \qquad \Sigma_k a_k = 1.$$

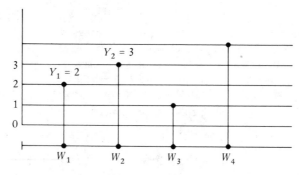

Figure 5.8 A marked Poisson process. $W_1, W_2, \ldots$ are the event times in a Poisson process of rate λ. The random variables $Y_1, Y_2, \ldots$ are the markings, assumed to be independent and identically distributed, and independent of the Poisson process.

In Figure 5.8, the original Poisson event times $W_1, W_2, \ldots$ are shown on the bottom axis. Then a point is placed in the (t, y) plane at (W_n, Y_n) for every n. For every integer $k = 0, 1, 2, \ldots$ one obtains a point process that corresponds to the times W_n for which $Y_n = k$. The same reasoning as in the zero-one case applies to imply that each of these processes is Poisson, the rate for the kth process being λa_k, and that processes for distinct values of k are independent.

To state the corresponding decomposition result when the values Y_1, $Y_2, \ldots$ are continuous random variables requires a higher level of sophistication, although the underlying ideas are basically the same. To set the stage for the formal statement, we first define what we mean by a nonhomogeneous Poisson point process in the plane, thus extending the homogeneous processes of the previous section. Let $\theta = \theta(x, y)$ be a nonnegative function defined on a region S in the (x, y) plane. For each subset A of S, let $\mu(A) = \int\int_A \theta(x, y)dxdy$ be the volume under $\theta(x, y)$ enclosed by A. A nonhomogeneous Poisson point process of intensity function $\theta(x, y)$ is a point process $\{N(A); A \subset S\}$ for which

(i) for each subset A of S, the random variable $N(A)$ has a Poisson distribution with mean $\mu(A)$; and
(ii) for disjoint subsets $A_1, \ldots, A_m$ of S, the random variables $N(A_1), \ldots, N(A_m)$ are independent.

It is easily seen that the homogeneous Poisson point process of intensity λ corresponds to the function $\theta(x, y)$ being constant, and $\theta(x, y) = \lambda$ for all x, y.

With this definition in hand, we state the appropriate decomposition result for general marked Poisson processes.

Theorem 5.7 *Let (W_1, Y_1), (W_2, Y_2), . . . be a marked Poisson process where $W_1, W_2, \ldots$ are the waiting times in a Poisson process of rate λ and $Y_1, Y_2, \ldots$ are independent identically distributed continuous random variables having probability density function $g(y)$. Then (W_1, Y_1), (W_2, Y_2), . . . form a two-dimensional nonhomogeneous Poisson point process in the (t, y) plane, where the mean number of points in a region A is given by*

$$\mu(A) = \int_A \int \lambda g(y)dy \, dt \tag{5.33}$$

Figure 5.9 diagrams the scene.

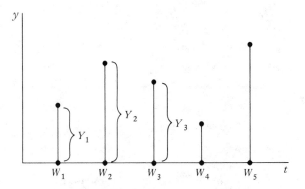

Figure 5.9 A marked Poisson process

Theorem 5.7 asserts that the numbers of points in disjoint intervals are independent random variables. For example, the waiting times corresponding to positive values Y_1, Y_2, . . . form a Poisson process, as do the times associated with negative values, and these two processes are independent.

Example *Crack Failure*   The following model is proposed to describe the failure time of a sheet or volume of material subjected to a constant stress σ. The failure time is viewed in two parts, crack initiation and crack propagation.

Crack initiation occurs according to a Poisson process whose rate per unit time and unit volume is a constant $\lambda_\sigma > 0$ depending on the stress level σ. Crack initiation per unit time then is a Poisson process of rate $\lambda_\sigma |V|$ where $|V|$ is the volume of material under consideration. We let W_1, W_2, . . . be the times of crack initiation.

Once begun, a crack grows at a random rate until reaching a critical size at which instant structural failure occurs. Let Y_k be the time to reach critical size for the kth crack. The cumulative distribution function $G_\sigma(y) = \Pr\{Y_k \leq y\}$ depends on the constant stress level σ.

We assume that crack initiations are sufficiently sparse as to make Y_1, Y_2, . . . independent random variables. That is, we do not allow two small cracks to join and form a larger one.

The structural failure time Z is the smallest of $W_1 + Y_1$, $W_2 + Y_2$, ... It is not necessarily the case that the first crack to begin will cause system failure. A later crack may grow to critical size faster.

In the (t, y) plane, the event $\{\min\{W_k + Y_k\} > z\}$ corresponds to no points falling in the triangle $\triangle = \{(t, y): t + y \leq z, t \geq 0, y \geq 0\}$, as shown in Figure 5.10.

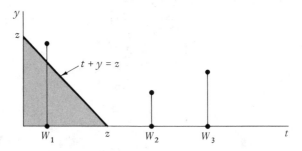

Figure 5.10   A crack failure model

The number of points $N(\triangle)$ falling in the triangle $\triangle$ has a Poisson distribution with mean $\mu(\triangle)$ given, according to (5.33), by

$$\mu(\Delta) = \int_{\Delta} \int \lambda_\sigma |V| ds\, g_\sigma(u) du$$

$$= \int_0^z \lambda_\sigma |V| \left\{ \int_0^{z-s} g_\sigma(u) du \right\} ds$$

$$= \lambda_\sigma |V| \int_0^z G_\sigma(z - s) ds$$

$$= \lambda_\sigma |V| \int_0^z G_\sigma(v) dv.$$

From this we obtain the cumulative distribution function for structural failure time,

$$\Pr\{Z \leq z\} = 1 - \Pr\{Z > z\} = 1 - \Pr\{N(\Delta) = 0\}$$
$$= 1 - \exp\{-\lambda_\sigma |V| \int_0^z G_\sigma(v) dv\}.$$

Observe the appearance of the so-called *size effect* in the model wherein the structure volume $|V|$ affects the structural failure time even at constant stress level σ. The parameter λ_σ and distribution function $G_\sigma(y)$ would require experimental determination.

Example *The Strength Versus Length of Filaments* It was noted that the logarithm of mean tensile strength of brittle fibers, such as boron filaments, in general varied linearly with the length of the filament, but that this relation did not hold for short filaments. It was suspected that the breakdown in the log linear relation might be due to testing or measurement problems, rather than being an inherent property of short filaments. Evidence supporting this idea was the observation that short filaments would break in the test clamps, rather than between them as desired, more often than would long filaments. Some means of correcting observed mean strengths to account for filaments breaking in, rather than between, the clamps was desired. It was decided to compute the ratio between the actual mean strength and an ideal mean strength, obtained under the assumption that there was no stress in the clamps, as a correction factor.

Since the molecular bonding strength is several orders of magnitude higher than generally observed strengths, it was felt that failure typically was caused by flaws. There are a number of different types of flaws, both internal flaws such as voids, inclusions, and weak grain boundaries, and external or surface flaws such as notches and cracks that cause stress concentrations. Let us suppose that flaws occur independently in a Poisson manner along the length of the filament. We let Y_k be the strength of the filament at the kth flaw and suppose Y_k has the cumulative distribution function $G(y)$, $y > 0$. We have plotted this information in Figure 5.11. The flaws reduce the strength. Opposing the strength is the stress in the filament. Ideally, the stress should be constant along the filament between the clamp faces and zero within the clamp. In practice the stress tapers off to zero over some positive length in the clamp. As a first approximation it is reasonable to assume that the stress decreases linearly. Let l be the length of the clamp and t the distance between the clamps, called the *gage length,* as illustrated in Figure 5.11.

The filament holds as long as the stress has not exceeded the strength as determined by the weakest flaw. That is, the filament will support a stress of y as long as no flaw points fall in the stress trapezoid of Figure 5.11. The number of points in this trapezoid has a Poisson distribution with mean $\mu(B) + 2\mu(A)$. In particular, no points fall there with probability $e^{-[\mu(B)+2\mu(A)]}$. If we let S be the strength of the filament, then

$$\Pr\{S > y\} = e^{-2\mu(A)-\mu(B)}.$$

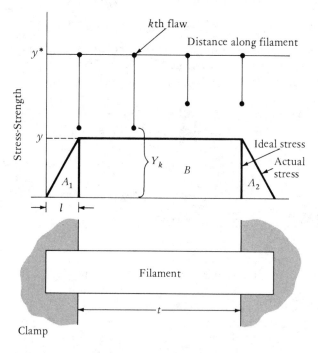

Figure 5.11   The stress versus strength of a filament under tension. Flaws reducing the strength of a filament below its theoretical maximum $y^\star$ are distributed randomly along its length. The stress in the filament is constant at the level y between clamps and tapers off to zero within the clamps. The filament fails if at any point along its length a flaw reduces its strength below its stress.

We compute

$$\mu(A) = \int_0^l G\left(\frac{xy}{l}\right)\lambda dx$$

and

$$\mu(B) = \lambda t G(y).$$

Finally, the mean strength of the filament is

$$E[S] = \int_0^\infty \Pr\{S > y\}dy = \int_0^\infty \exp\{-\lambda[tG(y) + 2\int_0^l G\left(\frac{xy}{l}\right)dx]\}dy.$$

For an ideal filament we use the same expression but with $l = 0$.

Chapter 6 | Continuous Time Markov Chains

6.1 Pure Birth Processes

In this chapter we present several important examples of continuous time, discrete state, Markov processes. Specifically, we deal here with a family of random variables $\{X(t); 0 \leq t < \infty\}$ where the possible values of $X(t)$ are the nonnegative integers. We shall restrict attention to the case where $\{X(t)\}$ is a Markov process with stationary transition probabilities. Thus, the transition probability function for $t > 0$,

$$P_{ij}(t) = \Pr\{X(t + u) = j | X(u) = i\}, \qquad i, j = 0, 1, 2, \ldots,$$

is independent of $u \geq 0$.

It is usually more natural in investigating particular stochastic models based on physical phenomena to prescribe the so-called infinitesimal probabilities relating to the process and then derive from them an explicit expression for the transition probability function. For the case at hand, we will postulate the form of $P_{ij}(h)$ for h small and, using the Markov property, we will derive a system of differential equations satisfied by $P_{ij}(t)$ for all $t > 0$. The solution of these equations under suitable boundary conditions gives $P_{ij}(t)$. We recall that the Poisson process introduced in Chapter 5 was in fact treated from just this point of view.

By way of introduction to the general pure birth process we review briefly the axioms characterizing the Poisson process.

6.1.1 Postulates for the Poisson Process

The Poisson process was considered in Chapter 5 where it was shown that it could be defined by a few simple postulates. In order to define more general processes of a similar kind, let us point out various further properties that the Poisson process possesses. In particular, it is a Markov process on the nonnegative integers that has the following properties:

(i) $\Pr\{X(t + h) - X(t) = 1 | X(t) = x\} = \lambda h + o(h)$ as $h \downarrow 0$

$$(x = 0, 1, 2, . . .).$$

The precise interpretation of (i) is the relationship

$$\lim_{h \to 0+} \frac{\Pr\{X(t + h) - X(t) = 1 | X(t) = x\}}{h} = \lambda.$$

The $o(h)$ symbol means that if we divide this term by h then the resulting value tends to zero as h tends to zero. Notice that the right-hand side is independent of x.

(ii) $\Pr\{X(t + h) - X(t) = 0 | X(t) = x\} = 1 - \lambda h + o(h)$ as $h \downarrow 0$.
(iii) $X(0) = 0$.

These properties are easily verified by direct computation, since the explicit formulas for all the relevant probabilities are available.

6.1.2 Pure Birth Process

A natural generalization of the Poisson process is to permit the chance of an event occurring at a given instant of time to depend upon the number of events that have already occurred. An example of this phenomenon is the reproduction of living organisms (and hence the name of the process), in which under certain conditions—sufficient food, no mortality, no migration, for example—the infinitesimal probability of a birth at a given instant is proportional (directly) to the population size at that time. This example is known as the Yule process and will be considered in detail later.

Consider a sequence of positive numbers, $\{\lambda_k\}$. We define a pure birth process as a Markov process satisfying the postulates:

(i) $\Pr\{X(t + h) - X(t) = 1 | X(t) = k\} = \lambda_k h + o_{1,k}(h)$, $(h \to 0+)$,
(ii) $\Pr\{X(t + h) - X(t) = 0 | X(t) = k\} = 1 - \lambda_k h + o_{2,k}(h)$, (6.1)
(iii) $\Pr\{X(t + h) - X(t) < 0 | X(t) = k\} = 0$, $(k \geq 0)$.

As a matter of convenience we often add the postulate

(iv) $X(0) = 0$.

With this postulate $X(t)$ does not denote the population size but, rather, the number of births in the time interval $(0, t]$.

Note that the left sides of (i) and (ii) are just $P_{k,k+1}(h)$ and $P_{k,k}(h)$, respectively (owing to stationarity), so that $o_{1,k}(h)$ and $o_{2,k}(h)$ do not depend upon t.

We define $P_n(t) = \Pr\{X(t) = n\}$, assuming $X(0) = 0$.

In exactly the same way as for the Poisson process, we may derive a system of differential equations satisfied by $P_n(t)$ for $t \geq 0$, namely

$$P_0'(t) = -\lambda_0 P_0(t),$$

$$P_n'(t) = -\lambda_n P_n(t) + \lambda_{n-1} P_{n-1}(t) \qquad \text{for} \quad n \geq 1,$$

$$(6.2)$$

with boundary conditions

$$P_0(0) = 1, \qquad P_n(0) = 0, \qquad n > 0.$$

Indeed, if $h > 0$, $n \geq 1$, then by invoking the law of total probability, the Markov property, and Postulate (iii) we obtain

$$P_n(t + h) = \sum_{k=0}^{\infty} P_k(t) \Pr\{X(t + h) = n | X(t) = k\}$$

$$= \sum_{k=0}^{\infty} P_k(t) \Pr\{X(t + h) - X(t) = n - k | X(t) = k\}$$

$$= \sum_{k=0}^{n} P_k(t) \Pr\{X(t + h) - X(t) = n - k | X(t) = k\}.$$

Now for $k = 0, 1, \ldots, n - 2$ we have

$$\Pr\{X(t + h) - X(t) = n - k | X(t) = k\}$$

$$\leq \Pr\{X(t + h) - X(t) \geq 2 | X(t) = k\}$$

$$= o_{1,k}(h) + o_{2,k}(h)$$

or

$$\Pr\{X(t + h) - X(t) = n - k | X(t) = k\} = o_{3,n,k}(h)$$

$$k = 0, \ldots, n - 2.$$

Thus

$$P_n(t + h) = P_n(t)[1 - \lambda_n h + o_{2,n}(h)] + P_{n-1}(t)[\lambda_{n-1}h + o_{1,n-1}(h)]$$

$$+ \sum_{k=0}^{n-2} P_k(t) o_{3,n,k}(h)$$

or

$$P_n(t + h) - P_n(t)$$

$$= P_n(t)[-\lambda_n h + o_{2,n}(h)] + P_{n-1}(t)[\lambda_{n-1}h + o_{1,n-1}(h)] + o_n(h), \quad (6.3)$$

where, clearly, $\lim_{h \downarrow 0} o_n(h)/h = 0$ uniformly in $t \geq 0$ since $o_n(h)$ is bounded by the finite sum $\sum_{k=0}^{n-2} o_{3,n,k}(h)$ which does not depend on t.

Dividing by h and passing to the limit $h \downarrow 0$, we validate the relations (6.2) where on the left side we should, to be precise, write the derivative from the right. With a little more care, however, we can derive the same relation involving the derivative from the left. In fact, from (6.3) we see at once that the $P_n(t)$ are continuous functions of t. Replacing t by $t - h$ in (6.3), dividing by h, and passing to the limit $h \downarrow 0$, we find that each $P_n(t)$ has a left derivative that also satisfies Equation (6.2).

The first equation of (6.2) can be solved immediately and yields

$$P_0(t) = \exp\{-\lambda_0 t\} \qquad \text{for} \quad t > 0. \tag{6.4}$$

Define S_k as the time between the kth and the $(k + 1)$st birth, so that

$$P_n(t) = \Pr\left\{ \sum_{i=0}^{n-1} S_i \le t < \sum_{i=0}^{n} S_i \right\}.$$

The random variables S_k are called the "sojourn times" between births, and

$$W_k = \sum_{i=0}^{k-1} S_i = \text{the time at which the } k\text{th birth occurs.}$$

We have already seen that $P_0(t) = \exp\{-\lambda_0 t\}$. Therefore,

$$\Pr\{S_0 \le t\} = 1 - \Pr\{X(t) = 0\} = 1 - \exp\{-\lambda_0 t\},$$

i.e., S_0 has an exponential distribution with parameter λ_0. It may be deduced from Postulates (i) through (iv) that S_k, $k > 0$, also has an exponential distribution with parameter λ_k and that the S_i's are mutually independent.

This description characterizes the pure birth process in terms of its sojourn times, in contrast to the infinitesimal description corresponding to (6.1).

To recursively solve the differential equations of (6.2) introduce $Q_n(t) = e^{\lambda_n t} P_n(t)$ for $n = 0, 1, \ldots$. Then

$$\begin{aligned} Q'_n(t) &= \lambda_n e^{\lambda_n t} P_n(t) + e^{\lambda_n t} P'_n(t) \\ &= e^{\lambda_n t} [\lambda_n P_n(t) + P'_n(t)] \\ &= e^{\lambda_n t} \lambda_{n-1} P_{n-1}(t) \qquad \text{[using (6.2)].} \end{aligned}$$

Integrating both sides of these equations and using the boundary condition $Q_n(0) = 0$ for $n \ge 1$ gives

$$Q_n(t) = \int_0^t e^{\lambda_n x} \lambda_{n-1} P_{n-1}(x)\, dx,$$

or

$$P_n(t) = \lambda_{n-1} e^{-\lambda_n t} \int_0^t e^{\lambda_n x} P_{n-1}(x)\, dx, \qquad n = 1, 2, \ldots . \tag{6.5}$$

It is now clear that all $P_k(t) \ge 0$, but there is still a possibility that

$$\sum_{n=0}^{\infty} P_n(t) < 1.$$

To secure the validity of the process, i.e., to assure that $\sum_{n=0}^{\infty} P_n(t) = 1$ for all t, we must restrict the λ_k according to the following

$$\sum_{n=0}^{\infty} P_n(t) = 1 \quad \text{if and only if} \quad \sum_{n=0}^{\infty} \frac{1}{\lambda_n} = \infty. \tag{6.6}$$

The intuitive argument for this result is as follows: The time S_k between consecutive births is exponentially distributed with a corresponding parameter λ_k. Therefore, the quantity $\sum_n 1/\lambda_n$ equals the expected time before the population becomes infinite. By comparison $1 - \sum_{n=0}^{\infty} P_n(t)$ is the probability that $X(t) = \infty$.

If $\sum_n \lambda_n^{-1} < \infty$ then the expected time for the population to become infinite is finite. It is then plausible that for all $t > 0$ the probabililty that $X(t) = \infty$ is positive.

When no two of the birth parameters $\lambda_0, \lambda_1, \ldots$ are equal, the differential equation (6.5) may be solved to give the explicit formula

$$P_0(t) = e^{-\lambda_0 t},$$

$$P_1(t) = \lambda_0\left(\frac{1}{\lambda_1 - \lambda_0} e^{-\lambda_0 t} + \frac{1}{\lambda_0 - \lambda_1} e^{-\lambda_1 t}\right) \tag{6.7}$$

and,

$$P_n(t) = \Pr\{X(t) = n \mid X(0) = 0\}$$

$$= \lambda_0 \cdots \lambda_{n-1}[B_{0,n} e^{-\lambda_0 t} + \cdots + B_{n,n} e^{-\lambda_n t}] \quad \text{for} \quad n > 1, \tag{6.8}$$

where

$$B_{0,n} = \frac{1}{(\lambda_1 - \lambda_0) \cdots (\lambda_n - \lambda_0)},$$

$$B_{k,n} = \frac{1}{(\lambda_0 - \lambda_k) \cdots (\lambda_{k-1} - \lambda_k)(\lambda_{k+1} - \lambda_k) \cdots (\lambda_n - \lambda_k)}$$

and $\qquad\qquad\qquad\qquad\qquad\qquad\qquad\qquad$ for $\quad 0 < k < n, \tag{6.9}$

$$B_{n,n} = \frac{1}{(\lambda_0 - \lambda_n) \cdots (\lambda_{n-1} - \lambda_n)}.$$

Because $\lambda_j \neq \lambda_k$ when $j \neq k$ by assumption, the denominator in (6.8) does not vanish, and $B_{k,n}$ is well defined.

We will verify that $P_1(t)$, as given by (6.7), satisfies (6.5). Equation (6.4) gives $P_0(t) = e^{-\lambda_0 t}$. We next substitute this in (6.5) when $n = 1$, thereby obtaining

$$P_1(t) = \lambda_0 e^{-\lambda_1 t} \int_0^t e^{\lambda_1 x} e^{-\lambda_0 x} dx$$

$$= \lambda_0 e^{-\lambda_1 t}(\lambda_0 - \lambda_1)^{-1}[1 - e^{-(\lambda_0 - \lambda_1)t}]$$

$$= \lambda_0 \left(\frac{1}{\lambda_1 - \lambda_0} e^{-\lambda_0 t} + \frac{1}{\lambda_0 - \lambda_1} e^{-\lambda_1 t} \right)$$

in agreement with (6.7).

The induction proof for a general n involves tedious and difficult algebra. The case $n = 2$ is suggested as a problem.

6.1.3 The Yule Process

The Yule process arises in physics and biology and describes the growth of a population in which each member has a probability $\beta h + o(h)$ of giving birth to a new member during an interval of time length h ($\beta > 0$). Assuming independence and no interaction among members of the population, the binomial theorem gives

$$\Pr\{X(t + h) - X(t) = 1 | X(t) = n\} = \binom{n}{1} [\beta h + o(h)][1 - \beta h + o(h)]^{n-1}$$
$$= n\beta h + o_n(h),$$

i.e., for the Yule process the infinitesimal parameters are $\lambda_n = n\beta$. In words, the total population birth rate is directly proportional to the population size, the proportionality constant being the individual birth rate β. As such, the Yule process forms a stochastic analog of the deterministic population growth model represented by the differential equation $dy/dt = \alpha y$. In the deterministic model, the rate dy/dt of population growth is directly proportional to population size y. In the stochastic model, the infinitesimal deterministic increase dy is replaced by the probability of a unit increase during the infinitesimal time interval dt. Similar connections between deterministic rates and birth (and death) parameters arise frequently in stochastic modeling. Examples abound in this chapter.

The system of equations (6.2) in the case that $X(0) = 1$ becomes

$$P_n'(t) = -\beta[nP_n(t) - (n - 1)P_{n-1}(t)], \qquad n = 1, 2, \ldots,$$

under the initial conditions

$$P_1(0) = 1, \qquad P_n(0) = 0, \qquad n = 2, 3, \ldots.$$

Its solution is

$$P_n(t) = e^{-\beta t}(1 - e^{-\beta t})^{n-1}, \qquad n \geq 1, \tag{6.10}$$

as may be verified directly. We recognize (6.10) as the geometric distribution (1.26) with $p = e^{-\beta t}$.

The general solution analogous to (6.8) but for pure birth processes starting from $X(0) = 1$ is

$$P_n(t) = \lambda_1 \cdots \lambda_{n-1}[B_{1,n}e^{-\lambda_1 t} + \cdots + B_{n,n}e^{-\lambda_n t}], \qquad n > 1. \quad (6.11)$$

When $\lambda_n = \beta n$, we will show that (6.11) reduces to the solution given in (6.10) for a Yule process with parameter β. Then

$$B_{1,n} = \frac{1}{(\lambda_2 - \lambda_1)(\lambda_3 - \lambda_1) \cdots (\lambda_n - \lambda_1)}$$

$$= \frac{1}{\beta^{n-1}1(2) \cdots (n - 1)}$$

$$= \frac{1}{\beta^{n-1}(n - 1)!},$$

$$B_{2,n} = \frac{1}{(\lambda_1 - \lambda_2)(\lambda_3 - \lambda_2) \cdots (\lambda_n - \lambda_2)}$$

$$= \frac{1}{\beta^{n-1}(-1)(1)(2) \cdots (n - 2)}$$

$$= \frac{-1}{\beta^{n-1}(n - 2)!}$$

and

$$B_{k,n} = \frac{1}{(\lambda_1 - \lambda_k) \cdots (\lambda_{k-1} - \lambda_k)(\lambda_{k+1} - \lambda_k) \cdots (\lambda_n - \lambda_k)}$$

$$= \frac{(-1)^{k-1}}{\beta^{n-1}(k - 1)!(n - k)!}$$

Thus, according to (6.11),

$$P_n(t) = \beta^{n-1}(n - 1)!(B_{1,n}e^{-\beta t} + \cdots + B_{n,n}e^{-n\beta t})$$

$$= \sum_{k=1}^{n} \frac{(n - 1)!}{(k - 1)!(n - k)!}(-1)^{k-1}e^{-k\beta t}$$

$$= e^{-\beta t}\sum_{j=0}^{n-1} \frac{(n - 1)!}{j!(n - 1 - j)!}(-e^{-\beta t})^j$$

$$= e^{-\beta t}(1 - e^{-\beta t})^{n-1} \qquad \text{[see (1.67), p. 42]}$$

which establishes (6.10).

Problems 6.1

1. Using Equation (6.10), calculate the mean and variance for the Yule process where $X(0) = 1$.

2. Let $N(t)$ be a pure birth process for which

   $\Pr\{$an event happens in $(t, t + h)|N(t)$ is odd$\} = \alpha h + o(h)$,

   $\Pr\{$an event happens in $(t, t + h)|N(t)$ is even$\} = \beta h + o(h)$,

   where $o(h)/h \to 0$ as $h \downarrow 0$. Take $N(0) = 0$. Find the following probabilities:

   $$P_0(t) = \Pr\{N(t) \text{ is even}\}; \qquad P_1(t) = \Pr\{N(t) \text{ is odd}\}.$$

   Hint: Derive the differential equations

   $$P_0'(t) = \alpha P_1(t) - \beta P_0(t) \quad \text{and} \quad P_1'(t) = -\alpha P_1(t) + \beta P_0(t)$$

   and solve them by using $P_0(t) + P_1(t) = 1$.

3. Under the conditions of Problem 2, determine $E[N(t)]$.

4. Consider a pure birth process on the states $0, 1, \ldots, N$ for which $\lambda_k = (N - k)\lambda$ for $k = 0, 1, \ldots, N$. Suppose that $X(0) = 0$. Determine $P_n(t) = \Pr\{X(t) = n\}$.

5. Beginning with $P_0(t) = e^{-\lambda_0 t}$ and using Equation (6.5), calculate $P_1(t)$, $P_2(t)$, and $P_3(t)$ and verify that these probabilities conform with Equation (6.7), assuming distinct birth parameters.

6. Verify that $P_2(t)$, as given by (6.8), satisfies (6.5), by following the calculations in the text that showed that $P_1(t)$ satisfies (6.5).

7. Operations 1, 2, and 3 are to be performed in succession on a major piece of equipment. Operation k takes a random duration S_k that is exponentially distributed with parameter λ_k for $k = 1, 2, 3$, and all operation times are independent. Let $X(t)$ denote the operation being performed at time t, with time $t = 0$ marking the start of the first operation. Suppose that $\lambda_1 = 5$, $\lambda_2 = 3$, and $\lambda_3 = 13$. Determine
   (a) $P_1(t) = \Pr\{X(t) = 1\}$.
   (b) $P_2(t) = \Pr\{X(t) = 2\}$.
   (c) $P_3(t) = \Pr\{X(t) = 3\}$.

6.2 Pure Death Processes

Complementing the increasing pure birth process is the decreasing pure death process. It moves successively through states $N, N - 1, \ldots, 2, 1$ and ultimately is absorbed in state 0 (extinction). The process is specified by the death parameters $\mu_k > 0$ for $k = 1, 2, \ldots, N$, where the sojourn time in state k is exponentially distributed with parameter μ_k, all sojourn times being independent. A typical sample path is depicted in Figure 6.1.

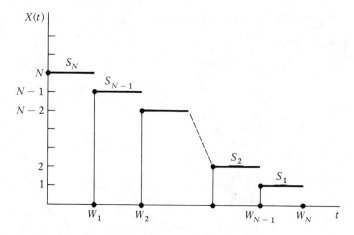

Figure 6.1 A typical sample path of a pure death process, show-ing the sojourn times $S_N, \ldots, S_1$ and the waiting times W_1, $W_2, \ldots, W_N$.

Alternatively, we have the infinitesimal description of a pure death pro-cess as a Markov process $X(t)$ whose state space is $0, 1, \ldots, N$ and for which

(i) $\Pr\{X(t + h) = k - 1 | X(t) = k\} = \mu_k h + o(h), \ k = 1, \ldots, N$
(ii) $\Pr\{X(t + h) = k | X(t) = k\} = 1 - \mu_k h + o(h), \ k = 1, \ldots, N$
(iii) $\Pr\{X(t + h) > k | X(t) = k\} = 0, \ k = 0, 1, \ldots, N.$ (6.12)

The parameter μ_k is the "death rate" operating or in effect while the process sojourns in state k. It is a common and useful convention to assign $\mu_0 = 0$.

When the death parameters $\mu_1, \mu_2, \ldots, \mu_N$ are distinct, that is $\mu_j \neq \mu_k$ if $j \neq k$, then we have the explicit transition probabilities

$$P_N(t) = e^{-\mu_N t},$$

and for $n < N$,

$$P_n(t) = \Pr\{X(t) = n | X(0) = N\}$$

$$= \mu_{n+1}\mu_{n+2} \cdots \mu_N [A_{n,n} e^{-\mu_n t} + \cdots + A_{N,n} e^{-\mu_N t}] \quad (6.13)$$

where

$$A_{k,n} = \frac{1}{(\mu_N - \mu_k) \cdots (\mu_{k+1} - \mu_k)(\mu_{k-1} - \mu_k) \cdots (\mu_n - \mu_k)}.$$

6.2.1 The Linear Death Process

As an example, consider a pure death process in which the death rates are proportional to population size. This process, which we will call the *linear death process,* complements the Yule or linear birth process. The parameters are $\mu_k = k\alpha$ where α is the individual death rate in the population. Then

$$
A_{n,n} = \frac{1}{(\mu_N - \mu_n)(\mu_{N-1} - \mu_n) \cdots (\mu_{n+1} - \mu_n)}
$$

$$
= \frac{1}{\alpha^{N-n-1}(N - n)(N - n - 1) \cdots (2)(1)},
$$

$$
A_{n+1,n} = \frac{1}{(\mu_N - \mu_{n+1}) \cdots (\mu_{n+2} - \mu_{n+1})(\mu_n - \mu_{n+1})}
$$

$$
= \frac{1}{\alpha^{N-n-1}(N - n - 1) \cdots (1)(-1)},
$$

$$
A_{k,n} = \frac{1}{(\mu_N - \mu_k) \cdots (\mu_{k+1} - \mu_k)(\mu_{k-1} - \mu_k) \cdots (\mu_n - \mu_k)}
$$

$$
= \frac{1}{\alpha^{N-n-1}(N - k) \cdots (1)(-1)(-2) \cdots (n - k)}
$$

$$
= \frac{1}{\alpha^{N-n-1}(-1)^{k-n}(N - k)!(k - n)!}.
$$

Then

$$
P_n(t) = \mu_{n+1}\mu_{n+2} \cdots \mu_N \sum_{k=n}^{N} A_{k,n} e^{-\mu_k t}
$$

$$
= \alpha^{N-n-1} \frac{N!}{n!} \sum_{k=n}^{N} \frac{e^{-k\alpha t}}{\alpha^{N-n-1}(-1)^{k-n}(N - k)!(k - n)!}
$$

$$
= \frac{N!}{n!} e^{-n\alpha t} \sum_{j=0}^{N-n} \frac{(-1)^j e^{-j\alpha t}}{(N - n - j)! j!}
$$

$$
= \frac{N!}{n!(N - n)!} e^{-n\alpha t} (1 - e^{-\alpha t})^{N-n}, \qquad n = 0, \ldots, N. \tag{6.14}
$$

Let T be the time of population extinction. Formally, $T = \min\{t \geq 0; X(t) = 0\}$. Then $T \leq t$ if and only if $X(t) = 0$, which leads to the cumulative distribution function of T via

$$
F_T(t) = \Pr\{T \leq t\} = \Pr\{X(t) = 0\}
$$
$$
= P_0(t) = (1 - e^{-\alpha t})^N, \qquad t \geq 0. \tag{6.15}
$$

The linear death process can be viewed in yet another way, a way that again confirms the intimate connection between the exponential distribution and a continuous time parameter Markov chain. Consider a population consisting of N individuals, each of whose lifetimes is an independent exponentially distributed random variable with parameter α. Let $X(t)$ be the number of survivors in this population at time t. Then $X(t)$ is the linear pure death process whose parameters are $\mu_k = k\alpha$ for $k = 0, 1, \ldots, N$. To help

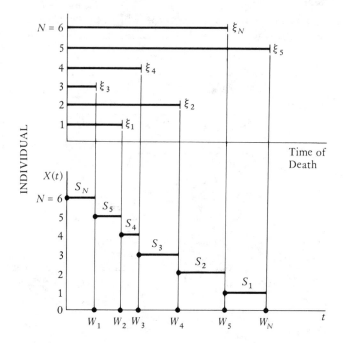

Figure 6.2 The linear death process. As depicted here, the third individual is the first to die, the first individual is the second to die, etc.

understand this connection, let $\xi_1, \xi_2, \ldots, \xi_N$ denote the times of death of the individuals labeled $1, 2, \ldots, N$, respectively. Figure 6.2 shows the relation between the individual lifetimes $\xi_1, \xi_2, \ldots, \xi_N$ and the death process $X(t)$.

The sojourn time in state N, denoted S_N, equals the time of the earliest death, or $S_N = \min\{\xi_1, \ldots, \xi_N\}$. Since the lifetimes are independent and have the same exponential distribution,

$$\begin{aligned}
\Pr\{S_N > t\} &= \Pr\{\min\{\xi_1, \ldots, \xi_N\} > t\} \\
&= \Pr\{\xi_1 > t, \ldots, \xi_N > t\} \\
&= [\Pr\{\xi_1 > t\}]^N \\
&= e^{-N\alpha t}.
\end{aligned}$$

That is, S_N has an exponential distribution with parameter $N\alpha$. Similar reasoning applies when there are k members alive in the population. The memoryless property of the exponential distribution implies that the remaining lifetime of each of these k individuals is exponentially distributed with parameter α. Then the sojourn time S_k is the minimum of these k remaining lifetimes and hence is exponentially distributed with parameter $k\alpha$. To give one more approach in terms of transition rates, each individual in the population has a constant death rate of α in the sense that

$$\Pr\{t < \xi_1 < t + h | t < \xi_1\} = \frac{\Pr\{t < \xi_1 < t + h\}}{\Pr\{t < \xi_1\}}$$

$$= \frac{e^{-\alpha t} - e^{-\alpha(t+h)}}{e^{-\alpha t}}$$

$$= 1 - e^{-\alpha h}$$

$$= \alpha h + o(h) \qquad \text{as} \quad h \downarrow 0.$$

If each of k individuals alive in the population at time t has a constant death rate of α, then the total population death rate should be $k\alpha$, directly proportional to the population size. This shortcut approach to specifying appropriate death parameters is a powerful and often used tool of stochastic modeling. The next example furnishes another illustration of its use.

6.2.2 Cable Failure Under Static Fatigue

A cable comprised of parallel fibers in tension is being designed to support a high altitude weather balloon. With a design load of 1000 kgs and a design lifetime of 100 years, how many fibers should be used in the cable?

The low weight, high strength fibers to be used are subject to *static fatigue,* or eventual failure when subjected to a constant load. The higher the constant load, the shorter the life, and experiments have established a linear plot on log-log axes between average failure time and load that is shown in Figure 6.3.

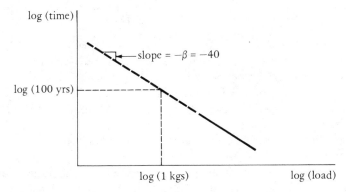

Figure 6.3 A linear relation between log mean failure time and log load.

The relation between mean life μ_T and load l that is illustrated in Figure 6.3 takes the analytic form

$$\log_{10}\mu_T = 2 - 40 \log_{10}l.$$

Were the cable to be designed on the basis of average life, to achieve the 100 year design target, each fiber should carry 1 kg. Since the total load is 1000 kgs, then $N = \frac{1000}{1} = 1000$ fibers should be used in the cable.

One might suppose that this large number ($N = 1000$) of fibers would justify designing the cable based on average fiber properties. We shall see that such reasoning is dangerously wrong.

Let us suppose, however, as is the case with many modern high performance structural materials, that there is a large amount of random scatter of individual fiber lifetimes about the mean. How does this randomness affect the design problem?

Some assumption must be made concerning the probability distribution governing individual fiber lifetimes. In practice, it is extremely difficult to gather sufficient data to determine this distribution with any degree of certainty. Most data do show, however, a significant degree of skewness or asymmetry. Because it qualitatively matches observed data, and because it leads to a pure death process model that is accessible to exhaustive analysis, we will assume that the probability distribution for the failure time T of a single fiber subjected to the time-varying tensile load $l(t)$ is given by

$$\Pr\{T \le t\} = 1 - \exp\{-\int_0^t K[l(s)]ds\}, \qquad t \ge 0.$$

This distribution corresponds to a *failure rate* or *hazard rate* of $r(t) = K[l(t)]$ wherein a single fiber, having not failed prior to time t and carrying the load $l(t)$, will fail during the interval $(t, t + \Delta t]$ with probability

$$\Pr\{t < T \le t + \Delta t | T > t\} = K[l(t)]\Delta t + o(\Delta t).$$

The function $K[l]$, called the *breakdown rule*, expresses how changes in load affect the failure probability. We are concerned with the *power law breakdown* rule in which $K[l] = l^\beta/A$ for some positive constants A and β. Assuming power law breakdown, under a constant load $l(t) = l$, the single fiber failure time is exponentially distributed with mean $\mu_T = E[T|l] = 1/K[l] = Al^{-\beta}$. A plot of mean failure time versus load is linear on log-log axes, matching the observed properties of our fiber type. For the design problem we have $\beta = 40$ and $A = 100$.

Now place N of these fibers in parallel and subject the resulting bundle or cable to a total load, constant in time, of NL, where L is the nominal load per fiber. What is the probability distribution of the time at which the cable fails? Since the fibers are in parallel, this system failure time equals the failure time of the last fiber.

Under the stated assumptions governing single fiber behavior, $X(t)$, the number of unfailed fibers in the cable at time t, evolves as a pure death process with parameters $\mu_k = kK[NL/k]$ for $k = 1, 2, \ldots, N$. Given $X(t) = k$ surviving fibers at time t, and assuming that the total bundle load NL is shared equally among them, then each carries load NL/k and has a corresponding failure rate of $K[NL/k]$. As there are k such survivors in the bundle, the bundle or system failure rate is $\mu_k = kK[NL/k]$ as claimed.

It was mentioned earlier that the system failure time was W_N, the waiting time to the Nth fiber failure. Then $\Pr\{W_N \le t\} = \Pr\{X(t) = 0\} = P_0(t)$ where $P_n(t)$ is given explicitly by (6.13) in terms of $\mu_1, \ldots, \mu_N$. Alternatively, we may bring to bear the sojourn time description of the pure death process and, following Figure 6.1, write

$$W_N = S_N + S_{N-1} + \cdots + S_1$$

where $S_N, S_{N-1}, \ldots, S_1$ are independent exponentially distributed random variables and S_k has parameter $\mu_k = kK[NL/k] = k(NL/k)^\beta/A$. The mean system failure time is readily computed to be

$$
\begin{aligned}
E[W_N] &= E[S_N] + \cdots + E[S_1] \\
&= AL^{-\beta} \sum_{k=1}^{N} \frac{1}{k}\left(\frac{k}{N}\right)^\beta \\
&= AL^{-\beta} \sum_{k=1}^{N} \left(\frac{k}{N}\right)^{\beta-1}\left(\frac{1}{N}\right).
\end{aligned}
\tag{6.16}
$$

The sum in the expression for $E[W_N]$ seems formidable at first glance, but a very close approximation is readily available when N is large. Figure 6.4 compares the sum to an integral.

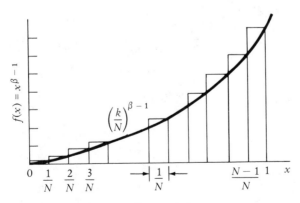

Figure 6.4 The sum $\sum_{k=1}^{N} (k/N)^{\beta-1}(1/N)$ is a Riemann approximation to $\int_0^1 x^{\beta-1}dx = 1/\beta$.

From Figure 6.4 we see that

$$\sum_{k=1}^{N}\left(\frac{k}{N}\right)^{\beta-1}\left(\frac{1}{N}\right) \cong \int_{0}^{1} x^{\beta-1} dx = \frac{1}{\beta}.$$

Indeed, we readily obtain

$$\frac{1}{\beta} = \int_{0}^{1} x^{\beta-1} dx \leq \sum_{k=1}^{N}\left(\frac{k}{N}\right)^{\beta-1}\left(\frac{1}{N}\right) \leq \int_{1/N}^{1+1/N} x^{\beta-1} dx$$

$$= \left(\frac{1}{\beta}\right)\left[\left(1 + \frac{1}{N}\right)^{\beta} - \left(\frac{1}{N}\right)^{\beta}\right].$$

When $N = 1000$ and $\beta = 40$, the numerical bounds are

$$\left(\frac{1}{40}\right) \leq \sum_{k=1}^{N}\left(\frac{k}{N}\right)^{\beta-1}\left(\frac{1}{N}\right) \leq \left(\frac{1}{40}\right)(1.0408),$$

which shows that the integral determines the sum to within about 4 percent. Substituting $1/\beta$ for the sum in (6.16) gives the average cable life

$$E[W_N] \cong \frac{A}{\beta L^{\beta}},$$

to be compared with the average fiber life of

$$\mu_T = \frac{A}{L^{\beta}}.$$

That is, a cable lasts only about $1/\beta$ as long as an average fiber under an equivalent load. With $A = 100$, $\beta = 40$, and $N = 1000$, the designed cable would last, on the average, $100/[40(1)^{40}] = 2.5$ years, far short of the desired life of 100 years. The cure is to increase the number of fibers in the cable, thereby decreasing the per fiber load. Increasing the number of fibers from N to N' decreases the nominal load per fiber from L to $L' = NL/N'$. To achieve parity in fiber-cable lifetimes we equate

$$\frac{A}{L^{\beta}} = \frac{A}{\beta(NL/N')^{\beta}}$$

or

$$N' = N\beta^{1/\beta}.$$

For the given data, this calls for $N' = 1000(40)^{1/40} = 1097$ fibers. That is, the design lifetime can be restored by increasing the number of fibers in the cable by about 10 percent.

Problems 6.2

1. Give the transition probabilities for the pure death process described by $X(0) = 3$, $\mu_3 = 1$, $\mu_2 = 2$, and $\mu_1 = 3$.

2. Consider a cable comprised of fibers following the breakdown rule $K[l] = \sinh(l) = \frac{1}{2}(e^l - e^{-l})$ for $l \geq 0$. Show that the mean cable life is given by

$$E[W_N] = \sum_{k=1}^{N} \{k \sinh(NL/k)\}^{-1} = \sum_{k=1}^{N} \left\{ \frac{k}{N} \sinh\left(\frac{L}{k/N}\right) \right\}^{-1} \left(\frac{1}{N}\right)$$

$$\cong \int_0^1 \{x \sinh(L/x)\}^{-1} dx.$$

3. Consider the linear death process (Section 6.2.1) in which $X(0) = N = 5$ and $\alpha = 2$. Determine $\Pr\{X(t) = 2\}$.

   Hint: Use Equation (6.14).

4. Let T be the time to extinction in the linear death process with parameters $X(0) = N$ and α (see Section 6.2.1).

   (a) Using the sojourn time viewpoint, show that

   $$E[T] = \frac{1}{\alpha}\left[\frac{1}{N} + \frac{1}{N-1} + \cdots + \frac{1}{1} \right].$$

   (b) Using the binomial expansion of $(1 - e^{-\alpha t})^N$ verify the result of (a) by using Equation (6.15) in

   $$E[T] = \int_0^\infty \Pr\{T > t\}dt = \int_0^\infty [1 - F_T(t)]dt.$$

6.3 Birth and Death Processes

An obvious generalization of the pure birth and pure death processes discussed in Sections 6.1 and 6.2 is to permit $X(t)$ both to increase and to decrease. Thus if at time t the process is in state n it may, after a random sojourn time, move to either of the neighboring states $n + 1$ or $n - 1$. The resulting *birth and death process* can then be regarded as the continuous time analog of a random walk (Section 3.5.3).

Birth and death processes form a powerful tool in the kit of the stochastic modeler. The richness of the birth and death parameters facilitates modeling a variety of phenomena. At the same time standard methods of analysis are available for determining numerous important quantities such as stationary distributions and mean first passage times. This section and later sections contain several examples of birth and death processes and illustrate how they are used to draw conclusions about phenomena in a variety of disciplines.

6.3.1 Postulates

As in the case of the pure birth processes we assume that $X(t)$ is a Markov process on the states 0, 1, 2, . . . and that its transition probabilities $P_{ij}(t)$ are stationary, i.e.,

$$P_{ij}(t) = \Pr\{X(t + s) = j | X(s) = i\} \qquad \text{for all} \quad s \geq 0.$$

In addition we assume that the $P_{ij}(t)$ satisfy

(1) $P_{i,i+1}(h) = \lambda_i h + o(h)$ as $h \downarrow 0, i \geq 0$
(2) $P_{i,i-1}(h) = \mu_i h + o(h)$ as $h \downarrow 0, i \geq 1$
(3) $P_{i,i}(h) = 1 - (\lambda_i + \mu_i)h + o(h)$ as $h \downarrow 0, i \geq 0$
(4) $P_{ij}(0) = \delta_{ij}$
(5) $\mu_0 = 0, \lambda_0 > 0, \mu_i, \lambda_i > 0, i = 1, 2, \ldots$

The $o(h)$ in each case may depend on i. The matrix

$$\mathbf{A} = \begin{Vmatrix} -\lambda_0 & \lambda_0 & 0 & 0 & \cdots \\ \mu_1 & -(\lambda_1 + \mu_1) & \lambda_1 & 0 & \cdots \\ 0 & \mu_2 & -(\lambda_2 + \mu_2) & \lambda_2 & \cdots \\ 0 & 0 & \mu_3 & -(\lambda_3 + \mu_3) & \cdots \\ \vdots & \vdots & \vdots & \vdots & \end{Vmatrix}$$

$$(6.17)$$

is called the *infinitesimal generator* of the process. The parameters λ_i and μ_i are called, respectively, the infinitesimal birth and death rates. In Postulates 1 and 2 we are assuming that if the process starts in state i, then in a small interval of time the probabilities of the population increasing or decreasing by 1 are essentially proportional to the length of the interval.

Since the $P_{ij}(t)$ are probabilities we have $P_{ij}(t) \geq 0$ and

$$\sum_{j=0}^{\infty} P_{ij}(t) \leq 1. \qquad (6.18)$$

Using the Markov property of the process we may also derive the so-called *Chapman-Kolmogorov equation*

$$P_{ij}(t + s) = \sum_{k=0}^{\infty} P_{ik}(t) P_{kj}(s). \qquad (6.19)$$

This equation states that in order to move from state i to state j in time $t + s$, $X(t)$ moves to some state k in time t and then from k to j in the remaining time s. This is the continuous time analog of formula (3.11).

So far we have mentioned only the transition probabilities $P_{ij}(t)$. In order to obtain the probability that $X(t) = n$ we must specify where the process starts or more generally the probability distribution for the initial state. We then have

$$\Pr\{X(t) = n\} = \sum_{i=0}^{\infty} q_i P_{in}(t),$$

where

$$q_i = \Pr\{X(0) = i\}.$$

6.3.2 Sojourn Times

With the aid of the preceding assumptions we may calculate the distribution of the random variable S_i which is the sojourn time of $X(t)$ in state i; that is, given that the process is in state i, what is the distribution of the time S_i until it first leaves state i? If we let

$$\Pr\{S_i \geq t\} = G_i(t)$$

it follows easily by the Markov property that as $h \downarrow 0$

$$G_i(t + h) = G_i(t)G_i(h) = G_i(t)[P_{ii}(h) + o(h)]$$
$$= G_i(t)[1 - (\lambda_i + \mu_i)h] + o(h)$$

or

$$\frac{G_i(t + h) - G_i(t)}{h} = -(\lambda_i + \mu_i)G_i(t) + o(1),$$

so that

$$G_i'(t) = -(\lambda_i + \mu_i)G_i(t). \tag{6.20}$$

If we use the conditions $G_i(0) = 1$ the solution of this equation is

$$G_i(t) = \exp[-(\lambda_i + \mu_i)t],$$

i.e., S_i follows an exponential distribution with mean $(\lambda_i + \mu_i)^{-1}$. The proof presented here is not quite complete, since we have used the intuitive relationship

$$G_i(h) = P_{ii}(h) + o(h)$$

without a formal proof.

According to Postulates 1 and 2, during a time duration of length h a transition occurs from state i to $i + 1$ with probability $\lambda_i h + o(h)$ and from state i to $i - 1$ with probability $\mu_i h + o(h)$. It follows intuitively that, given that a transition occurs at time t, the probability that this transition is to state $i + 1$ is $\lambda_i/(\lambda_i + \mu_i)$ and to state $i - 1$ is $\mu_i/(\lambda_i + \mu_i)$. The rigorous demonstration of this result is beyond the scope of this book.

It leads to an important characterization of a birth and death process, however, wherein the description of the motion of $X(t)$ is as follows: The process sojourns in a given state i for a random length of time whose distribution function is an exponential distribution with parameter $(\lambda_i + \mu_i)$. When leaving state i the process enters either state $i + 1$ or state $i - 1$ with probabilities $\lambda_i/(\lambda_i + \mu_i)$ and $\mu_i/(\lambda_i + \mu_i)$, respectively. The motion is analogous to that of a random walk except that transitions occur at random times rather than at fixed time periods.

The traditional procedure for constructing birth and death processes is to prescribe the birth and death parameters $\{\lambda_i, \mu_i\}_{i=0}^{\infty}$ and to build the path structure by utilizing the preceding description concerning the waiting times and the conditional transition probabilities of the various states. We determine realizations of the process as follows. Suppose $X(0) = i$; the particle spends a random length of time, exponentially distributed with parameter $(\lambda_i + \mu_i)$, in state i and subsequently moves with probability $\lambda_i/(\lambda_i + \mu_i)$ to state $i + 1$ and with probability $\mu_i/(\lambda_i + \mu_i)$ to state $i - 1$. Next the particle sojourns a random length of time in the new state and then moves to one of its neighboring states, and so on. More specifically, we observe a value t_1 from the exponential distribution with parameter $(\lambda_i + \mu_i)$ that fixes the initial sojourn time in state i. Then we toss a coin with probability of heads $p_i = \lambda_i/(\lambda_i + \mu_i)$. If heads (tails) appears we move the particle to state $i + 1$ $(i - 1)$. In state $i + 1$ we observe a value t_2 from the exponential distribution with parameter $(\lambda_{i+1} + \mu_{i+1})$ that fixes the sojourn time in the second state visited. If the particle at the first transition enters state $i - 1$, the subsequent sojourn time t_2' is an observation from the exponential distribution with parameter $(\lambda_{i-1} + \mu_{i-1})$. After the second wait is completed, a binomial trial is performed that chooses the next state to be visited, and the process continues in the same way.

A typical outcome of these sampling procedures determines a realization of the process. Its form could be

$$
X(t) = \begin{cases}
i, & 0 < t < t_1, \\
i + 1, & t_1 < t < t_1 + t_2, \\
i, & t_1 + t_2 < t < t_1 + t_2 + t_3, \\
\vdots & \\
\vdots &
\end{cases}
$$

Thus by sampling from exponential and binomial distributions appropriately, we construct typical sample paths of the process. Now it is possible to assign to this set of paths (realizations of the process) a probability measure in a consistent way so that $P_{ij}(t)$ is determined satisfying (6.18) and (6.19). This result is rather deep and its rigorous discussion is beyond the level of this book. The process obtained in this manner is called the minimal process associated with the infinitesimal matrix **A** defined in (6.17).

The above construction of the minimal process is fundamental since the infinitesimal parameters need not determine a unique stochastic process obeying (6.18), (6.19), and Postulates 1 through 5 of page 226. In fact there could be several Markov processes that possess the same infinitesimal generator. Fortunately, such complications do not arise in the modeling of common phenomena. In the special case of birth and death processes for which $\lambda_0 > 0$, a sufficient condition that there exists a unique Markov process with transition probability function $P_{ij}(t)$ for which the infinitesimal relations (6.18) and (6.19) hold is that

$$\sum_{n=0}^{\infty} \frac{1}{\lambda_n \theta_n} \sum_{k=0}^{n} \theta_k = \infty, \tag{6.21}$$

where

$$\theta_0 = 1, \qquad \theta_n = \frac{\lambda_0 \lambda_1 \cdots \lambda_{n-1}}{\mu_1 \mu_2 \cdots \mu_n}, \qquad n = 1, 2, \ldots.$$

In most practical examples of birth and death processes the condition (6.21) is met and the birth and death process associated with the prescribed parameters is uniquely determined.

6.3.3 Differential Equations of Birth and Death Processes

As in the case of the pure birth and pure death processes the transition probabilities $P_{ij}(t)$ satisfy a system of differential equations known as the backward Kolmogorov differential equations. These are given by

$$P'_{0j}(t) = -\lambda_0 P_{0j}(t) + \lambda_0 P_{1j}(t), \tag{6.22}$$

$$P'_{ij}(t) = \mu_i P_{i-1,j}(t) - (\lambda_i + \mu_i) P_{ij}(t) + \lambda_i P_{i+1,j}(t), \qquad i \geq 1,$$

and the boundary condition $P_{ij}(0) = \delta_{ij}$.

To derive these we have from Equation (6.19)

$$P_{ij}(t + h) = \sum_{k=0}^{\infty} P_{ik}(h) P_{kj}(t) \tag{6.23}$$

$$= P_{i,i-1}(h) P_{i-1,j}(t) + P_{i,i}(h) P_{ij}(t) + P_{i,i+1}(h) P_{i+1,j}(t)$$

$$+ \Sigma'_k P_{ik}(h) P_{kj}(t),$$

where the last summation is over all $k \neq i - 1, i, i + 1$. Using Postulates 1, 2, and 3 of Section 6.3.1 we obtain

$$\Sigma'_k P_{ik}(h) P_{kj}(t) \leq \Sigma'_k P_{ik}(h)$$

$$= 1 - [P_{i,i}(h) + P_{i,i-1}(h) + P_{i,i+1}(h)]$$

$$= 1 - [1 - (\lambda_i + \mu_i)h + o(h) + \mu_i h + o(h) + \lambda_i h + o(h)]$$

$$= o(h),$$

so that

$$P_{ij}(t + h) = \mu_i h P_{i-1,j}(t) + [1 - (\lambda_i + \mu_i)h] P_{ij}(t) + \lambda_i h P_{i+1,j}(t) + o(h).$$

Transposing the term $P_{ij}(t)$ to the left-hand side and dividing the equation by h, we obtain, after letting $h \downarrow 0$,

$$P'_{ij}(t) = \mu_i P_{i-1,j}(t) - (\lambda_i + \mu_i) P_{ij}(t) + \lambda_i P_{i+1,j}(t).$$

The backward equations are deduced by decomposing the time interval $(0, t + h)$, where h is positive and small, into the two periods

$$(0, h), (h, t + h),$$

and examining the transition in each period separately. In this sense the backward equations result from a "first step analysis," the first step being over the short time interval of duration h.

A different result arises from a "last step analysis" which proceeds by splitting the time interval $(0, t + h)$ into the two periods

$$(0, t), (t, t + h)$$

and adapting the preceding reasoning. From this viewpoint, under more stringent conditions, we can derive a further system of differential equations

$$P'_{i0}(t) = -\lambda_0 P_{i,0}(t) + \mu_1 P_{i,1}(t), \tag{6.24}$$

$$P'_{ij}(t) = \lambda_{j-1} P_{i,j-1}(t) - (\lambda_j + \mu_j) P_{ij}(t) + \mu_{j+1} P_{i,j+1}(t), \qquad j \geq 1,$$

with the same initial condition $P_{ij}(0) = \delta_{ij}$. These are known as the forward Kolmogorov differential equations. To derive these equations we interchange t and h in Equation (6.23), and under stronger assumptions in addition to Postulates 1, 2, and 3 it can be shown that the last term is again $o(h)$. The remainder of the argument is the same as before. The usefulness of the differential equations will become apparent in the examples that we study in the next section.

A sufficient condition that (6.24) hold is that $[P_{kj}(h)]/h = o(1)$ for $k \neq j$, $j - 1, j + 1$ where the $o(1)$ term apart from tending to zero is uniformly bounded with respect to k for fixed j as $h \to 0$. In this case it can be proved that $\sum'_k P_{ik}(t) P_{kj}(h) = o(h)$.

Example *Linear Growth with Immigration*   A birth and death process is called a linear growth process if $\lambda_n = \lambda n + a$ and $\mu_n = \mu n$ with $\lambda > 0$, $\mu > 0$, and $a > 0$. Such processes occur naturally in the study of biological reproduction and population growth. If the state n describes the current population size, then the average instantaneous rate of growth is $\lambda n + a$. Similarly, the probability of the state of the process decreasing by one after the elapse of a small duration h of time is $\mu n h + o(h)$. The factor λn represents the natural growth of the population owing to its current size while the second factor a may be interpreted as the infinitesimal rate of increase of the population due to an external source such as immigration. The component μn which gives the mean infinitesimal death rate of the present population possesses the obvious interpretation.

If we substitute the above values of λ_n and μ_n in (6.24) we obtain

$$P'_{i0}(t) = -a P_{i0}(t) + \mu P_{i1}(t),$$

$$P'_{ij}(t) = [\lambda(j - 1) + a] P_{i,j-1}(t) - [(\lambda + \mu)j + a] P_{ij}(t)$$
$$+ \mu(j + 1) P_{i,j+1}(t), \qquad j \geq 1.$$

Now if we multiply the *j*th equation by *j* and sum, it follows that the expected value

$$E[X(t)] = M(t) = \sum_{j=1}^{\infty} jP_{ij}(t)$$

satisfies the differential equation

$$M'(t) = a + (\lambda - \mu)M(t),$$

with initial condition $M(0) = i$, if $X(0) = i$. The solution of this equation is

$$M(t) = at + i \quad \text{if } \lambda = \mu,$$

and

$$M(t) = \frac{a}{\lambda - \mu} \{e^{(\lambda - \mu)t} - 1\} + ie^{(\lambda - \mu)t} \quad \text{if } \lambda \neq \mu. \qquad (6.25)$$

The second moment or variance may be calculated in a similar way. It is interesting to note that $M(t) \to \infty$ as $t \to \infty$ if $\lambda \geq \mu$, while if $\lambda < \mu$ the mean population size for large *t* is approximately

$$\frac{a}{\mu - \lambda}.$$

These results suggest that in the second case, wherein $\lambda < \mu$, the population stabilizes in the long run in some form of statistical equilibrium. Indeed it can be shown that a limiting probability distribution $\{\pi_j\}$ exists for which $\lim_{t \to \infty} P_{ij}(t) = \pi_j, j = 0, 1, \ldots$. Such limiting distributions for general birth and death processes are the subject of the next section.

Problems 6.3

1. Patients arrive at a hospital emergency room according to a Poisson process of rate λ. The patients are treated by a single doctor on a first come, first served basis. The doctor treats patients more quickly when the number of patients waiting is higher. An industrial engineering time study suggests that the mean patient treatment time when there are *k* patients in the system is of the form $m_k = \alpha - \beta k/(k + 1)$, where α and β are constants with $\alpha > \beta > 0$. Let $N(t)$ be the number of patients in the system at time *t* (waiting and being treated). Argue that $N(t)$ might be modeled as a birth and death process with parameters $\lambda_k = \lambda$ for $k = 0, 1, \ldots$ and $\mu_k = 1/m_k$ for $k = 1, 2, \ldots$. State explicitly any necessary assumptions.

2. Collards were planted equally spaced in a single row in order to provide an experimental setup for observing the chaotic movements of the flea beetle (P. cruciferae). A beetle at position *k* in the row remains on that plant for a random length of time having mean m_k (which varies with

the "quality" of the plant) and then is equally likely to move right $(k + 1)$ or left $(k - 1)$. Model the position of the beetle at time t as a birth and death process having parameters $\lambda_k = \mu_k = 1/(2m_k)$ for $k = 1, 2, \ldots, N - 1$ where the plants are numbered $0, 1, \ldots, N$. What assumptions might be plausible at the ends 0 and N?

6.4 The Limiting Behavior of Birth and Death Processes

For a general birth and death process that has no absorbing states, it can be proved that the limits

$$\lim_{t \to \infty} P_{ij}(t) = \pi_j \geq 0 \qquad (6.26)$$

exist and are independent of the initial state i. It may happen that $\pi_j = 0$ for all states j. When the limits π_j are strictly positive, however, and satisfy

$$\sum_{j=0}^{\infty} \pi_j = 1, \qquad (6.27)$$

they form a probability distribution that is called, naturally enough, the *limiting distribution* of the process. The limiting distribution is also a *stationary distribution* in that

$$\pi_j = \sum_{i=0}^{\infty} \pi_i P_{ij}(t), \qquad (6.28)$$

which tells us that if the process starts in state i with probability π_i, then at any time t it will be in state i with the same probability π_i. The proof of (6.28) follows from (6.19) and (6.26) if we let $t \to \infty$ and use the fact that $\sum_{i=0}^{\infty} \pi_i = 1$.

The general importance of birth and death processes as models derives in large part from the availability of standard formulas for determining if a limiting distribution exists and what its values are when it does. These formulas derive from the forward equations

$$P'_{i,0}(t) = -\lambda_0 P_{i,0}(t) + \mu_1 P_{i,1}(t), \qquad (6.29)$$

$$P'_{i,j}(t) = \lambda_{j-1} P_{i,j-1}(t) - (\lambda_j + \mu_j) P_{ij}(t) + \mu_{j+1} P_{i,j+1}(t), \qquad j \geq 1,$$

with the initial condition $P_{ij}(0) = \delta_{ij}$. Now pass to the limit as $t \to \infty$ in (6.29) and observe first that the limit of the right side of (6.29) exists according to (6.26). Therefore the limit of the left side, the derivatives $P'_{ij}(t)$, exists as well. Since the probabilities are converging to a constant, the limit of these derivatives must be zero. In summary, passing to the limit in (6.29) produces

$$0 = -\lambda_0 \pi_0 + \mu_1 \pi_1$$

$$0 = \lambda_{j-1} \pi_{j-1} - (\lambda_j + \mu_j) \pi_j + \mu_{j+1} \pi_{j+1}, \qquad j \geq 1. \qquad (6.30)$$

The solution to (6.30) is obtained by induction. Letting

$$\theta_0 = 1 \quad \text{and} \quad \theta_j = \frac{\lambda_0 \lambda_1 \cdots \lambda_{j-1}}{\mu_1 \mu_2 \cdots \mu_j} \quad \text{for} \quad j \geq 1, \qquad (6.31)$$

we have $\pi_1 = \lambda_0 \pi_0 / \mu_1 = \theta_1 \pi_0$. Then, assuming that $\pi_k = \theta_k \pi_0$ for $k = 1, \ldots, j$, we obtain

$$
\begin{aligned}
\mu_{j+1} \pi_{j+1} &= (\lambda_j + \mu_j)\theta_j \pi_0 - \lambda_{j-1}\theta_{j-1}\pi_0 \\
&= \lambda_j \theta_j \pi_0 + (\mu_j \theta_j - \lambda_{j-1}\theta_{j-1})\pi_0 \\
&= \lambda_j \theta_j \pi_0,
\end{aligned}
$$

and finally

$$\pi_{j+1} = \theta_{j+1}\pi_0.$$

In order that the sequence $\{\pi_j\}$ define a distribution we must have $\Sigma_j \pi_j = 1$. If $\Sigma \theta_j < \infty$ then we may sum the following

$$
\begin{array}{rcl}
\pi_0 &=& \theta_0 \ \pi_0 \\
\pi_1 &=& \theta_1 \ \pi_0 \\
\pi_2 &=& \theta_2 \ \pi_0 \\
& \vdots & \\
& \vdots & \\
\hline
1 &=& (\Sigma \ \theta_k)\pi_0
\end{array}
$$

to see that $\pi_0 = 1/\Sigma_{k=0}^{\infty} \theta_k$ and then

$$\pi_j = \theta_j \pi_0 = \frac{\theta_j}{\Sigma_{k=0}^{\infty} \theta_k} \quad \text{for} \quad j = 0, 1, \ldots. \qquad (6.32)$$

If $\Sigma \theta_k = \infty$, then necessarily $\pi_0 = 0$, and then $\pi_j = \theta_j \pi_0 = 0$ for all j, and there is no limiting distribution ($\lim_{t \to \infty} P_{ij}(t) = 0$ for all j).

Example *Linear Growth with Immigration* As described on page 230, this process has birth parameters $\lambda_n = a + \lambda n$ and death parameters $\mu_n = \mu n$ for $n = 0, 1, \ldots$ where $\lambda > 0$ is the individual birth rate, $a > 0$ is the rate of immigration into the population, and $\mu > 0$ is the individual death rate.

Suppose $\lambda < \mu$. It was shown in Section 6.3 that the population mean $M(t)$ converges to $a/(\mu - \lambda)$ as $t \to \infty$. Here we will determine the limiting distribution of the process under the same condition $\lambda < \mu$.

Then $\theta_0 = 1$, $\theta_1 = a/\mu$, $\theta_2 = a(a + \lambda)/\mu(2\mu)$, $\theta_3 = a(a + \lambda)(a + 2\lambda)/\mu(2\mu)(3\mu)$ and, in general,

$$\theta_k = \frac{a(a + \lambda) \cdots [a + (k - 1)\lambda]}{\mu^k (k)!}$$

$$= \frac{(a/\lambda)[(a/\lambda) + 1] \cdots [(a/\lambda) + k - 1]}{k!} \left(\frac{\lambda}{\mu}\right)^k$$

$$= \left(\begin{matrix} (a/\lambda) + k - 1 \\ k \end{matrix}\right) \left(\frac{\lambda}{\mu}\right)^k.$$

Now use the infinite binomial formula

$$(1 - x)^{-N} = \sum_{k=0}^{\infty} \left(\begin{matrix} N + k - 1 \\ k \end{matrix}\right) x^k \qquad \text{for} \quad |x| < 1,$$

to determine that

$$\sum_{k=0}^{\infty} \theta_k = \sum_{k=0}^{\infty} \left(\begin{matrix} (a/\lambda) + k - 1 \\ k \end{matrix}\right) \left(\frac{\lambda}{\mu}\right)^k = \left(1 - \frac{\lambda}{\mu}\right)^{-(a/\lambda)}$$

when $\lambda < \mu$. Thus $\pi_0 = (1 - \lambda/\mu)^{a/\lambda}$ and

$$\pi_k = \left(\frac{\lambda}{\mu}\right)^k \frac{(a/\lambda)[(a/\lambda) + 1] \cdots [(a/\lambda) + k - 1]}{k!} (1 - \lambda/\mu)^{a/\lambda}$$

$$\text{for} \quad k \geq 1.$$

Example *Repairman Models* A system is composed of N machines. At most $M \leq N$ can be operating at any one time; the rest are "spares." When a machine is operating, it operates a random length of time until failure. Suppose this failure time is exponentially distributed with parameter μ.

When a machine fails, it undergoes repair. At most R machines can be "in repair" at any one time. The repair time is exponentially distributed with parameter λ. Thus a machine can be in any of four states: (i) operating, (ii) "up," but not operating, i.e., a spare, (iii) in repair, (iv) waiting for repair. There are a total of N machines in the system. At most M can be operating. At most R can be in repair.

The action is diagrammed in Figure 6.5.

Let $X(t)$ be the number of machines "up" at time t, either operating or spare. Then, (we assume) the number operating is $\min\{X(t), M\}$ and the number of spares is $\max\{0, X(t) - M\}$. Let $Y(t) = N - X(t)$ be the number of machines "down." Then the number in repair is $\min\{Y(t), R\}$ and the number waiting for repair is $\max\{0, Y(t) - R\}$. The foregoing formulas permit us to determine the number of machines in any category, once $X(t)$ is known.

Then $X(t)$ is a finite state birth and death process* with parameters

---

*The definition of birth and death processes was given for an infinite number of states. The adjustments in the definitions and analyses for the case of a finite number of states is straightforward and even simpler and left to the reader.

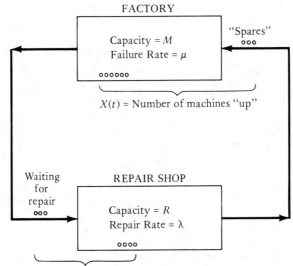

FACTORY

Capacity = M
Failure Rate = μ

"Spares"

$X(t)$ = Number of machines "up"

Waiting for repair

REPAIR SHOP

Capacity = R
Repair Rate = λ

$Y(t)$ = Number of machines "down"

Figure 6.5 Repairman model

$$\lambda_n = \lambda \times \min\{N - n, R\}$$

$$= \begin{cases} \lambda R & \text{for} \quad n = 0, 1, \ldots, N - R, \\ \lambda(N - n) & \text{for} \quad n = N - R + 1, \ldots, N, \end{cases}$$

and

$$\mu_n = \mu \times \min\{n, M\} = \begin{cases} \mu n & \text{for} \quad n = 0, 1, \ldots, M, \\ \mu M & \text{for} \quad n = M + 1, \ldots, N. \end{cases}$$

It is now a routine task to determine the limiting probability distribution for any values of λ, μ, N, M, and R. (See Problems 1 and 2 at the end of this section.) In terms of the limiting probabilities $\pi_0, \pi_1, \ldots, \pi_N$, some quantities of interest are

Average Machines Operating $= \pi_1 + 2\pi_2 + \cdots + M\pi_M$
$$+ M(\pi_{M+1} + \cdots + \pi_N).$$

Long Run Utilization $= \dfrac{\text{Average Machines Operating}}{\text{Capacity}}$

$$= \dfrac{\pi_1 + 2\pi_2 + \cdots + M\pi_M}{M}$$
$$+ (\pi_{M+1} + \cdots + \pi_N)$$

Average Idle Repair Capacity $= 1\pi_{N-R+1} + 2\pi_{N-R+2} + \cdots + R\pi_N.$

These and other similar quantities can be used to evaluate the desirability of adding additional repair capability, additional spare machines, and other possible improvements.

The stationary distribution assumes quite simple forms in certain special cases. For example, consider the special case in which $M = N = R$. The situation arises, for instance, when each machine's operator becomes its repairman upon its failure. Then $\lambda_n = \lambda(N - n)$ and $\mu_n = \mu n$ for $n = 0,$ $1, \ldots, N,$ and following (6.31), we determine $\theta_0 = 1,$ $\theta_1 = \lambda N/\mu,$ $\theta_2 = (\lambda N)\lambda(N - 1)/\mu(2\mu),$ and, in general

$$\theta_k = \frac{N(N-1)\cdots(N-k+1)}{(1)(2)\cdots(k)}\left(\frac{\lambda}{\mu}\right)^k = \binom{N}{k}\left(\frac{\lambda}{\mu}\right)^k.$$

The binomial formula $(1 + x)^N = \sum_{k=0}^{N}\binom{N}{k}x^k$ applies to yield

$$\sum_{k=0}^{N}\theta_k = \sum_{k=0}^{N}\binom{N}{k}\left(\frac{\lambda}{\mu}\right)^k = \left(1 + \frac{\lambda}{\mu}\right)^N.$$

Thus $\pi_0 = [1 + (\lambda/\mu)]^{-N} = [\mu/(\lambda + \mu)]^N$ and

$$\pi_k = \binom{N}{k}\left(\frac{\lambda}{\mu}\right)^k[\mu/(\lambda + \mu)]^N$$

$$= \binom{N}{k}\left(\frac{\lambda}{\lambda + \mu}\right)^k\left(\frac{\mu}{\lambda + \mu}\right)^{N-k}. \tag{6.33}$$

We recognize (6.33) as the familiar binomial distribution.

Example *Logistic Process* Suppose we consider a population whose size $X(t)$ ranges between two fixed integers N and M $(N < M)$ for all $t \geq 0$. We assume that the birth and death rates per individual at time t are given by

$$\lambda = \alpha(M - X(t)) \quad \text{and} \quad \mu = \beta(X(t) - N),$$

and that the individual members of the population act independently of each other. The resulting birth and death rates for the population then become

$$\lambda_n = \alpha n(M - n) \quad \text{and} \quad \mu_n = \beta n(n - N).$$

To see this we observe that if the population size $X(t)$ is n, then each of the n individuals has an infinitesimal birth rate λ so that $\lambda_n = \alpha n(M - n)$. The same rationale applies in the interpretation of the μ_n.

Under such conditions one would expect the process to fluctuate between the two constants N and M, since, for example, if $X(t)$ is near M, the death rate is high and the birth rate low and then $X(t)$ will tend toward N. Ultimately the process should display stationary fluctuations between the two limits N and M.

The stationary distribution in this case is

$$\pi_{N+m} = \frac{c}{N+m}\binom{M-N}{m}\left(\frac{\alpha}{\beta}\right)^m, \qquad m = 0, 1, 2, \ldots, M - N,$$

where c is an appropriate constant determined so that $\Sigma_m \, \pi_{N+m} = 1$. To see this we observe that

$$
\begin{aligned}
\theta_{N+m} \\
&= \frac{\lambda_N \lambda_{N+1} \cdots \lambda_{N+m-1}}{\mu_{N+1}\mu_{N+2} \cdots \mu_{N+m}} \\
&= \frac{\alpha^m N\,(N+1)\cdots(N+m-1)(M-N)\cdots(M-N-m+1)}{\beta^m(N+1)\cdots(N+m)m!} \\
&= \frac{N}{N+m}\binom{M-N}{m}\left(\frac{\alpha}{\beta}\right)^m.
\end{aligned}
$$

Example *Some Genetic Models* Consider a population consisting of N individuals which are either of gene type a or gene type A. The state of the process $X(t)$ represents the number of a-individuals at time t. We assume that the probability that the state changes during the time interval $(t, t + h)$ is $\lambda h + o(h)$ independent of the values of $X(t)$ and that the probability of two or more changes occurring in a time interval h is $o(h)$.

The changes in the population structure are effected as follows. An individual is to be replaced by another chosen randomly from the population; i.e., if $X(t) = j$ then an a-type is selected to be replaced with probability j/N and an A-type with probability $1 - j/N$. We refer to this stage as death. Next, birth takes place by the following rule. Another selection is made randomly from the population to determine the type of the new individual replacing the one that died. The model introduces mutation pressures that admit the possibility that the type of the new individual may be altered upon birth. Specifically, let γ_1 denote the probability that an a-type mutates to an A-type, and let γ_2 denote the probability of an A-type mutating to an a-type.

The probability that the new individual added to the population is of type a is

$$\frac{j}{N}(1 - \gamma_1) + \left(1 - \frac{j}{N}\right)\gamma_2. \tag{6.34}$$

We deduce this formula as follows: The probability that we select an a-type and that no mutation occurs is $(j/N)(1 - \gamma_1)$. Moreover, the final type may be an a-type if we select an A-type that subsequently mutates into an a-type. The probability of this contingency is $(1 - j/N)\gamma_2$. The combination of these two possibilities gives (6.34).

We assert that the conditional probability that $X(t+) - X(t) = 1$, when a change of state occurs, is

$$\left(1 - \frac{j}{N}\right)\left[\frac{j}{N}(1 - \gamma_1) + \left(1 - \frac{j}{N}\right)\gamma_2\right] \qquad \text{where} \quad X(t) = j. \quad (6.35)$$

In fact, the a-type population size can increase only if an A-type dies (is replaced). This probability is $1 - (j/N)$. The second factor is the probability that the new individual is of type a as in (6.34).

In a similar way we find that the conditional probability that $X(t+) - X(t) = -1$ when a change of state occurs is

$$\frac{j}{N}\left[\left(1 - \frac{j}{N}\right)(1 - \gamma_2) + \frac{j}{N}\gamma_1\right] \qquad \text{where} \quad X(t) = j.$$

The stochastic process described is thus a birth and death process with a finite number of states whose infinitesimal birth and death rates are

$$\lambda_j = \lambda\left(1 - \frac{j}{N}\right)\left[\frac{j}{N}(1 - \gamma_1) + \left(1 - \frac{j}{N}\right)\gamma_2\right]$$

and

$$\mu_j = \lambda\frac{j}{N}\left[\frac{j}{N}\gamma_1 + \left(1 - \frac{j}{N}\right)(1 - \gamma_2)\right],$$

respectively corresponding to an a-type population size j, $0 \le j \le N$.

Although these parameters seem rather complicated, it is interesting to see what happens to the stationary measure $\{\pi_k\}_{k=0}^{N}$ if we let the population size $N \to \infty$ and the probabilities of mutation per individual γ_1 and γ_2 tend to zero in such a way that $\gamma_1 N \to \kappa_1$ and $\gamma_2 N \to \kappa_2$, where $0 < \kappa_1, \kappa_2 < \infty$. At the same time we shall transform the state of the process to the interval $[0, 1]$ by defining new states j/N, i.e., the fraction of a-types in the population. To examine the stationary density at a fixed fraction x, where $0 < x < 1$, we shall evaluate π_k as $k \to \infty$ in such a way that $k = [xN]$, where $[xN]$ is the greatest integer less than or equal to xN.

Keeping these relations in mind we write

$$\lambda_j = \frac{\lambda(N - j)}{N^2}(1 - \gamma_1 - \gamma_2)j\left(1 + \frac{a}{j}\right) \qquad \text{where} \quad a = \frac{N\gamma_2}{1 - \gamma_1 - \gamma_2},$$

and

$$\mu_j = \frac{\lambda(N - j)}{N^2}(1 - \gamma_1 - \gamma_2)j\left(1 + \frac{b}{N - j}\right)$$

$$\text{where} \quad b = \frac{N\gamma_1}{1 - \gamma_1 - \gamma_2}.$$

Then

$$\log \theta_k = \sum_{j=0}^{k-1} \log \lambda_j - \sum_{j=1}^{k} \log \mu_j$$

$$= \sum_{j=1}^{k-1} \log\left(1 + \frac{a}{j}\right) - \sum_{j=1}^{k-1} \log\left(1 + \frac{b}{N-j}\right) + \log Na$$

$$- \log(N-k)k\left(1 + \frac{b}{N-k}\right).$$

Now using the expansion

$$\log(1 + x) = x - \frac{x^2}{2} + \frac{x^3}{3} - \cdots, \qquad |x| < 1,$$

it is possible to write

$$\sum_{j=1}^{k-1} \log\left(1 + \frac{a}{j}\right) = a\sum_{j=1}^{k-1} \frac{1}{j} + c_k,$$

where c_k approaches a finite limit as $k \to \infty$. Therefore, using the relation

$$\sum_{j=1}^{k-1} \frac{1}{j} \sim \log k \qquad \text{as} \quad k \to \infty,$$

we have

$$\sum_{j=1}^{k-1} \log\left(1 + \frac{a}{j}\right) \sim \log k^a + c_k \qquad \text{as} \quad k \to \infty.$$

In a similar way we obtain

$$\sum_{j=1}^{k-1} \log\left(1 + \frac{b}{N-j}\right) \sim \log \frac{N^b}{(N-k)^b} + d_k \qquad \text{as} \quad k \to \infty,$$

where d_k approaches a finite limit as $k \to \infty$. Using the above relations we have

$$\log \theta_k \sim \log\left(C_k \frac{k^a(N-k)^b Na}{N^b(N-k)k}\right) \qquad \text{as} \quad k \to \infty, \qquad (6.36)$$

where $\log C_k = c_k + d_k$, which approaches a limit, say C, as $k \to \infty$. Notice that $a \to \kappa_2$ and $b \to \kappa_1$ as $N \to \infty$. Since $k = [Nx]$ we have, for $N \to \infty$,

$$\theta_k \sim C\kappa_2 N^{\kappa_2 - 1} x^{\kappa_2 - 1}(1 - x)^{\kappa_1 - 1}.$$

Now from (6.36) we have

$$\theta_k \sim aC_k k^{a-1}\left(1 - \frac{k}{N}\right)^{b-1}.$$

Therefore

$$\frac{1}{Na}\sum_{k=0}^{N-1}\theta_k \sim \frac{a}{N}\sum_{k=0}^{N-1}C_k\left(\frac{k}{N}\right)^{a-1}\left(1-\frac{k}{N}\right)^{b-1}.$$

Since $C_k \to C$ as k tends to ∞ we recognize the right side as the Riemann sum approximation of

$$\kappa_2 C\int_0^1 x^{\kappa_2-1}(1-x)^{\kappa_1-1}dx.$$

Thus

$$\sum_{i=0}^{N}\theta_i \sim N^{\kappa_2}\kappa_2 C\int_0^1 x^{\kappa_2-1}(1-x)^{\kappa_1-1}dx,$$

so that the resulting density on $[0, 1]$ is

$$\frac{\theta_k}{\Sigma\,\theta_i} \sim \frac{1}{N}\frac{x^{\kappa_2-1}(1-x)^{\kappa_1-1}}{\int_0^1 x^{\kappa_2-1}(1-x)^{\kappa_1-1}dx} = \frac{x^{\kappa_2-1}(1-x)^{\kappa_1-1}dx}{\int_0^1 x^{\kappa_2-1}(1-x)^{\kappa_1-1}dx},$$

since $dx \sim 1/N$. This is a beta distribution with parameters κ_1 and κ_2.

Problems 6.4

1. For the repairman model of the Example on page 234, suppose that $M = N = 5$, $R = 1$, $\lambda = 2$, and $\mu = 1$. Using the limiting distribution for the system, determine
   (a) The average number of machines operating.
   (b) The equipment utilization.
   (c) The average idle repair capacity.
   How do these system performance measures change if a second repairman is added?

2. A system consists of three machines and two repairmen. At most two machines can operate at any time. The amount of time that an operating machine works before breaking down is exponentially distributed with mean 5. The amount of time that it takes a single repairman to fix a machine is exponentially distributed with mean 4. Only one repairman can work on a failed machine at any given time. Let $X(t)$ be the number of machines in operating condition at time t.
   (a) Calculate the long run probability distribution for $X(t)$.
   (b) If an operating machine produces 100 units of output per hour, what is the long run output per hour of the system?

3. Determine the stationary distribution, when it exists, for a birth and death process having constant parameters $\lambda_n = \lambda$ for $n = 0, 1, \ldots$ and $\mu_n = \mu$ for $n = 1, 2, \ldots$.

4. Consider the birth and death parameters $\lambda_n = \theta < 1$, and $\mu_n = n/(n + 1)$ for $n = 0, 1, \ldots$. Determine the stationary distribution.

5. A birth and death process has parameters $\lambda_n = \lambda$ and $\mu_n = n\mu$, for $n = 0, 1, \ldots$. Determine the stationary distribution.

6.5 Birth and Death Processes with Absorbing States

Birth and death processes in which $\lambda_0 = 0$ arise frequently and are correspondingly important. For these processes, the zero state is an absorbing state. A central example is the linear growth birth and death process without immigration (cf. page 230). In this case $\lambda_n = n\lambda$ and $\mu_n = n\mu$. Since growth of the population results exclusively from the existing population, it is clear that when the population size becomes zero it remains zero thereafter, i.e., 0 is an absorbing state.

6.5.1 Probability of Absorption into State 0

It is of interest to compute the probability of absorption into state 0 starting from state i $(i \geq 1)$. This is not, a priori, a certain event since conceivably the particle (i.e., state variable) may wander forever among the states $(1, 2, \ldots)$ or possibly drift to infinity.

Let u_i $(i = 1, 2, \ldots)$ denote the probability of absorption into state 0 from the initial state i. We can write a recursion formula for u_i by considering the possible states after the first transition. We know that the first transition entails the movements

$$i \rightarrow i + 1 \quad \text{with probability} \quad \frac{\lambda_i}{\mu_i + \lambda_i},$$

$$i \rightarrow i - 1 \quad \text{with probability} \quad \frac{\mu_i}{\mu_i + \lambda_i}.$$

Invoking the familiar first step analysis we directly obtain

$$u_i = \frac{\lambda_i}{\mu_i + \lambda_i} u_{i+1} + \frac{\mu_i}{\mu_i + \lambda_i} u_{i-1}, \qquad i \geq 1, \tag{6.37}$$

where $u_0 = 1$.

Another method for deriving (6.37) is to consider the "embedded random walk" associated with a given birth and death process. Specifically we examine the birth and death process only at the transition times. The discrete time Markov chain generated in this manner is denoted by $\{Y_n\}_{n=0}^{\infty}$, where $Y_0 = X_0$ is the initial state and Y_n $(n \geq 1)$ is the state at the nth transition. Obviously, the transition probability matrix has the form

$$\mathbf{P} = \begin{Vmatrix} 1 & 0 & 0 & 0 & \cdot & \cdot & \cdot \\ q_1 & 0 & p_1 & 0 & \cdot & \cdot & \cdot \\ 0 & q_2 & 0 & p_2 & & \cdot & \cdot \\ \cdot & & \cdot & & & & \\ \cdot & & \cdot & & & & \\ \cdot & & \cdot & & & & \end{Vmatrix},$$

where

$$p_i = \frac{\lambda_i}{\lambda_i + \mu_i} = 1 - q_i \qquad \text{for} \quad i \geq 1.$$

The probability of absorption into state 0 for the embedded random walk is the same as for the birth and death process since both processes execute the same transitions. A closely related problem (Gambler's ruin) for a random walk was examined in Section 3.6.1.

We turn to the task of solving (6.37) subject to the conditions $u_0 = 1$ and $0 \leq u_i \leq 1$ ($i \geq 1$). Rewriting (6.37) we have

$$(u_{i+1} - u_i) = \frac{\mu_i}{\lambda_i} (u_i - u_{i-1}), \qquad i \geq 1.$$

Defining $v_i = u_{i+1} - u_i$, we obtain

$$v_i = \frac{\mu_i}{\lambda_i} v_{i-1}, \qquad i \geq 1.$$

Iteration of the last relation yields the formula $v_i = \rho_i v_0$, where

$$\rho_0 = 1 \quad \text{and} \quad \rho_i = \frac{\mu_1 \mu_2 \cdots \mu_i}{\lambda_1 \lambda_2 \cdots \lambda_i} \qquad \text{for} \quad i \geq 1,$$

and with $u_{i+1} - u_i = v_i$, then

$$u_{i+1} - u_i = v_i = \rho_i v_0 = \rho_i(u_1 - u_0) = \rho_i(u_1 - 1) \qquad \text{for} \quad i \geq 1.$$

Summing these last equations from $i = 1$ to $i = m - 1$ we have

$$u_m - u_1 = (u_1 - 1) \sum_{i=1}^{m-1} \rho_i, \qquad m > 1. \tag{6.38}$$

Since u_m, by its very meaning, is bounded by 1 we see that if

$$\sum_{i=1}^{\infty} \rho_i = \infty \tag{6.39}$$

then necessarily $u_1 = 1$ and $u_m = 1$ for all $m \geq 2$. In other words, if (6.39) holds then ultimate absorption into state 0 is certain from any initial state.

Suppose $0 < u_1 < 1$; then, of course,

$$\sum_{i=1}^{\infty} \rho_i < \infty.$$

Obviously, u_m is decreasing in m since passing from state m to state 0 requires entering the intermediate states in the intervening time. Furthermore, it can be shown that $u_m \to 0$ as $m \to \infty$. Now letting $m \to \infty$ in (6.38) permits us to solve for u_1; thus

$$u_1 = \frac{\sum_{i=1}^{\infty} \rho_i}{1 + \sum_{i=1}^{\infty} \rho_i}$$

and then from (6.38) we obtain

$$u_m = \frac{\sum_{i=m}^{\infty} \rho_i}{1 + \sum_{i=1}^{\infty} \rho_i}, \qquad m \ge 1.$$

6.5.2 Mean Time Until Absorption

Consider the problem of determining the mean time until absorption, starting from state m.

We assume that condition (6.39) holds so that absorption is certain. Notice that we cannot reduce our problem to a consideration of the embedded random walk since the actual time spent in each state is relevant for the calculation of the mean absorption time.

Let w_i be the mean absorption time starting from state i (this could be infinite). Considering the possible states following the first transition, instituting a first step analysis, and recalling the fact that the mean waiting time in state i is $(\lambda_i + \mu_i)^{-1}$ (it is actually exponentially distributed with parameter $\lambda_i + \mu_i$), we deduce the recursion relation

$$w_i = \frac{1}{\lambda_i + \mu_i} + \frac{\lambda_i}{\lambda_i + \mu_i} w_{i+1} + \frac{\mu_i}{\lambda_i + \mu_i} w_{i-1}, \qquad i \ge 1, \quad (6.40)$$

and where $w_0 = 0$. Letting $z_i = w_i - w_{i+1}$ and rearranging (6.40) leads to

$$z_i = \frac{1}{\lambda_i} + \frac{\mu_i}{\lambda_i} z_{i-1}, \qquad i \ge 1. \tag{6.41}$$

Iterating this relation gives

$$z_1 = \frac{1}{\lambda_1} + \frac{\mu_1}{\lambda_1} z_0,$$

$$z_2 = \frac{1}{\lambda_2} + \frac{\mu_2}{\lambda_2} z_1 = \frac{1}{\lambda_2} + \frac{\mu_2}{\lambda_2 \lambda_1} + \frac{\mu_2 \mu_1}{\lambda_2 \lambda_1} z_0,$$

$$z_3 = \frac{1}{\lambda_3} + \frac{\mu_3}{\lambda_3 \lambda_2} + \frac{\mu_3 \mu_2}{\lambda_3 \lambda_2 \lambda_1} + \frac{\mu_3 \mu_2 \mu_1}{\lambda_3 \lambda_2 \lambda_1} z_0,$$

and finally

$$z_m = \sum_{i=1}^{m} \frac{1}{\lambda_i} \prod_{j=i+1}^{m} \frac{\mu_j}{\lambda_j} + \left(\prod_{j=1}^{m} \frac{\mu_j}{\lambda_j} \right) z_0.$$

(The product $\prod_{m+1}^{m} \mu_j/\lambda_j$ is interpreted as 1.) Using the notation

$$\rho_0 = 1 \quad \text{and} \quad \rho_i = \frac{\mu_1 \mu_2 \cdots \mu_i}{\lambda_1 \lambda_2 \cdots \lambda_i}, \qquad i \geq 1,$$

the expression for z_m becomes

$$z_m = \sum_{i=1}^{m} \frac{1}{\lambda_i} \frac{\rho_m}{\rho_i} + \rho_m z_0,$$

or, since $z_m = w_m - w_{m+1}$ and $z_0 = w_0 - w_1 = -w_1$, then

$$\frac{1}{\rho_m} (w_m - w_{m+1}) = \sum_{i=1}^{m} \frac{1}{\lambda_i \rho_i} - w_1. \qquad (6.42)$$

If $\sum_{i=1}^{\infty} (1/\lambda_i \rho_i) = \infty$, then inspection of (6.42) reveals that necessarily $w_1 = \infty$. Indeed, it is probabilistically evident that $w_m < w_{m+1}$ for all m and this property would be violated for m large if we assume to the contrary that w_1 is finite.

Now suppose $\sum_{i=1}^{\infty} (1/\lambda_i \rho_i) < \infty$; then letting $m \to \infty$ in (6.42) gives

$$w_1 = \sum_{i=1}^{\infty} \frac{1}{\lambda_i \rho_i} - \lim_{m \to \infty} \frac{1}{\rho_m} (w_m - w_{m+1}).$$

It is more involved but still possible to prove that

$$\lim_{m \to \infty} \frac{1}{\rho_m} (w_m - w_{m+1}) = 0,$$

and then

$$w_1 = \sum_{i=1}^{\infty} \frac{1}{\lambda_i \rho_i}.$$

We summarize the discussion of this section in the following theorem:

Theorem 6.1 *Consider a birth and death process with birth and death parameters λ_n and μ_n, $n \geq 1$, where $\lambda_0 = 0$ so that 0 is an absorbing state.*

The probability of absorption into state 0 from the initial state m is

$$u_m = \begin{cases} \dfrac{\sum_{i=m}^{u} \rho_i}{1 + \sum_{i=1}^{u} \rho_i} & \text{if } \sum_{i=1}^{\infty} \rho_i < \infty, \\[4mm] 1 & \text{if } \sum_{i=1}^{\infty} \rho_i = \infty. \end{cases} \qquad (6.43)$$

The mean time to absorption is

$$
w_m = \begin{cases} \infty & \text{if } \sum_{i=1}^{\infty} \dfrac{1}{\lambda_i \rho_i} = \infty, \\[4mm] \sum_{i=1}^{\infty} \dfrac{1}{\lambda_i \rho_i} + \sum_{k=1}^{m-1} \rho_k \sum_{j=k+1}^{\infty} \dfrac{1}{\lambda_j \rho_j} & \text{if } \sum_{i=1}^{\infty} \dfrac{1}{\lambda_i \rho_i} < \infty, \end{cases} \tag{6.44}
$$

where $\rho_0 = 1$ *and* $\rho_i = (\mu_1 \mu_2 \ldots \mu_i)/(\lambda_1 \lambda_2 \ldots \lambda_i).$

Example *Population Processes* Consider the linear growth birth and death process without immigration (cf. page 230) for which $\mu_n = n\mu$ and $\lambda_n = n\lambda$, $n = 0, 1, \ldots$. During a short time interval of length h, a *single individual* in the population dies with probability $\mu h + o(h)$ and gives birth to a new individual with probability $\lambda h + o(h)$, and thus $\mu > 0$ and $\lambda > 0$ represent the *individual* death and birth rates, respectively.

Substitution of $a = 0$ and $i = m$ in Equation (6.25) determines the mean population size at time t for a population starting with $X(0) = m$ individuals. This mean population size is $M(t) = me^{(\lambda - \mu)t}$, exhibiting exponential growth or decay according as $\lambda > \mu$ or $\lambda < \mu$.

Let us now examine the extinction phenomenon and determine the probability that the population eventually dies out. This phenomenon corresponds to absorption in state 0 for the birth and death process.

When $\lambda_n = n\lambda$ and $\mu_n = n\mu$, a direct calculation yields $\rho_i = (\mu/\lambda)^i$ and then

$$
\sum_{i=m}^{\infty} \rho_i = \sum_{i=m}^{\infty} (\mu/\lambda)^i = \begin{cases} \dfrac{(\mu/\lambda)^m}{1 - (\mu/\lambda)} & \text{when } \lambda > \mu, \\[4mm] \infty & \text{when } \lambda \leq \mu. \end{cases}
$$

From Theorem 6.1, the probability of eventual extinction starting with m individuals is

$$
\Pr\{\text{Extinction}|X(0) = m\} = \begin{cases} (\mu/\lambda)^m & \text{when } \lambda > \mu, \\[2mm] 1 & \text{when } \lambda \leq \mu. \end{cases} \tag{6.45}
$$

When $\lambda = \mu$, the process is sure to vanish eventually. Yet in this case the mean population size remains constant at the initial population level. Similar situations where mean values do not adequately describe population behavior frequently arise when stochastic elements are present.

We turn attention to the mean time to extinction assuming extinction is certain, that is, when $\lambda \leq \mu$. For a population starting with a single individual, then, from (6.44) with $m = 1$ we determine this mean time to be

$$\sum_{i=1}^{\infty} \frac{1}{\lambda_i \rho_i} = \frac{1}{\lambda} \sum_{i=1}^{\infty} \frac{1}{i} \left(\frac{\lambda}{\mu}\right)^i$$

$$= \frac{1}{\lambda} \sum_{i=1}^{\infty} \int_0^{(\lambda/\mu)} x^{i-1} dx$$

$$= \frac{1}{\lambda} \int_0^{(\lambda/\mu)} \sum_{i=1}^{\infty} x^{i-1} dx$$

$$= \frac{1}{\lambda} \int_0^{(\lambda/\mu)} \frac{dx}{(1-x)}$$

$$= -\frac{1}{\lambda} \ln(1-x) \Big|_0^{(\lambda/\mu)}$$

$$= \begin{cases} \dfrac{1}{\lambda} \ln\left(\dfrac{\mu}{\mu - \lambda}\right) & \text{when } \mu > \lambda, \\[2mm] \infty & \text{when } \mu = \lambda. \end{cases} \qquad (6.46)$$

When the birth rate λ exceeds the death rate μ, a linear growth birth and death process can, with strictly positive probability, grow without limit. In contrast, many natural populations exhibit density dependent behavior wherein the individual birth rates decrease or the individual death rates increase or both changes occur as the population grows. These changes are ascribed to factors including limited food supplies, increased predation, crowding, and limiting nesting sites. Accordingly, we introduce a notion of environmental *carrying capacity K,* an upper bound that the population size cannot exceed.

Since all individuals have a chance of dying, with a finite carrying capacity, all populations will eventually become extinct. Our measure of population fitness will be the mean time to extinction, and it is of interest to population ecologists studying colonization phenomena to examine how the capacity K, the birth rate λ, and the death rate μ affect this mean population lifetime.

The model should have the properties of exponential growth (on the average) for small populations, as well as the ceiling K beyond which the population cannot grow. There are several ways of approaching the population size K and staying there at equilibrium. Since all such models give more or less the same qualitative results, we stipulate the simplest model in which the birth parameters are

$$\lambda_n = \begin{cases} n\lambda & \text{for } n = 0, 1, \ldots, K-1 \\[2mm] 0 & \text{for } n \geq K. \end{cases}$$

Theorem 6.1 yields w_1, the mean time to population extinction starting with a single individual, as given by

$$w_1 = \sum_{i=1}^{\infty} \frac{1}{\lambda_i \rho_i} = \sum_{i=1}^{\infty} \frac{\lambda_1 \lambda_2 \cdots \lambda_{i-1}}{\mu_1 \mu_2 \cdots \mu_i} = \frac{1}{\mu} \sum_{i=1}^{K} \frac{1}{i} \left(\frac{\lambda}{\mu}\right)^{i-1}. \tag{6.47}$$

Equation (6.47) isolates the distinct factors influencing the mean time to population extinction. The first factor is $1/\mu$, the mean lifetime of an individual since μ is the individual death rate. Thus, the sum in (6.47) represents the mean *generations* or mean lifespans to population extinction, a dimensionless quantity which we denote by

$$M_g = \mu w_1 = \sum_{i=1}^{K} \frac{1}{i} \theta^{i-1} \qquad \text{where} \quad \theta = \frac{\lambda}{\mu}. \tag{6.48}$$

Next we examine the influence of the birth-death or reproduction ratio $\theta = \lambda/\mu$ and the carrying capacity K on the mean time to extinction. Since λ represents the individual birth rate and $1/\mu$ is the mean life of a single member in the population, we may interpret the reproduction ratio $\theta = \lambda(1/\mu)$ as the mean number of offspring of an arbitrary individual in the population. Accordingly, we might expect significantly different behavior when $\theta < 1$ as opposed to when $\theta > 1$, and this is indeed the case. A carrying capacity of $K = 100$ is small. When K is of the order of 100 or more, we have the following accurate approximations, their derivations being sketched in Problems 1 and 2 at the end of this section:

$$M_g \cong \begin{cases} \frac{1}{\theta} \ln\left(\frac{1}{1-\theta}\right) & \text{for} \quad \theta < 1, \\ .5772157 + \ln K & \text{for} \quad \theta = 1, \\ \frac{1}{K}\left(\frac{\theta^K}{\theta - 1}\right) & \text{for} \quad \theta > 1. \end{cases} \tag{6.49}$$

The contrast between $\theta < 1$ and $\theta > 1$ is vivid. When $\theta < 1$, the mean generations to extinction M_g is almost independent of the carrying capacity K and approaches the asymptotic value $\theta^{-1} \ln(1 - \theta)^{-1}$ quite rapidly. When $\theta > 1$, the mean generations to extinction M_g grows exponentially in K. Some calculations based on (6.49) are given in Table 6.1.

Table 6.1 **Mean generations to extinction for a population starting with a single parent and where θ is the reproduction rate and K is the environmental capacity.**

K	$\theta = .8$	$\theta = 1$	$\theta = 1.2$
10	1.96	2.88	3.10
100	2.01	5.18	4140899
1000	2.01	7.48	7.59×10^{76}

Example *Sterile Male Insect Control*   The screwworm fly, a cattle pest in warm climates, was eliminated from the southeastern United States by the release into the environment of sterilized adult male screwworm flies. When these males, artificially sterilized by radiation, mate with native females, there are no offspring, and in this manner part of the reproductive capacity of the natural population is nullified by their presence. If the sterile males are sufficiently plentiful so as to cause even a small decline in the population level, then this decline accelerates in succeeding generations even if the number of sterile males is maintained at approximately the same level, because the ratio of sterile to fertile males will increase as the natural population drops. Because of this compounding effect, if the sterile male control method works at all, it works to such an extent as to drive the native population to extinction in the area in which it is applied.

Recently, a multibillion dollar effort involving the sterile male technique has been proposed for the control of the cotton boll weevil. In this instance, it was felt that a pretreatment with a pesticide could reduce the natural population size to a level such that the sterile male technique would become effective. Let us examine this assumption, first with a deterministic model, and then in a stochastic setting.

For both models we suppose that sexes are present in equal numbers, that sterile and fertile males are equally competitive, and that a constant number S of sterile males is present in each generation. In the deterministic case, if N_0 fertile males are in the parent generation and the N_0 fertile females choose mates equally likely from the entire male population, then the fraction $N_0/(N_0 + S)$ of these matings will be with fertile males and will produce offspring. Letting θ denote the number of offspring of either sex in a fertile mating, we calculate the size N, of the next generation according to

$$N_1 = \theta N_0 \left(\frac{N_0}{N_0 + S} \right). \tag{6.50}$$

For a numerical example, suppose that there are $N_0 = 100$ fertile males and an equal number of fertile females in the parent generation of the native population, and that $S = 100$ sterile male insects are released. If $\theta = 4$, meaning that a fertile mating produces four males and four females for the succeeding generation, then the number of either sex in the first generation is

$$N_1 = 4(100) \left(\frac{100}{100 + 100} \right) = 200;$$

the population has increased and the sterile male control method has failed.

On the other hand, if a pesticide can be used to reduce the initial population size to $N_0 = 20$, or 20 percent of its former level, and $S = 100$ sterile males are released, then

Table 6.2 **The trend of an insect population subject to sterile male releases.**

Generation	Number of Insects Natural Population	Number of Sterile Insects	Ratio Sterile to Fertile	Number of Progeny
Parent	20	100	5:1	13.33
F_1	13.33	100	7.5:1	6.27
F_2	6.27	100	16:1	1.48
F_3	1.48	100	67.5:1	.09
F_4	.09	100	1156:1	—

$$N_1 = 4(20)\left(\frac{20}{20 + 100}\right) = 13.33$$

and the population is declining. The succeeding population sizes are given in Table 6.2. With the pretreatment, the population becomes extinct by the fourth generation.

Often deterministic or average value models will adequately describe the evolution of large populations. But extinction is a small population phenomenon, and even in the presence of significant long term trends, small populations are strongly influenced by the chance fluctuations that determine which of extinction or recolonization will occur. This fact motivates us to examine a stochastic model of the evolution of a population in the presence of sterile males. The factors in our model are

λ, the individual birth rate;

μ, the individual death rate;

$\theta = \lambda/\mu$, the mean offspring per individual;

K, the carrying capacity of the environment;

S, the constant number of sterile males in the population; and

m, the initial population size.

We assume that both sexes are present in equal numbers in the natural population, and that $X(t)$, the number of either sex present at time t evolves as a birth and death process with parameters

$$\lambda_n = \begin{cases} \lambda n\left(\dfrac{n}{n + S}\right) & \text{if } 0 \leq n < K, \\ 0 & \text{for } n \geq K, \end{cases}$$

and   (6.51)

$$\mu_n = \mu n \quad \text{for } n = 0, 1, \ldots .$$

This is the colonization model of the "Population Processes" Example, modified in analogy with (6.50) by including in the birth rate, the factor $n/(n + S)$ to represent the probability that a given mating will be fertile.

To calculate the mean time to extinction w_m as given in (6.44), we first use (6.51) to determine

$$\rho_k = \frac{\mu_1 \mu_2 \cdots \mu_k}{\lambda_1 \lambda_2 \cdots \lambda_k} = \left(\frac{\mu}{\lambda}\right)^k \frac{(k + S)!}{k! S!} \qquad \text{for} \quad k = 1, \ldots, K - 1,$$

$$\rho_0 = 1 \quad \text{and} \quad \rho_K = \infty \quad \text{or} \quad 1/\rho_K = 0,$$

and then substitute these expressions for ρ_k into (6.44) to obtain

$$\begin{aligned}
w_m &= \sum_{j=1}^{K} \frac{1}{\lambda_j \rho_j} + \sum_{k=1}^{m-1} \rho_k \sum_{j=k+1}^{K} \frac{1}{\lambda_j \rho_j} \\
&= \sum_{k=0}^{m-1} \rho_k \sum_{j=k+1}^{K} \frac{1}{\lambda_j \rho_j} = \sum_{k=0}^{m-1} \rho_k \sum_{j=k+1}^{K} \frac{1}{\mu_j \rho_{j-1}} \\
&= \frac{1}{\mu} \left\{ \sum_{k=0}^{m-1} \sum_{j=k}^{K-1} \frac{1}{j + 1} \; \theta^{j-k} \frac{j!(S + k)!}{k!(S + j)!} \right\}.
\end{aligned} \qquad (6.52)$$

Because of the factorials, Equation (6.52) presents numerical difficulties when direct computations are attempted. A simple iterative scheme works to provide accurate and effective computation, however. We let

$$\alpha_k = \sum_{j=k}^{K-1} \frac{1}{j + 1} \; \theta^{j-k} \frac{j!(S + k)!}{k!(S + j)!}$$

so that $w_m = (\alpha_0 + \ldots + \alpha_{m-1})/\mu$. But it is easily verified that

$$\alpha_{k-1} = \frac{1}{k} + \theta \left(\frac{k}{S + k}\right) \alpha_k.$$

Beginning with $\alpha_K = 0$, one successively computes $\alpha_{K-1}, \alpha_{K-2}, \ldots,$ α_0, and then $w_m = (\alpha_0 + \ldots + \alpha_{m-1})/\mu$.

Using this method, we have computed the mean generations to extinction in the stochastic model for comparison with the deterministic model as given in Table 6.2. Table 6.3 lists the mean generations to extinction for various initial population sizes m when $K = S = 100$, $\lambda = 4$, and $\mu = 1$ so that $\theta = 4$. Instead of the four generations to extinction as predicted by the deterministic model when $m = 20$, we now estimate that the population will persist for over 8 billion generations!

What is the explanation for the dramatic difference between the predictions of the deterministic model and the predictions of the stochastic model? The stochastic model allows the small but positive probability that the population will not die out but will recolonize and return to a higher level near the environmental capacity K, and then persist for an enormous length of time.

Table 6.3 **The mean
lifespans to extinction in a
birth and death model of a
population containing a
constant number $S = 100$ of
sterile males.**

Initial Population Size	Mean Lifespans to Extinction
20	8,101,227,748
10	4,306,531
5	3,822
4	566
3	65
2	6.3
1	1.2

While both models are qualitative, the practical implications cannot be dismissed. In any large scale control effort, a wide range of habitats and microenvironments is bound to be encountered. The stochastic model suggests the likely possibility that some subpopulation in some pocket might persist and later recolonize the entire area. A sterile male program that depends on a pretreatment with an insecticide for its success is chancy at best.

Problems 6.5

1. Assuming $\theta < 1$, verify the following steps in the approximation to M_g, the mean generation to extinction as given in (6.48):

$$M_g = \sum_{i=1}^{K} \frac{1}{i} \theta^{i-1} = \theta^{-1} \sum_{i=1}^{K} \int_0^\theta x^{i-1} dx$$

$$= \theta^{-1} \int_0^\theta \frac{1 - x^K}{1 - x} dx = \theta^{-1} \int_0^\theta \frac{dx}{1 - x} - \theta^{-1} \int_0^\theta \frac{x^K}{1 - x} dx$$

$$= \frac{1}{\theta} \ln \frac{1}{1 - \theta} - \theta^{-1} \int_0^\theta x^K (1 + x + x^2 + \cdots) dx$$

$$= \frac{1}{\theta} \ln \frac{1}{1 - \theta} - \frac{1}{\theta} \left(\frac{\theta^{K+1}}{K + 1} + \frac{\theta^{K+2}}{K + 2} + \cdots \right)$$

$$= \frac{1}{\theta} \ln \frac{1}{1 - \theta} - \frac{\theta^K}{K + 1} \left(1 + \frac{K + 1}{K + 2} \theta + \frac{K + 1}{K + 3} \theta^2 + \cdots \right)$$

$$\cong \frac{1}{\theta} \ln \frac{1}{1 - \theta} - \frac{\theta^K}{(K + 1)(1 - \theta)}.$$

2. Assume that $\theta > 1$ and verify the following steps in the approximation to M_g, the mean generation to extinction as given in (6.48):

$$M_g = \sum_{i=1}^{K} \frac{1}{i} \theta^{i-1} = \theta^K \sum_{i=1}^{K} \frac{1}{i} \theta^{K-i+1}$$

$$= \theta^K \sum_{j=1}^{K} \frac{1}{K - j + 1} \left(\frac{1}{\theta} \right)^j$$

$$= \frac{\theta^{K-1}}{K} \left[1 + \frac{K}{K - 1} \left(\frac{1}{\theta} \right) + \frac{K}{K - 2} \left(\frac{1}{\theta} \right)^2 + \cdots + \frac{K}{1} \left(\frac{1}{\theta} \right)^{K-1} \right]$$

$$\cong \frac{\theta^{K-1}}{K} \left[\frac{1}{1 - (1/\theta)} \right] = \frac{\theta^K}{K(\theta - 1)}.$$

3. Consider the sterile male control model as described in the Example entitled "Sterile Male Insect Control" and let u_m be the probability that the population becomes extinct before growing to size K starting with $X(0) = m$ individuals. Show that

$$u_m = \frac{\sum_{i=0}^{m-1} \rho_i}{\sum_{i=0}^{K-1} \rho_i} \quad \text{for} \quad m = 1, \ldots, K,$$

where $\rho_i = \theta^{-i} \dfrac{(S + i)!}{i! S!}$.

4. Consider a birth and death process on the states $0, 1, \ldots, 5$ with parameters

$$\lambda_0 = \mu_0 = \lambda_5 = \mu_5 = 0$$

$$\lambda_1 = 1, \quad \lambda_2 = 2, \quad \lambda_3 = 3, \quad \lambda_4 = 4$$

$$\mu_1 = 4, \quad \mu_2 = 3, \quad \mu_3 = 2, \quad \mu_2 = 1.$$

Note that 0 and 5 are absorbing states. Suppose the process begins in state $X(0) = 2$.

(a) What is the probability of eventual absorption in state 0?

(b) What is the mean time to absorption?

6.6 Finite State Continuous Time Markov Chains

A continuous time Markov chain $X(t)$ $(t > 0)$ is a Markov process on the states 0, 1, 2, ... We assume as usual that the transition probabilities are stationary, i.e.,

$$P_{ij}(t) = \Pr\{X(t + s) = j \mid X(s) = i\}. \tag{6.53}$$

In this section we consider only the case where the state space S is finite, labeled as $\{0, 1, 2, . . . , N\}$.

The Markov property asserts that $P_{ij}(t)$ satisfies

(a) $P_{ij}(t) \geq 0,$

(b) $\sum_{j=0}^{N} P_{ij}(t) = 1,$ $i, j = 0, 1, . . . , N,$ and

(c) $P_{ik}(s + t) = \sum_{j=0}^{N} P_{ij}(s) P_{jk}(t)$ for $t, s \geq 0$
(Chapman-Kolmogorov relation),

and we postulate in addition that

(d) $\lim_{t \to 0+} P_{ij}(t) = \begin{cases} 1, & i = j, \\ 0, & i \neq j. \end{cases}$

If $\mathbf{P}(t)$ denotes the matrix $\|P_{ij}(t)\|_{i,j=0}^{N}$ then property (c) can be written compactly in matrix notation as

$$\mathbf{P}(t + s) = \mathbf{P}(t)\mathbf{P}(s), \qquad t, s \geq 0. \tag{6.54}$$

Property (d) asserts that $\mathbf{P}(t)$ is continuous at $t = 0$ since the fact $\mathbf{P}(0) = \mathbf{I}$ (= identity matrix) is implied by (6.54). It follows simply from (6.54) that $\mathbf{P}(t)$ is continuous for all $t > 0$. In fact if $s = h > 0$ in (6.54), then because of (d) we have

$$\lim_{h \to 0+} \mathbf{P}(t + h) = \mathbf{P}(t) \lim_{h \to 0+} \mathbf{P}(h) = \mathbf{P}(t)\mathbf{I} = \mathbf{P}(t). \tag{6.55}$$

On the other hand, for $t > 0$ and $0 < h < t$ we write (6.54) in the form

$$\mathbf{P}(t) = \mathbf{P}(t - h)\mathbf{P}(h). \tag{6.56}$$

But $\mathbf{P}(h)$ is near the identity when h is sufficiently small and so $\mathbf{P}(h)^{-1}$ [the inverse of $\mathbf{P}(h)$] exists and also approaches the identity $\mathbf{I}$. Therefore

$$\mathbf{P}(t) = \mathbf{P}(t) \lim_{h \to 0+} (\mathbf{P}(h))^{-1} = \lim_{h \to 0+} \mathbf{P}(t - h). \tag{6.57}$$

The limit relations (6.55) and (6.57) together show that $\mathbf{P}(t)$ is continuous.

Actually, $\mathbf{P}(t)$ is not only continuous but also differentiable in that the limits

$$\lim_{h \to 0+} \frac{1 - P_{ii}(h)}{h} = q_i,$$

$$\lim_{h \to 0+} \frac{P_{ij}(h)}{h} = q_{ij}, \qquad i \neq j$$

(6.58)

exist, where $0 \leq q_{ij} < \infty$ $(i \neq j)$ and $0 \leq q_i < \infty$. Starting with the relation

$$1 = P_{ii}(h) + \sum_{j=0, j \neq i}^{N} P_{ij}(h),$$

dividing by h, and letting h decrease to zero yields directly the relation

$$q_i = \sum_{j=0, j \neq i}^{N} q_{ij}.$$

The rates q_i and q_{ij} furnish an infinitesimal description of the process with

$$\Pr\{X(t + h) = j | X(t) = i\} = q_{ij}h + o(h) \qquad \text{for} \quad i \neq j,$$

$$\Pr\{X(t + h) = i | X(t) = i\} = 1 - q_i h + o(h).$$

In contrast to the infinitesimal description, the sojourn description of the process proceeds as follows: Starting in state i, the process sojourns there for a duration that is exponentially distributed with parameter q_i. The process then jumps to state $j \neq i$ with probability $p_{ij} = q_{ij}/q_i$; the sojourn time in state j is exponentially distributed with parameter q_j, and so on. The sequence of states visited by the process, denoted by $\xi_0, \xi_1, \ldots$, is a Markov chain with discrete parameter, called the *embedded Markov chain*. Conditioned on the state sequence $\xi_0, \xi_1, \ldots$, the successive sojourn times $S_0, S_1, \ldots$ are independent exponentially distributed random variables with parameters $q_{\xi_0}, q_{\xi_1}, \ldots$, respectively.

Assuming that (6.58) has been verified we now derive an explicit expression for $P_{ij}(t)$ in terms of the infinitesimal matrix

$$\mathbf{A} = \begin{Vmatrix} -q_0 & q_{01} & \cdots & q_{0N} \\ q_{10} & -q_1 & & q_{1N} \\ \cdot & & & \\ \cdot & & & \\ \cdot & & & \\ q_{N0} & q_{N1} & \cdots & -q_N \end{Vmatrix}.$$

The limit relations (6.58) can be expressed concisely in matrix form:

$$\lim_{h \to 0+} \frac{\mathbf{P}(h) - \mathbf{I}}{h} = \mathbf{A}, \qquad (6.59)$$

which shows that $\mathbf{A}$ is the matrix derivative of $\mathbf{P}(t)$ at $t = 0$. Formally, $\mathbf{A} = \mathbf{P}'(0)$.

With the aid of (6.59) and referring to (6.54) we have

$$\frac{\mathbf{P}(t + h) - \mathbf{P}(t)}{h} = \frac{\mathbf{P}(t)[\mathbf{P}(h) - \mathbf{I}]}{h} = \frac{\mathbf{P}(h) - \mathbf{I}}{h} \mathbf{P}(t). \qquad (6.60)$$

The limit on the right exists and this leads to the matrix differential equation

$$\mathbf{P}'(t) = \mathbf{P}(t)\mathbf{A} = \mathbf{A}\mathbf{P}(t), \qquad (6.61)$$

where $\mathbf{P}'(t)$ denotes the matrix whose elements are $P'_{ij}(t) = dP_{ij}(t)/dt$. The existence of $P'_{ij}(t)$ is an obvious consequence of (6.59) and (6.60).

Example *The Two State Markov Chain* Consider a Markov chain $\{X(t)\}$ with states $\{0, 1\}$ whose infinitesimal matrix is

$$\mathbf{A} = \begin{matrix} & 0 & 1 \\ 0 & \\ 1 & \end{matrix} \begin{Vmatrix} -\lambda & \lambda \\ \mu & -\mu \end{Vmatrix}$$

The process alternates between states 0 and 1. The sojourn times in state 0 are independent and exponentially distributed with parameter λ. Those in state 1 are independent and exponentially distributed with parameter μ.

In the special case, the matrix differential equation (6.61) becomes

$$\begin{Vmatrix} P'_{00}(t) & P'_{01}(t) \\ P'_{10}(t) & P'_{11}(t) \end{Vmatrix} = \begin{Vmatrix} P_{00}(t) & P_{01}(t) \\ P_{10}(t) & P_{11}(t) \end{Vmatrix} \times \begin{Vmatrix} -\lambda & \lambda \\ \mu & -\mu \end{Vmatrix},$$

the first element of which is

$$P'_{00}(t) = -\lambda P_{00}(t) + \mu P_{01}(t). \qquad (6.62)$$

Now $P_{01}(t) = 1 - P_{00}(t)$, which placed in (6.62) gives

$$P'_{00}(t) = \mu - (\lambda + \mu)P_{00}(t).$$

Let $Q_{00}(t) = e^{(\lambda+\mu)t}P_{00}(t)$. Then

$$\begin{aligned} \frac{dQ_{00}(t)}{dt} &= e^{(\lambda+\mu)t}P'_{00}(t) + (\lambda + \mu)e^{(\lambda+\mu)t}P_{00}(t) \\ &= e^{(\lambda+\mu)t}[P'_{00}(t) + (\lambda + \mu)P_{00}(t)] \\ &= \mu e^{(\lambda+\mu)t} \end{aligned}$$

which can be integrated immediately to yield

$$\begin{aligned} Q_{00}(t) &= \mu \int e^{(\lambda+\mu)t}dt + C \\ &= \left(\frac{\mu}{\lambda + \mu}\right)e^{(\lambda+\mu)t} + C. \end{aligned}$$

The initial condition $Q_{00}(0) = 1$ determines the constant of integration to be $C = \lambda/(\lambda + \mu)$. Thus

$$Q_{00}(t) = e^{(\lambda+\mu)t}P_{00}(t) = \left(\frac{\mu}{\lambda + \mu}\right)e^{(\lambda+\mu)t} + \left(\frac{\lambda}{\lambda + \mu}\right)$$

and

$$P_{00}(t) = \frac{\mu}{\lambda + \mu} + \frac{\lambda}{\lambda + \mu} e^{-(\lambda+\mu)t}. \tag{6.63}$$

Since $P_{01}(t) = 1 - P_{00}(t)$, we have

$$P_{01}(t) = \frac{\lambda}{\lambda + \mu} - \frac{\lambda}{\lambda + \mu} e^{-(\lambda+\mu)t}, \tag{6.64}$$

and, by symmetry,

$$P_{11}(t) = \frac{\lambda}{\lambda + \mu} + \frac{\mu}{\lambda + \mu} e^{-(\lambda+\mu)t}, \tag{6.65}$$

$$P_{10}(t) = \frac{\mu}{\lambda + \mu} - \frac{\mu}{\lambda + \mu} e^{-(\lambda+\mu)t} \tag{6.66}$$

Returning to the general Markov chain on states $\{0, 1, \ldots, N\}$, the differential equations (6.61) under the initial condition $\mathbf{P}(0) = \mathbf{I}$ can be solved by standard methods to yield the formula

$$\mathbf{P}(t) = e^{\mathbf{A}t} = \mathbf{I} + \sum_{n=1}^{\infty} \frac{\mathbf{A}^n t^n}{n!}. \tag{6.67}$$

When the Markov chain is irreducible (all states communicate) then $P_{ij}(t) > 0$ for $i, j = 0, 1, \ldots, N$ and $\lim_{t\to\infty} P_{ij}(t) = \pi_j > 0$ exists independently of the initial state i. The limiting distribution may be found by passing to the limit in (6.61), noting that $\lim_{t\to\infty} \mathbf{P}'(t) = 0$. The resulting equations for $\boldsymbol{\pi} = (\pi_0, \pi_1, \ldots, \pi_N)$ are

$$0 = \boldsymbol{\pi}\mathbf{A} = (\pi_0, \pi_1, \ldots, \pi_N) \begin{Vmatrix} -q_0 & q_{01} & \cdots & q_{0N} \\ q_{10} & -q_1 & \cdots & q_{1N} \\ \cdot & \cdot & & \cdot \\ \cdot & \cdot & & \cdot \\ \cdot & \cdot & & \cdot \\ q_{N0} & q_{N1} & \cdots & -q_N \end{Vmatrix},$$

which is the same as

$$\pi_j q_j = \sum_{i \neq j} \pi_i q_{ij}, \qquad j = 0, 1, \ldots, N. \tag{6.68}$$

Equation (6.68) together with

$$\pi_0 + \pi_1 + \cdots + \pi_N = 1 \tag{6.69}$$

determines the limiting distribution.

Equation (6.68) has a mass balance interpretation that aids us in understanding it. The left side $\pi_j q_j$ represents the long run rate at which particles executing the Markov process leave state j. This rate must equal the long run rate at which particles arrive at state j if equilibrium is to be maintained. Such arriving particles must come from some state $i \neq j$, and a particle moves from state $i \neq j$ to state j at rate q_{ij}. Therefore, the right side $\Sigma_{i \neq j} \, \pi_i q_{ij}$ represents the total rate of arriving particles.

Example   *Industrial Mobility and the Peter Principle*   Let us suppose that a draftsman position at a large engineering firm can be occupied by a worker at any of three levels: T = Trainee, J = Junior Draftsman, and S = Senior Draftsman. Let $X(t)$ denote the level of the person in the position at time t, and suppose that $X(t)$ evolves as a Markov chain whose infinitesimal matrix is

$$
\mathbf{A} =
\begin{array}{c}
T \\ J \\ S
\end{array}
\begin{array}{c}
\begin{array}{ccc}
T & J & S
\end{array} \\
\left\|
\begin{array}{ccc}
-a_T & a_T & 0 \\
a_{JT} & -a_J & a_{JS} \\
a_S & 0 & -a_S
\end{array}
\right\|
\end{array}
$$

Thus a Trainee stays at that rank for an exponentially distributed time having parameter a_T and then becomes a Junior Draftsman. A Junior Draftsman stays at that level for an exponentially distributed length of time having parameter $a_J = a_{JT} + a_{JS}$. Then the Junior Draftsman leaves the position and is replaced by a Trainee with probability a_{JT}/a_J, or is promoted to a Senior Draftsman with probability a_{JS}/a_J, and so on.

Alternatively, we may describe the model by specifying the movements during short time intervals according to

$$\Pr\{X(t + h) = J | X(t) = T\} = a_T h + o(h),$$

$$\Pr\{X(t + h) = T | X(t) = J\} = a_{JT} h + o(h),$$

$$\Pr\{X(t + h) = S | X(t) = J\} = a_{JS} h + o(h),$$

$$\Pr\{X(t + h) = T | X(0) = S\} = a_S h + o(h),$$

and

$$\Pr\{X(t + h) = i | X(t) = i\} = 1 - a_i h + o(h) \qquad \text{for} \quad i = T, J, S.$$

The equations for the equilibrium distribution (π_T, π_J, π_S) are, according to (6.68),

$$a_T \pi_T = a_{JT} \pi_J + a_S \pi_S$$

$$a_J \pi_J = a_T \pi_T$$

$$a_S \pi_S = a_{JS} \pi_J,$$

$$1 = \pi_T + \pi_J + \pi_S,$$

and the solution is

$$\pi_T = \frac{a_S a_J}{a_S a_J + a_S a_T + a_T a_{JS}}$$

$$\pi_J = \frac{a_S a_T}{a_S a_J + a_S a_T + a_T a_{JS}}$$

$$\pi_S = \frac{a_T a_{JS}}{a_S a_J + a_S a_T + a_T a_{JS}}.$$

Let us consider a numerical example for comparison with an alternative model to be developed later. We suppose that the mean time in the three states are

State	Mean Time
T	.1
J	.2
S	1.0

and that a Junior Draftsman leaves and is replaced by a trainee with probability $\frac{2}{5}$ and is promoted to a Senior Draftsman with probability $\frac{3}{5}$. These suppositions lead to the prescription $a_T = 10$, $a_{JT} = 2$, $a_{JS} = 3$, $a_S = 1$. The equilibrium probabilities are

$$\pi_T = \frac{1(5)}{1(5) + 1(10) + 10(3)} = \frac{5}{45} = .11,$$

$$\pi_J = \frac{10}{45} = .22,$$

and

$$\pi_S = \frac{30}{45} = .67.$$

But the duration that people spend in any given position is not exponentially distributed in general. A bimodal distribution is often observed in which many people leave rather quickly, while others persist for a substantial time. A possible explanation for this phenomenon is found in the "Peter Principle" which asserts that a worker is promoted until first reaching a position in which he or she is incompetent. When this happens, the worker stays in that job until retirement. Let us modify the industrial mobility model to accommodate the Peter Principle by considering two types of Junior Draftsmen, *Competent* and *Incompetent*. We suppose that a fraction p of trainees are Competent, and $q = 1 - p$ are Incompetent. We assume that a competent Junior Draftsman stays at that level for an exponentially distributed duration with parameter a_C and then is promoted to Senior Draftsman. Finally, an incompetent Junior Draftsman stays in that position until retirement, an exponentially distributed sojourn with parameter a_I, and then he or she is replaced by a trainee. The relevant infinitesimal matrix is given by

$$
\mathbf{A} = \begin{array}{c} \\ T \\ I \\ C \\ S \end{array}
\begin{array}{c} \\ \left\| \begin{array}{cccc} & & & \\ -a_T & qa_T & pa_T & \\ a_I & -a_I & & \\ & & -a_C & a_C \\ a_S & & & -a_S \end{array} \right\| \end{array}
$$

with column labels $T \quad I \quad C \quad S$.

The duration in the Junior Draftsman position now follows a probability law that is a mixture of exponential densities. To compare this model with the previous model, suppose that $p = \frac{3}{5}$, $q = \frac{2}{5}$, $a_I = 2.86$, and $a_C = 10$. These numbers were chosen so as to make the mean duration as a Junior Draftsman,

$$
p\left(\frac{1}{a_C}\right) + q\left(\frac{1}{a_I}\right) = (\tfrac{3}{5})(.10) + (\tfrac{2}{5})(.35) = .20,
$$

the same as in the previous calculations. The probability density of this duration is

$$
f(t) = \tfrac{3}{5}(10)e^{-10t} + \tfrac{2}{5}(2.86)e^{-2.86t} \qquad \text{for} \quad t \geq 0.
$$

This density is plotted in Figure 6.6 for comparison with the exponential density $g(t) = 5e^{-5t}$, which has the same mean. The bimodal tendency is indicated in that $f(t) > g(t)$ when t is near zero and when t is very large.

With the numbers as given and $a_T = 10$ and $a_S = 1$ as before, the stationary distribution $(\pi_T, \pi_I, \pi_C, \pi_S)$ is found by solving

$$
\begin{aligned}
10\pi_T &= 2.86\pi_I \quad\quad\quad 1\pi_S \\
2.86\pi_I &= 4\pi_T \\
10\pi_C &= 6\pi_T \\
1\pi_S &= 10\pi_C \\
1 &= \pi_T + \pi_I + \pi_C + \pi_S.
\end{aligned}
$$

The solution is

$$
\pi_T = .111, \qquad \pi_I = .155,
$$
$$
\pi_S = .667, \qquad \pi_C = .067.
$$

Let us make two observations before leaving this example. First, the limiting probabilities π_T, π_S, and $\pi_J = \pi_I + \pi_C$ agree between the two

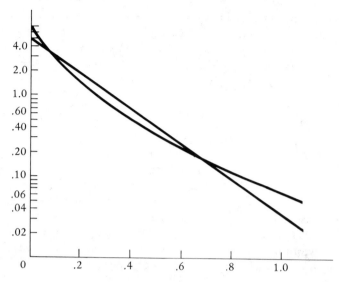

Figure 6.6 The exponential density (straight line) versus the mixed exponential density (curved line). Both distributions have the same mean. A logarithmic scale was used to accentuate the differences.

models. This is a common occurrence in stochastic modeling wherein the limiting behavior of a process is rather insensitive to certain details of the model and depends only on the first moments or means. When this happens, the model assumptions can be chosen for their mathematical convenience with no loss.

The second observation is specific to the Peter Principle. We have assumed that $p = \frac{3}{5}$ of Trainees are Competent Junior Draftsmen and only $q = \frac{2}{5}$ are Incompetent. Yet in the long run, a Junior Draftsman is found to be Incompetent with probability $\pi_I/(\pi_I + \pi_C) = .155/(.155 + .067) = .70!$

Example   *Redundancy and the Burn-in Phenomenon*   An airlines reservation system has two computers, one on-line and one backup. The operating computer fails after an exponentially distributed duration having parameter μ and is replaced by the standby. There is one repair facility and repair times are exponentially distributed with parameter λ. Let $X(t)$ be the number of computers in operating condition at time t. Then $X(t)$ is a Markov chain whose infinitesimal matrix is

$$
\mathbf{A} = \begin{array}{c} 0 \\ 1 \\ 2 \end{array} \left\| \begin{array}{ccc} 0 & 1 & 2 \\ -\lambda & \lambda & 0 \\ \mu & -(\lambda + \mu) & \lambda \\ 0 & \mu & -\mu \end{array} \right\|.
$$

The stationary distribution (π_0, π_1, π_2) satisfies

$$\lambda\pi_0 = \mu\pi_1$$
$$(\lambda + \mu)\pi_1 = \lambda\pi_0 + \mu\pi_2$$
$$\mu\pi_2 = \lambda\pi_1$$
$$1 = \pi_0 + \pi_1 + \pi_2,$$

and the solution is

$$\pi_0 = \frac{1}{1 + (\lambda/\mu) + (\lambda/\mu)^2},$$

$$\pi_1 = \frac{\lambda/\mu}{1 + (\lambda/\mu) + (\lambda/\mu)^2},$$

$$\pi_2 = \frac{(\lambda/\mu)^2}{1 + (\lambda/\mu) + (\lambda/\mu)^2}.$$

The reliability, or probability that at least one computer is operating, is $1 - \pi_0 = \pi_1 + \pi_2$.

Often in practice the assumption of exponentially distributed operating times is not realistic because of the so-called *burn-in phenomenon*. This idea is best explained in terms of the *hazard* rate $r(t)$ associated with a probability density function $f(t)$ of a nonnegative failure time T. Recall that $r(t)\Delta t$ measures the conditional probability that the item fails in the next time interval $(t, t + \Delta t)$ given that it has survived up to time t, and therefore we have

$$r(t) = \frac{f(t)}{1 - F(t)} \qquad \text{for} \quad t \geq 0$$

where $F(t)$ is the cumulative distribution function associated with the probability density function $f(t)$.

A constant hazard rate $r(t) = \lambda$ for all t corresponds to the exponential density function $f(t) = \lambda e^{-\lambda t}$ for $t \geq 0$. The burn-in phenomenon is described by a hazard rate that is initially high and then decays to a constant level, where it persists, possibly later to rise again (aging). It corresponds to a situation in which a newly manufactured or newly repaired item has a significant probability of failing early in its use. If the item survives this test period, however, it then operates in an exponential or memoryless manner. The early failures might correspond to incorrect manufacture or faulty repair, or might be a property of the materials used.

Anyone familiar with automobile repairs has experienced the burn-in phenomenon.

One of many possible ways to model the burn-in phenomenon is to use a mixture of exponential densities

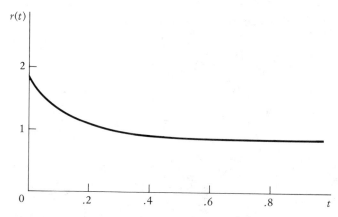

Figure 6.7 The hazard rate corresponding to the density given in (6.70). The higher hazard rate at the initial t values represents the burn-in phenomenon.

$$f(t) = p\alpha e^{-\alpha t} + q\beta e^{-\beta t}, \qquad t \geq 0 \qquad (6.70)$$

where $0 < p = 1 - q < 1$ and α, β are positive. The density function for which $p = .1$, $\alpha = 10$, $q = .9$, and $\beta = .909 \ldots = 1/1.1$ has mean one. Its hazard rate is plotted in Figure 6.7 where the burn-in higher initial level is evident.

We may incorporate the burn-in phenomenon corresponding to the mixed exponential density (6.70) by expanding the state space according to the following table:

Notation	State
0	Both computers down
1_A	One operating computer, current up time has parameter α
1_B	One operating computer, current up time has parameter β
2_A	Two operating computers, current up time has parameter α
2_B	Two operating computers, current up time has parameter β

Equation (6.70) corresponds to a probability p that a computer beginning operation will have an exponentially distributed up time with parameter α, and a probability q that the parameter is β. Accordingly we have the infinitesimal matrix

$$
\mathbf{A} =
\begin{array}{c}
0 \\ 1_A \\ 1_B \\ 2_A \\ 2_B
\end{array}
\begin{array}{ccccc}
\quad 0 \quad & \quad 1_A \quad & \quad 1_B \quad & \quad 2_A \quad & \quad 2_B \quad \\
-\lambda & p\lambda & q\lambda & & \\
\alpha & -(\lambda + \alpha) & & \lambda & \\
\beta & & -(\lambda + \beta) & & \lambda \\
& p\alpha & q\alpha & -\alpha & \\
& p\beta & q\beta & & -\beta
\end{array}.
$$

The stationary distribution can be determined in the usual way by applying (6.69).

Problems 6.6

1. Let $X_1(t)$ and $X_2(t)$ be independent two state Markov chains having the same infinitesimal matrix

$$
\mathbf{A} =
\begin{array}{c}
0 \\ 1
\end{array}
\begin{array}{cc}
\quad 0 \quad & \quad 1 \quad \\
-\lambda & \lambda \\
\mu & -\mu
\end{array}
$$

   Argue that $Z(t) = X_1(t) + X_2(t)$ is a Markov chain on the state space $S = \{0, 1, 2\}$ and determine the transition probability matrix $\mathbf{P}(t)$ for $Z(t)$.

2. Let $X_1(t), X_2(t), \ldots, X_N(t)$ be independent two state Markov chains having the same infinitesimal matrix

$$
\mathbf{A} =
\begin{array}{c}
0 \\ 1
\end{array}
\begin{array}{cc}
\quad 0 \quad & \quad 1 \quad \\
-\lambda & \lambda \\
\mu & -\mu
\end{array}
$$

   Determine the infinitesimal matrix for the Markov chain $Z(t) = X_1(t) + \ldots + X_N(t)$.

3. A system consists of two units, both of which may operate simultaneously, and a single repair facility. The probability that an operating system will fail in a short time interval of length Δt is $\mu(\Delta t) + o(\Delta t)$. Repair times are exponentially distributed, but the parameter depends on whether the failure was *regular* or *severe*. The fraction of regular failures is p and corresponding exponential parameter is α. The fraction of severe failure is $q = 1 - p$ and the exponential parameter is $\beta < \alpha$.

   Model the system as a continuous time Markov chain by taking as states the pairs (x, y) where $x = 0, 1, 2$ is the number of units operating and $y = 0, 1, 2$ is the number of units undergoing repair for a severe failure. The possible states are $(2, 0), (1, 0), (1, 1), (0, 0), (0, 1),$

and $(0, 2)$. Specify the infinitesimal matrix $\mathbf{A}$. Assume that the units enter the repair shop on a first come, first served basis.

6.7 Set Valued Processes★

In physics, engineering, sociology, and biology, Markov processes arise whose values are subsets of a given finite set. In a typical application in physics, the process value $X(t)$ at time t may indicate the set of particles possessing a certain magnetism or "spin." In sociology, the process may track the set of people-pairs having a specified relation. These processes are finite state Markov chains, and in this sense, their theory was given in the previous section. The special nature of these set valued processes, however, often allows special techniques to be brought to bear. We illustrate with a pure death process having distinguishable organisms.

To recapitulate Section 6.2, a pure death process starting at N is a Markov process $\{X(t); t \geq 0\}$ on the state space $\{0, 1, \ldots, N\}$ whose transitions are always to the next lower state. The process is described by the death parameters $\mu_1, \mu_2, \ldots, \mu_N$. In applications where $X(t)$ is the number of living organisms in a population at time t, then conditional on $X(t) = k$, a death occurs during the interval $(t, t + \Delta t]$ with probability $\mu_k \Delta t + o(\Delta t)$, and otherwise the population remains at the same level. Death processes always traverse the same sequence of states, eventually being absorbed in the state 0. The process is completely specified by the values of the N exponentially distributed sojourn times before extinction.

Now we wish to model a population in which the death rates depend not only on the number of living organisms, but on exactly which organisms are alive. Accordingly, we define a death process with N distinguishable organisms as a Markov process $\{X(t); t \geq 0\}$ on the state space Ξ whose elements consist of all subsets of $N = \{1, 2, \ldots, N\}$, and whose probability law allows only jumps to smaller adjacent states. A typical sample path is depicted in Figure 6.8.

Let A, B, etc. denote subsets of $\{1, 2, \ldots, N\}$, or possible states of the process. The infinitesimal probability rates are denoted $q(A, B)$ with the interpretation

$$\Pr\{X(t + h) = B | X(t) = A\} = q(A, B)h + o(h) \qquad \text{for} \quad A \neq B. \qquad (6.71)$$

For the pure death process with distinguishable organisms, the infinitesimal elements satisfy

---

★This section contains material of a more difficult level. It is not prerequisite to what follows.

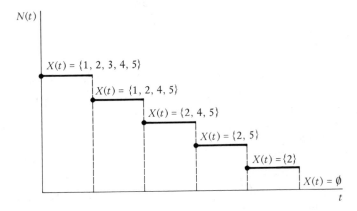

Figure 6.8 A pure death process with distinguished organisms. In the sample depicted, $N = 5$ and the organisms "die" in the following sequence: 3, 1, 4, 5, 2.

$$q(A, B) \geq 0 \quad \text{if} \quad B \subset A \quad \text{and} \quad {}^{\#}(A\backslash B) = 1,$$

$$q(A, A) = -\sum_{B}\{q(A, B); B \subset A, {}^{\#}(A\backslash B) = 1\}$$

and

$$q(A, B) = 0, \quad \text{otherwise.}$$

Here we use the notation ${}^{\#}(A\backslash B)$ for the number of elements in A but not in B. We assume that $q(A, A) < 0$ for all nonempty sets A. Thus the only absorbing state for $\{X(t)\}$ is the empty set $\emptyset$. It is convenient to write $q(A) = -q(A, A) > 0$.

In many applications, the state $X(t)$ identifies the set of surviving organisms at time t. If $A = B \cup \{j\}$ with j not an element of B, then $q(A, B)$ represents the death rate of organism j in the subset A.

To view the process from a different perspective, partition the state space according to

$$\Xi = \Xi_0 \cup \Xi_1 \cup \cdots \cup \Xi_N,$$

where Ξ_k contains exactly those subsets of $\{1, 2, \ldots, N\}$ having k elements. The number of sets in Ξ_k is ${}^{\#}(\Xi_k) = \binom{N}{k}$, where $\binom{N}{k} = N!/k!(N - k)!$. The process begins in the single state in Ξ_N, i.e., in $A_N = \{1, 2, \ldots, N\}$. Its sojourn time there is exponentially distributed with parameter $q(A_N)$. It then moves, independently of the sojourn time in A_N, to one of the N states in Ξ_{N-1}, occupying A_{N-1} in Ξ_{N-1} with probability $p(A_N, A_{N-1}) = q(A_N, A_{N-1})/q(A_N)$, and there its sojourn time is exponentially distributed with parameter $q(A_{N-1})$, independent of the past history of the process. It moves to a state A_{N-2} in Ξ_{N-2} with probability $p(A_{N-1}, A_{N-2}) = $

$q(A_{N-1})$, $q(A_{N-1}, A_{N-2})/q(A_{N-1})$, and so on, ultimately arriving in the unique state $A_0 = \emptyset$ in Ξ_0.

A *path* from A_i in Ξ_i to A_j in Ξ_j, with $i \geq j$, is an ordered $(i - j + 1)$-tuple $(A_i, A_{i-1}, \ldots, A_j)$ of states A_k in Ξ_k for $k = j, j + 1, \ldots, i$, such that $A_{k+1} \supset A_k$ for $k = j, j + 1, \ldots, i - 1$. From our description of the Markov process, the probability that the sequence of states visited by $\{X(t)\}$ is a particular path $(A_N, A_{N-1}, \ldots, A_0)$ is given by $p(A_N, A_{N-1}, \ldots, A_0) = p(A_N, A_{N-1})p(A_{N-1}, A_{N-2}) \cdots p(A_1, A_0)$ where $p(A, B) = q(A, B)/q(A)$. Given such a path, the successive sojourn times in the states, denoted by $S(A_N), S(A_{N-1}), \ldots, S(A_1)$, are conditionally independent random variables and are exponentially distributed with parameters $q(A_N), q(A_{N-1}), \ldots, q(A_1)$, respectively. As already mentioned, $A_0 = \emptyset$ is an absorbing state.

The preceding description furnishes many formulas for a number of quantities of interest. For example, let ν represent the mean time to population extinction. Along the path $A_N, A_{N-1}, \ldots, A_1, A_0$ this mean time is $q(A_N)^{-1} + q(A_{N-1})^{-1} + \ldots + q(A_1)^{-1}$. Since this path has probability $p(A_N, A_{N-1})p(A_{N-1}, A_{N-2}) \cdots p(A_1, A_0)$, where $p(A, B) = q(A, B)/q(A)$, we obtain

$$\nu = \sum_{\text{All paths}} p(A_N, A_{N-1}) \cdots p(A_1, A_0) \left[\frac{1}{q(A_N)} + \cdots + \frac{1}{q(A_1)} \right]. \quad (6.72)$$

While explicit, this formula quickly becomes infeasible because of the large number $N!$ of paths. The modern trend in such situations is to replace the explicit formula with algorithms suitable for digital computation. Accordingly, let $\nu(A)$ be the mean time to extinction starting from state $A \subset \{1, 2, \ldots, N\}$. Let A be a state in Ξ_k ($^\#(A) = k$) and suppose we have determined $\nu(B)$ for all states B in Ξ_{k-1}. A first step analysis readily yields the following recursion:

$$\nu(A) = \frac{1}{q(A)} + \sum_{B \text{ in } \Xi_{k-1}} \frac{q(A, B)}{q(A)} \nu(B). \quad (6.73)$$

The convention $\nu(A_0) = \nu(\emptyset) = 0$ for $A_0 = \emptyset$ in Ξ_0 starts the recursion. To determine $\nu = \nu(A_N)$, one such computation as in (6.73) is done for each of the $2^N - 1$ nonempty states in the system. This is a considerable reduction over the $N!$ terms in the sum given in (6.72).

A similar, but more complex, recursion can be developed for determining the exact distribution function of the time to extinction. We defer its presentation until after the following example, at which point the algorithm can be demonstrated numerically.

Example *A Fiber Bundle with Local Load Sharing* Section 6.2.2 was a model of a cable or bundle comprised of several fibers arranged in parallel and supporting a tensile load. The model assumed that a single nonfailed fiber, carrying load $l(t)$ at time t, fails in the interval $(t, t + h]$ with probability $K[l(t)]h + o(h)$ where $K(l) = l^\beta/A$ for positive constants β and A which are properties of the fiber material and size.

An important assumption in the earlier model was that all unfailed fibers shared the total bundle load NL equally. If at time t there were k surviving fibers, then each carried a load of NL/k. Since all unfailed fibers carried identical loads, it was possible to model the system as a pure death process.

Where interfiber friction is present, equal load sharing may not be a valid assumption. Where fibers are slightly twisted, or embedded in a ductile matrix thereby forming a composite system, friction tends to concentrate the effect of a break onto the fibers nearest to the broken element. In other words, the equal load sharing assumption is replaced by a local load sharing assumption which assigns different loads to nonfailed fibers depending on their proximity to broken fibers.

We introduce a rather severe form of stress concentration called *local load sharing* (LLS). According to this rule, the N fibers are viewed as being arranged in a circle. Whenever a fiber fails, it shifts its load to the nearest unfailed fibers on either side. Specifically, if an unbroken fiber is surrounded by r broken fibers, then the load on the fiber is given by

$$l = \begin{cases} (1 + r/2)L & \text{if } r = 0, 1, \ldots, N - 2, \\ NL & \text{if } r = N - 1, \end{cases}$$

where the total bundle load is NL.

Figure 6.9 shows all distinct configurations of five fibers. (All other configurations are rotations or reflections of these.)

An unfailed fiber carrying load l fails in an interval of length h with probability $K(l)h + o(h)$, where $K(l) = l^\beta/A$. Suppose $A = B \cup \{j\}$ where j is an element not in B. If r denotes the number of failed fibers adjacent to j in A, then j carries load $(1 + r/2)L$ and thus has failure rate $K[(1 + r/2)L]$. This, of course, equals $q(A, B)$. A sample is given in Figure 6.10.

Figure 6.11 gives all distinct sets of survivors together with the transition rates between them.

Referring to Table 6.4 and the algorithm represented by Equation (6.73), we determine the mean time to extinction. Starting from state 7, this mean time to extinction is exactly the mean sojourn time in that state, or $v(7) = 1/3125 = .0003$. State 6 leads only to state 7, whence $v(6) = 1/195.3125 + v(7) = .0054$, and similarly for state 5, $v(5) = .0054$. State 4 leads to state 5 or 6, whence $v(4) = 1/47.1875 + .6781\, v(5) + .3219\, v(6) = .0266$. For state 3, $v(3) = .9846\, v(5) + .0154\, v(6) = .0208$. State 2 leads to 3 and 4, whence $v(2) = 1/17.1875 + .8836\, v(3) + .1164\, v(4) = .1777$. Since state 1 always leads to state 2, $v(1) = 1/5 + v(2) = .3777$. This completes the computation.

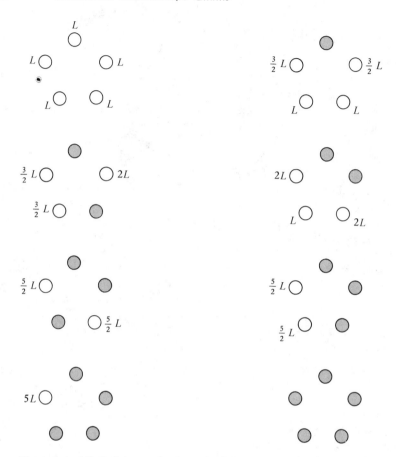

Figure 6.9 All distinct configurations of five fibers are given, and the loads carried by unfailed fibers under local load sharing are shown.

A more complex version of the same idea yields the exact cumulative distribution function for the time to extinction. We begin by describing the general algorithm for the distribution function of the time to traverse a network.

For a state A, a subset of $\{1, 2, \ldots, N\}$, let F_A be the distribution function of the time to extinction starting from A, and let G be the exponential distribution function with parameter $q(A)$. Using the independence of the sojourn time in A and the time to extinction after leaving A, we obtain

$$F_A = G \star \Sigma\{p(A, B)F_B; \quad B \subset A, {}^\#(A\backslash B) = 1\}, \qquad (6.74)$$

where $\star$ denotes convolution and $p(A, B) = q(A, B)/q(A)$.

We impose the condition that the sojourn parameters $q(A_N)$, $q(A_{N-1})$,

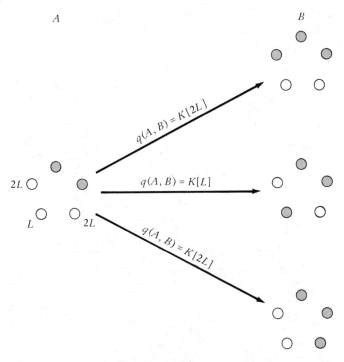

Figure 6.10   The transition rates from a set A to each of three possible successor sets for the fiber bundle model under local load sharing.

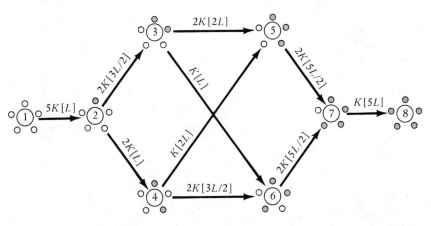

Figure 6.11   The transition rates between distinct sets of survivors in a fiber bundle model with local load sharing.

Table 6.4 **The sojourn parameters and transition probabilities for the set valued Markov process depicted in Figure 6.11.**

From State	Sojourn Parameter	To State	Rate $q(A, B)$	Probability $q(A, B)/q(A)$
1	$\lambda_1 = 5$	2	5	1.
2	$\lambda_2 = 17.1875$	3	15.1875	.8836
2		4	2	.1164
3	$\lambda_3 = 65$	5	64	.9846
3		6	1	.0154
4	$\lambda_4 = 47.1875$	5	32	.6781
4		6	15.1875	.3219
5	$\lambda_5 = 195.3125$	7	195.3125	1.
6	$\lambda_5 = 195.3125$	7	195.3125	1.
7	$\lambda_6 = 3125$	8	3125	1.

. . . , $q(A_1)$ along any path $A_N, A_{N-1}, \ldots, A_1$ are distinct, no two being the same. With this condition, we will show that the distribution function for the time to extinction from any state has a certain parametric form, and then (6.74) leads to a recursion that determines the coefficients in the parametric form.

Consider the set of sojourn parameters $q(A)$ for A, a nonempty subset of $\{1, 2, \ldots, N\}$. Suppose there are M distinct elements in this set and let $\lambda_1, \lambda_2, \ldots, \lambda_M$ denote these values in some order. For any state A, then F_A is a mixture of exponential functions with parameters in the set of λ_i's. The distribution function F_A has the form

$$F_A(t) = 1 - \sum_{m=1}^{M} c_{Am} e^{-\lambda_m t} \qquad \text{for} \quad t \geq 0, \qquad (6.75)$$

for some set of constants $c_{A1}, \ldots, c_{AM}$, some of which may be zero. Suppose that A is in Ξ_j, for j in $\{1, 2, \ldots, N\}$, and suppose that we have determined the distributions of time to extinction from all B in Ξ_{j-1}, i.e., the set of constants $\{c_{Bm}; B \in \Xi_{j-1}, m = 1, 2, \ldots, M\}$ is known. Let i be the index such that $q(A) = \lambda_i$, and note that $c_{Bi} = 0$ for all B in Ξ_{j-1} with $B \subset A$, since λ_i cannot appear again on a path from A. Then for $t \geq 0$, we obtain from (6.74)

$$F_A(t) = \Sigma\{[q(A, B)/q(A)](F_B \star G)(t); B \subset A, {}^{\#}(A\backslash B) = 1\}.$$

For such a state B, we have

$$(F_B \star G)(t) = \int_0^t \left\{1 - \sum_{\substack{m=1 \\ m \neq i}}^M c_{Bm} e^{-\lambda_m(t-y)}\right\} \lambda_i e^{-\lambda_i y} dy$$

$$= 1 - \sum_{\substack{m=1 \\ m \neq i}}^M c_{Bm}\left(\frac{\lambda_i}{\lambda_i - \lambda_m}\right) e^{-\lambda_m t}$$

$$- \left\{1 - \sum_{\substack{m=1 \\ m \neq i}}^M c_{Bm}\left(\frac{\lambda_i}{\lambda_i - \lambda_m}\right)\right\} e^{-\lambda_i t}.$$

Weighting this by $p(A, B) = q(A, B)/q(A)$ and summing over B, we can determine the constants $\{c_{Am}; m = 1, 2, \ldots, M\}$ in terms of $\{c_{Bm}; B \in \Xi_{j-1}, m = 1, 2, \ldots, M\}$, which gives us the following recursion:

Step 1 For each A in Ξ_1, if $q(A) = \lambda_i$, set $c_{Ai} = 1$, and $c_{Am} = 0$ for $m \neq i$. Then set $j = 2$.

Step 2 For each A in Ξ_j, if $q(A) = \lambda_i$, set

$$c_{Am} = \left\{\frac{1}{\lambda_i - \lambda_m}\right\}\Sigma\{c_{Bm}q(A, B); B \subset A, {}^{\#}(A\backslash B) = 1\} \qquad \text{for} \quad m \neq i,$$

and

$$c_{Ai} = 1 - \sum_{\substack{m=1 \\ m \neq i}}^M \left\{\frac{1}{\lambda_i - \lambda_m}\right\}\Sigma\{c_{Bm}q(A, B); B \subset A, {}^{\#}(A\backslash B) = 1\},$$

$$= 1 - \sum_{\substack{m=1 \\ m \neq i}}^M c_{Am}.$$

If $j = N$, STOP. Otherwise, add 1 to j, and return to Step 2.

We illustrate the procedure with the data given in Figure 6.11 and Table 6.4. Beginning with state 7, and $\lambda_6 = 3125$ and $c_{76} = 1$, we next consider states 5 and 6, which have the same data: $\lambda_5 = 195.3125$. According to Step 2 of the algorithm,

$$c_{66} = \frac{1}{\lambda_5 - \lambda_6}[c_{76}q(6, 7)] = \frac{1}{195.3125 - 3125}(195.3125)$$

$$= -0.06667$$

and

$$c_{65} = 1 - c_{66} = 1.06667.$$

At this stage we have determined that the cumulative distribution function of the time to extinction starting from state 6 is

$$F_6(t) = 1 - 1.06667e^{-195.3125t} + 0.06667e^{-3125t}, \qquad t \geq 0.$$

Since state 5 behaves exactly like state 6, then $c_{55} = 1.06667$ and $c_{56} = -0.06667$.

We continue:

$$c_{46} = \frac{1}{\lambda_4 - \lambda_6} [c_{56}q(4, 5) + c_{66}q(4, 6)]$$

$$= \frac{1}{47.1875 - 3125} [(-0.06667)(32) + (-0.06667)(15.1875)]$$

$$= 0.001022;$$

$$c_{45} = \frac{1}{\lambda_4 - \lambda_5} [c_{55}q(4, 5) + c_{65}q(4, 6)]$$

$$= \frac{1}{47.1875 - 195.3125} [(1.06667)(32) + (1.06667)(15.1875)]$$

$$= -0.339803;$$

$$c_{44} = 1 - c_{45} - c_{46} = 1.338781;$$

$$c_{36} = \frac{1}{\lambda_3 - \lambda_6} [c_{56}q(3, 5) + c_{66}q(3, 6)]$$

$$= \frac{1}{65 - 3125} [(-0.06667)(64) + (-0.06667)(1)]$$

$$= 0.001416;$$

$$c_{35} = \frac{1}{\lambda_3 - \lambda_5} [c_{55}q(3, 5) + c_{65}q(3, 6)]$$

$$= \frac{1}{65 - 195.3125} [(1.06667)(64) + (1.06667)(1)]$$

$$= -0.532054;$$

$$c_{33} = 1 - c_{36} - c_{35} = 1.530638.$$

Continuing these computations leads to the coefficients in Table 6.5. From this table we determine that the distribution function of the time to extinction starting with five individuals is

$$F_1(t) = 1 - 0.000000e^{-3125t} + 0.001292e^{-195.3125t}$$

$$- 0.010578e^{-47.1875t} - 0.040517e^{-65t}$$

$$+ 0.626167e^{-17.1875t} - 1.576364e^{-5t}.$$

Table 6.5 The coefficients c_{Am} in the distribution function $F_A(t)$ for the time to system failure starting from state A. The formula is $F_A(t) = \Sigma_m c_{Am} \exp\{-\lambda_m t\}$.

		m					
		1	2	3	4	5	6
	λ_m	5	17.1875	65	47.1875	195.3125	3125
A	1	1.576364	−.626167	.040517	.010578	−.001292	.000000
	2		1.526282	−.486220	−.089252	.049180	−.000008
	3			1.530638		−.532054	.001416
	4				1.338781	−.339803	.001022
	5					1.066667	−.066667
	6					1.066667	−.066667
	7						1.000000

Chapter 7 | Renewal Phenomena

7.1 Definition of a Renewal Process and Related Concepts

Renewal theory began with the study of stochastic systems whose evolution through time was interspersed with renewals or regeneration times when, in a statistical sense, the process began anew. Today, the subject is viewed as the study of general functions of independent, identically distributed, nonnegative random variables representing the successive intervals between renewals. The results are applicable in a wide variety of both theoretical and practical probability models.

A *renewal (counting) process* $\{N(t), t \geq 0\}$ is a nonnegative integer-valued stochastic process that registers the successive occurrences of an event during the time interval $(0, t]$, where the time durations between consecutive events are *positive, independent, identically distributed* random variables. Let the successive occurrence times between events be $\{X_k\}_{k=1}^{\infty}$ (often representing the lifetimes of some units successively placed into service) such that X_i is the elapsed time from the $(i - 1)$st event until the occurrence of the ith event. We write

$$F(x) = \Pr\{X_k \leq x\}, \qquad k = 1, 2, 3, \ldots,$$

for the common probability distribution of $X_1, X_2, \ldots$. A basic stipulation for renewal processes is $F(0) = 0$, signifying that $X_1, X_2, \ldots$ are positive random variables. We refer to

$$W_n = X_1 + X_2 + \cdots + X_n, \qquad n \geq 1$$

$$(W_0 = 0, \text{ by convention}) \quad (7.1)$$

as the *waiting time* until the occurrence of the nth event.

The relation between the interoccurrence times $\{X_k\}$ and the renewal counting process $\{N(t), t \geq 0\}$ is depicted in Figure 7.1. Note formally that

$$N(t) = \text{number of indices } n \text{ for which } 0 < W_n \leq t. \qquad (7.2)$$

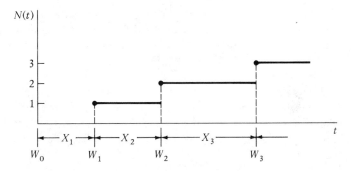

Figure 7.1 The relation between the interoccurrence times X_k and the renewal counting process $N(t)$.

In common practice the counting process $\{N(t), t \geq 0\}$ and the partial sum process $\{W_n, n \geq 0\}$ are interchangeably called the "renewal process." The prototypical renewal model involves successive replacements of light bulbs. A bulb is installed for service at time $W_0 = 0$, fails at time $W_1 = X_1$, and is then exchanged for a fresh bulb. The second bulb fails at time $W_2 = X_1 + X_2$ and is replaced by a third bulb. In general, the nth bulb burns out at time $W_n = X_1 + \ldots + X_n$ and is immediately replaced, and the process continues. It is natural to assume that the successive lifetimes are statistically independent, with probabilistically identical characteristics in that

$$\Pr\{X_k \leq x\} = F(x) \qquad \text{for} \quad k = 1, 2, \ldots .$$

In this process $N(t)$ records the number of light-bulb replacements up to time t.

The principal objective of renewal theory is to derive properties of certain random variables associated with $\{N(t)\}$ and $\{W_n\}$ from knowledge of the interoccurrence distribution F. For example, it is of significance and relevance to compute the expected number of renewals for the time duration $(0, t]$:

$$E[N(t)] = M(t)$$

is called the *renewal function*. For this end, several pertinent relationships and formulas are worth recording. In principle, the probability law of $W_n = X_1 + \ldots + X_n$ can be calculated in accordance with the convolution formula

$$\Pr\{W_n \leq x\} = F_n(x),$$

where $F_1(x) = F(x)$ is assumed known or prescribed, and then

$$F_n(x) = \int_0^\infty F_{n-1}(x - y)dF(y) = \int_0^x F_{n-1}(x - y)dF(y).$$

Such convolution formulas were reviewed in Section 1.2.5.

The fundamental connecting link between the waiting time process $\{W_n\}$ and the renewal counting process $\{N(t)\}$ is the observation that

$$N(t) \geq k \qquad \text{if and only if} \quad W_k \leq t. \tag{7.3}$$

In words, Equation (7.3) asserts that the number of renewals up to time t is at least k if and only if the kth renewal occurred on or before time t. Since this equivalence is the basis for much that follows, the reader should verify instances of it by referring to Figure 7.1.

It follows from (7.3) that

$$\Pr\{N(t) \geq k\} = \Pr\{W_k \leq t\}$$
$$= F_k(t), \qquad t \geq 0, \quad k = 1, 2, \ldots, \tag{7.4}$$

and consequently

$$\Pr\{N(t) = k\} = \Pr\{N(t) \geq k\} - \Pr\{N(t) \geq k + 1\}$$
$$= F_k(t) - F_{k+1}(t), \qquad t \geq 0, \quad k = 1, 2, \ldots. \tag{7.5}$$

For the renewal function $M(t) = E[N(t)]$ we sum the tail probabilities in the manner $E[N(t)] = \sum_{k=1}^\infty \Pr\{N(t) \geq k\}$, as derived in Equation (1.49) and then use (7.4) to obtain

$$M(t) = E[N(t)] = \sum_{k=1}^\infty \Pr\{N(t) \geq k\}$$
$$= \sum_{k=1}^\infty \Pr\{W_k \leq t\} = \sum_{k=1}^\infty F_k(t). \tag{7.6}$$

There are a number of other random variables of interest in renewal theory. Three of these are the *excess life* (also called the excess random variable), the *current life* (also called the age random variable), and the *total life*, defined, respectively, by

$$\gamma_t = W_{N(t)+1} - t \qquad \text{(excess or residual lifetime)}$$
$$\delta_t = t - W_{N(t)} \qquad \text{(current life or age random variable)}$$
$$\beta_t = \gamma_t + \delta_t \qquad \text{(total life).}$$

A pictorial description of these random variables is given in Figure 7.2.

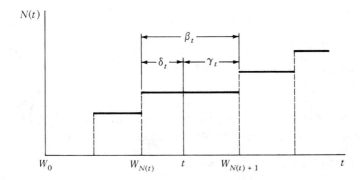

Figure 7.2 The excess life γ_t, the current life δ_t, and the total life β_t.

Problems 7.1

1. Which of the following are true statements?
   (a) $N(t) < k$ if and only if $W_k > t$.
   (b) $N(t) \leq k$ if and only if $W_k \geq t$.
   (c) $N(t) > k$ if and only if $W_k < t$.

2. Consider a renewal process for which the lifetimes $X_1, X_2, \ldots$ are discrete random variables having the Poisson distribution with mean λ. That is

$$\Pr\{X_k = n\} = \frac{e^{-\lambda}\lambda^n}{n!} \quad \text{for} \quad n = 0, 1, \ldots.$$

   (a) What is the distribution of the waiting time W_k?
   (b) Determine $\Pr\{N(t) = k\}$.

3. Let γ_t be the excess life and δ_t be the age in a renewal process having interoccurrence distribution function $F(x)$. Determine the conditional probability $\Pr\{\gamma_t > y | \delta_t = x\}$ and the conditional mean $E[\gamma_t | \delta_t = x]$.

7.2 Some Examples of Renewal Processes

The listing that follows suggests the wide scope and diverse contexts in which renewal processes arise. Several of the examples will be studied in more detail in later sections.

Poisson Processes
A Poisson process $\{N(t), t \geq 0\}$ with parameter λ is a renewal counting process having the exponential interoccurrence distribution

$$F(x) = 1 - e^{-\lambda x}, \qquad x \geq 0,$$

as established in Theorem 5.4. This particular renewal process possesses a host of special features highlighted later in Section 7.3.

Counter Processes

The times between successive electrical impulses or signals impinging on a recording device (counter) are often assumed to form a renewal process. Most physically realizable counters lock for some duration immediately upon registering an impulse and will not record impulses arriving during this dead period. Impulses are recorded only when the counter is free (i.e., unlocked). Under quite reasonable assumptions, the sequence of events of the times of recorded impulses forms a renewal process, but it should be emphasized that the renewal process of recorded impulses is a secondary renewal process derived from the original renewal process comprised of the totality of all arriving impulses.

Traffic Flow

The distances between successive cars on an indefinitely long single-lane highway are often assumed to form a renewal process. So also are the time durations between consecutive cars passing a fixed location.

Renewal Processes Associated with Queues

In a single-server queueing process there are imbedded many natural renewal processes. We cite two examples:

(i) If customer arrival times form a renewal process, then the times of the starts of successive busy periods generate a second renewal process.

(ii) For the situation in which the input process (the arrival pattern of customers) is Poisson, the successive moments when the server passes from a busy to a free state determine a renewal process.

Inventory Systems

In the analysis of most inventory processes it is customary to assume that the pattern of demands forms a renewal process. Most of the standard inventory policies induce renewal sequences, e.g., the times of replenishment of stock.

Renewal Processes in Markov Chains

Let $Z_0, Z_1, \ldots$ be a recurrent Markov chain. Suppose $Z_0 = i$ and consider the times (elapsed number of generations) between successive visits to state i. Specifically, let $W_0 = 0$ and

$$W_1 = \min\{n > 0 : Z_n = i\},$$

and

$$W_{k+1} = \min\{n > W_k : Z_n = i\} \qquad k = 1, 2, \ldots$$

Since each of these times is computed from the same starting state i, the Markov property guarantees that $X_k = W_k - W_{k-1}$ are independent and identically distributed and thus $\{X_k\}$ generates a renewal process.

Natural Embedded Renewal Processes

Natural embedded renewal processes can be found in many diverse fields of applied probability including branching processes, insurance risk models, phenomena of population growth, evolutionary genetic mechanisms, engineering systems, econometric structures, and elsewhere.

7.2.1 Block Replacement

Consider a light bulb whose life, measured in discrete units, is a random variable X where $\Pr\{X = k\} = p_k$ for $k = 1, 2, \ldots$. Assuming that one starts with a fresh bulb and that each bulb is replaced by a new one when it burns out, let $M(n) = E[N(n)]$ be the expected number of replacements up to time n.

Because of economies of scale, in a large building such as a factory or office it is often cheaper, on a per bulb basis, to replace all the bulbs, failed or not, than it is to replace a single bulb. A *block replacement policy* attempts to take advantage of this reduced cost by fixing a block period K and then replacing bulbs as they fail during periods $1, 2, \ldots, K - 1$, and replacing all bulbs, failed or not, in period K. This strategy is also known as "group relamping." If c_1 is the per bulb block replacement cost and c_2 is the per bulb failure replacement cost ($c_1 < c_2$), then the mean total cost during the block replacement cycle is $c_1 + c_2 M(K - 1)$ where $M(K - 1) = E[N(K - 1)]$ is the mean number of failure replacements. Since the block replacement cycle consists of K periods, the mean total cost per bulb per unit time is

$$\theta(K) = \frac{c_1 + c_2 M(K - 1)}{K}.$$

If we can determine the renewal function $M(n)$ from the life distribution $\{p_k\}$, then we can choose the block period $K = K^\star$ so as to minimize the cost rate $\theta(K)$. Of course, this cost must be compared to the cost of replacing only upon failure.

The renewal function $M(n)$, or expected number of replacements up to time n, solves the equation

$$M(n) = F_X(n) + \sum_{k=1}^{n-1} p_k M(n - k) \qquad \text{for} \quad n = 1, 2, \ldots$$

To derive this equation, condition on the life X_1 of the first bulb. If it fails after time n, there are no replacements during periods $[1, 2, \ldots, n]$. On the other hand, if it fails at time $k < n$, then we have its failure plus, on the average, $M(n - k)$ additional replacements during the interval $[k + 1, k +$

2, . . . , n]. Using the law of total probability to sum these contributions we obtain

$$M(n) = \sum_{k=n+1}^{\infty} p_k(0) + \sum_{k=1}^{n} p_k[1 + M(n - k)]$$

$$= F_X(n) + \sum_{k=1}^{n-1} p_k M(n - k) \qquad [\text{because } M(0) = 0]$$

as asserted.

Thus we determine

$$M(1) = F_X(1)$$

$$M(2) = F_X(2) + p_1 M(1)$$

$$M(3) = F_X(3) + p_1 M(2) + p_2 M(1)$$

and so on.

To consider a numerical example suppose that

$$p_1 = .1, \quad p_2 = .4, \quad p_3 = .3, \quad \text{and} \quad p_4 = .2,$$

and

$$c_1 = 2 \quad \text{and} \quad c_2 = 3.$$

Then

$$M(1) = p_1 = .1,$$

$$M(2) = (p_1 + p_2) + p_1 M(1) = (.1 + .4) + .1(.1) = .51,$$

$$M(3) = (p_1 + p_2 + p_3) + p_1 M(2) + p_2 M(1)$$

$$= (.1 + .4 + .3) + .1(.51) + .4(.1) = .891,$$

$$M(4) = (p_1 + p_2 + p_3 + p_4) + p_1 M(3) + p_2 M(2) + p_3 M(1)$$

$$= 1 + .1(.891) + .4(.51) + .3(.1) = 1.3231.$$

The average costs are

Block Period K	Cost $= \dfrac{c_1 + c_2 M(K - 1)}{K} = \theta(K)$
1	2.00000
2	1.15000
3	1.17667
4	1.16825
5	1.19386

The minimum cost block period is $K^\star = 2$.

We wish to elicit one more insight from this example. Forgetting about block replacement, we continue to calculate

$$M(5) = 1.6617$$

$$M(6) = 2.0647$$

$$M(7) = 2.4463$$

$$M(8) = 2.8336$$

$$M(9) = 3.2136$$

$$M(10) = 3.6016.$$

Let u_n be the probability that a replacement occurs in period n. Then $M(n) = M(n - 1) + u_n$ asserts that the mean replacements up to time n is the mean replacements up to time $n - 1$ plus the probability that a replacement occurs in period n. We calculate

n	$u_n = M(n) - M(n - 1)$
1	.1000
2	.4100
3	.3810
4	.4321
5	.3386
6	.4030
7	.3816
8	.3873
9	.3800
10	.3880

The probability of a replacement in period n seems to be converging. This is indeed the case, and the limit is the reciprocal of the mean bulb lifetime

$$\frac{1}{E[X_1]} = \frac{1}{.1(1) + .4(2) + .3(3) + .2(4)}$$

$$= .3846 \ldots$$

This calculation makes sense. If a light bulb lasts, on the average, $E[X_1]$ time units, then the probability it will need to be replaced in any period should approximate $1/E[X_1]$. Actually, the relationship is not so simple as just stated. Further discussion takes place in Section 7.4.

Problems 7.2

1. Let $X_1, X_2, \ldots$ be the interoccurrence times in a renewal process. Suppose $\Pr\{X_k = 1\} = p$ and $\Pr\{X_k = 2\} = q = 1 - p$. Verify that

$$M(n) = E[N(n)] = \frac{n}{1 + q} - \frac{q^2}{(1 + q)^2} + \frac{q^{n+2}}{(1 + q)^2}$$

for $n = 2, 4, 6, \ldots$.

2. Calculate the mean number of renewals $M(n) = E[N(n)]$ for the renewal process having interoccurrence distribution

$$p_1 = .4, \quad p_2 = .1, \quad p_3 = .3, \quad p_4 = .2$$

for $n = 1, 2, \ldots, 10$. Also calculate $u_n = M(n) - M(n - 1)$.

3. For the block replacement example of this section for which $p_1 = .1$, $p_2 = .4$, $p_3 = .3$, and $p_4 = .2$, suppose the costs are $c_1 = 4$ and $c_2 = 5$. Determine the minimal cost block period $K^\star$ and the cost of replacing upon failure alone.

4. Determine $M(n)$ when the interoccurrence times have the geometric distribution

$$\Pr\{X_1 = k\} = p_k = \beta(1 - \beta)^{k-1} \qquad \text{for} \quad k = 1, 2, \ldots,$$

where $0 < \beta < 1$.

7.3 The Poisson Process Viewed as a Renewal Process

As mentioned earlier, the Poisson process with parameter λ is a renewal process whose interoccurrence times have the exponential distribution $F(x) = 1 - e^{-\lambda x}$, $x \geq 0$. The memoryless property of the exponential distribution (see Section 1.4.2, Section 1.5.2, and Chapter 5) serves decisively in yielding the explicit computation of a number of functionals of the Poisson renewal process.

The Renewal Function
Since $N(t)$ has a Poisson distribution, then

$$\Pr\{N(t) = k\} = \frac{(\lambda t)^k e^{-\lambda t}}{k!}, \qquad k = 0, 1, \ldots,$$

and

$$M(t) = E[N(t)] = \lambda t.$$

Excess Life

Observe that the excess life at time t exceeds x if and only if there are no renewals in the interval $(t, t + x]$ (Figure 7.3). This event has the same probability as that of no renewals in the interval $(0, x]$, since a Poisson process has stationary independent increments. In formal terms, we have

$$\Pr\{\gamma_t > x\} = \Pr\{N(t + x) - N(t) = 0\} \tag{7.7}$$
$$= \Pr\{N(x) = 0\} = e^{-\lambda x}.$$

Thus, in a Poisson process, the excess life possesses the same exponential distribution

$$\Pr\{\gamma_t \leq x\} = 1 - e^{-\lambda x}, \qquad x \geq 0 \tag{7.8}$$

as every life, another manifestation of the memoryless property of the exponential distribution.

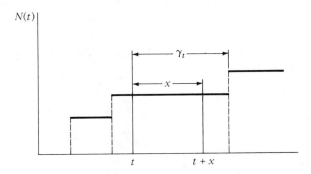

Figure 7.3   The excess life γ_t exceeds x if and only if there are no renewals in the interval $(t, t + x]$.

Current Life

The current life δ_t, of course, cannot exceed t, while for $x < t$ the current life exceeds x if and only if there are no renewals in $(t - x, t]$, which again has probability $e^{-\lambda x}$. Thus the current life follows the truncated exponential distribution

$$\Pr\{\delta_t \leq x\} = \begin{cases} 1 - e^{-\lambda x} & \text{for } 0 \leq x < t, \\ 1 & \text{for } t \leq x. \end{cases} \tag{7.9}$$

Mean Total Life

Using the evaluation of Section 1.5.1 for the mean of a nonnegative random variable, we have

$$E[\beta_t] = E[\gamma_t] + E[\delta_t]$$

$$= \frac{1}{\lambda} + \int_0^t \Pr\{\delta_t > x\}dx$$

$$= \frac{1}{\lambda} + \int_0^t e^{-\lambda x}dx$$

$$= \frac{1}{\lambda} + \frac{1}{\lambda}\left(1 - e^{-\lambda t}\right).$$

Observe that the mean total life is significantly larger than the mean life $1/\lambda = E[X_k]$ of any particular renewal interval. A more striking expression of this phenomenon is revealed when t is large, where the process has been in operation for a long duration. Then the mean total life $E[\beta_t]$ is approximately twice the mean life. These facts appear at first paradoxical.

Let us reexamine the manner of the definition of the total life β_t with a view to explaining on an intuitive basis the seeming discrepancy. First, an arbitrary time point t is fixed. Then β_t measures the length of the renewal interval containing the point t. Such a procedure will tend with higher likelihood to favor a lengthy renewal interval rather than one of short duration. The phenomenon is known as length-biased sampling and occurs, well disguised, in a number of sampling situations.

Joint Distribution of γ_t and δ_t

The joint distribution of γ_t and δ_t is determined in the same manner as the marginals. In fact, for any $x > 0$ and $0 < y < t$, the event $\{\gamma_t > x, \delta_t > y\}$ occurs if and only if there are no renewals in the interval $(t - y, t + x]$, which has probability $e^{-\lambda(x+y)}$. Thus

$$\Pr\{\gamma_t > x, \delta_t > y\} = \begin{cases} e^{-\lambda(x+y)} & \text{if } x > 0, \quad 0 < y < t, \\ 0 & \text{if } y \geq t. \end{cases} \tag{7.10}$$

For the Poisson process, observe that γ_t and δ_t are independent, since their joint distribution factors as the product of their marginal distributions.

Problems 7.3

1. Particles arrive at a counter according to a Poisson process of rate λ. An arriving particle is recorded with probability p and lost with probability $1 - p$ independently of the other particles. Show that the sequence of recorded particles is a Poisson process of rate λp.

2. Pulses arrive at a counter according to a Poisson process of rate λ. All physically realizable counters are imperfect, incapable of detecting all

signals that enter their detection chambers. After a particle or signal arrives, a counter must recuperate or renew itself in preparation for the next arrival. Signals arriving during the readjustment period, called dead time or locked time, are lost. We must distinguish between the arriving particles and the recorded particles. The experimenter observes only the particles recorded; from this observation he desires to infer the properties of the arrival process.

Suppose that each arriving pulse locks the counter for a fixed time τ. Determine the probability $p(t)$ that the counter is free at time t.

3. This problem is designed to aid in the understanding of length-biased sampling. Let X be a uniformly distributed random variable on $[0, 1]$. Then X divides $[0, 1]$ into the subintervals $[0, X]$ and $(X, 1]$. By symmetry, each subinterval has mean length $\frac{1}{2}$. Now pick one of these subintervals at random in the following way: Let Y be independent of X and uniformly distributed on $[0, 1]$, and pick the subinterval $[0, X]$ or $(X, 1]$ that Y falls in. Let L be the length of the subinterval so chosen. Formally

$$L = \begin{cases} X & \text{if } Y \leq X, \\ 1 - X & \text{if } Y > X. \end{cases}$$

Determine the mean of L.

7.4 The Asymptotic Behavior of Renewal Processes

A large number of the functionals that have explicit expressions for Poisson renewal processes are far more difficult to compute for other renewal processes. There are, however, many simple formulas that describe the asymptotic behavior, for large values of t, of a general renewal process. We summarize some of these asymptotic results in this section.

7.4.1 The Elementary Renewal Theorem

The Poisson process is the only renewal process (in continuous time) whose renewal function $M(t) = E[N(t)]$ is exactly linear. All renewal functions are asymptotically linear, however, in the sense that

$$\lim_{t \to \infty} \frac{M(t)}{t} = \lim_{t \to \infty} \frac{E[N(t)]}{t} = \frac{1}{\mu}, \tag{7.11}$$

where $\mu = E[X_k]$ is the mean interoccurrence time. This fundamental result, known as *the elementary renewal theorem*, is undoubtedly the most important result concerning renewal phenomena. It is invoked repeatedly to compute functionals describing the long run behavior of stochastic models having renewal processes associated with them.

The elementary renewal theorem (7.11) holds even when the interoccurrence times have infinite mean, and then $\lim_{t\to\infty} M(t)/t = 1/\infty = 0$.

The elementary renewal theorem is so intuitively plausible that it has often been viewed as obvious. The left side, $\lim_{t\to\infty} M(t)/t$, describes the long run mean number of renewals or replacements per unit time. The right side, $1/\mu$, is the reciprocal of the mean life of a component. Isn't it obvious that, if a component lasts, on the average, μ time units, then, in the long run these components will be replaced at the rate of $1/\mu$ per unit time? However plausible and convincing this argument may be, it is not obvious, and to establish the elementary renewal theorem requires several steps of mathematical analysis, beginning with the Law of Large Numbers. As our main concern is stochastic modeling, we omit this derivation, as well as the derivations of the other asymptotic results summarized in this section in order to give more space to their application.

Example *Age Replacement Policies* Let X_1, X_2, . . . represent the lifetimes of items (light bulbs, transistor cards, machines, etc.) that are successively placed in service, the next item commencing service immediately following the failure of the previous one. We stipulate that $\{X_k\}$ are independent and identically distributed positive random variables with finite mean $\mu = E[X_k]$. The elementary renewal theorem tells us to expect to replace items over the long run at a mean rate of $1/\mu$ per unit time.

In the long run, any replacement strategy that substitutes items prior to their failure will use more than $1/\mu$ items per unit time. Nonetheless, where there is some benefit in avoiding failure in service, and where units deteriorate, in some sense, with age, there may be an economic or reliability advantage in considering alternative replacement strategies. Telephone or utility poles serve as good illustrations of this concept. Clearly it is disadvantageous to allow these poles to fail in service because of the damage to the wires they carry, the damage to adjoining property, overtime wages paid for emergency replacements, and revenue lost while service is down. Therefore an attempt is usually made to replace older utility poles before they fail. Other instances of planned replacement occur in preventative maintenance strategies for aircraft, where "time" is now measured by operating hours.

An age replacement policy calls for replacing an item upon its failure or upon its reaching age T, whichever occurs first. Arguing intuitively, we would expect that the long run fraction of failure replacements, items that fail before age T, will be $F(T)$, and the corresponding fraction of (conceivably less expensive) planned replacements will be $1 - F(T)$. A renewal interval for this modified age replacement policy obviously follows a distribution law

$$F_T(x) = \begin{cases} F(x) & \text{for} \quad x < T, \\ 1 & \text{for} \quad x \geq T, \end{cases}$$

and the mean renewal duration is

$$\mu_T = \int_0^\infty \{1 - F_T(x)\}dx = \int_0^T \{1 - F(x)\}dx < \mu.$$

The elementary renewal theorem indicates that the long run mean replacement rate under age replacement is increased to $1/\mu_T$.

Now, let $Y_1, Y_2, \ldots$ denote the times between actual successive failures. The random variable Y_1 is composed of a random number of time periods of length T (corresponding to replacements not associated with failures), plus a last time period in which the distribution is that of a failure conditioned on failure before age T; that is, Y_1 has the distribution of $NT + Z$, where

$$\Pr\{N \geq k\} = \{1 - F(T)\}^k, \qquad k = 0, 1, \ldots,$$

and

$$\Pr\{Z \leq z\} = \frac{F(z)}{F(T)}, \qquad 0 \leq z \leq T.$$

Hence,

$$E[Y_1] = \frac{1}{F(T)} \left\{ T[1 - F(T)] + \int_0^T (F(T) - F(x))dx \right\}$$

$$= \frac{1}{F(T)} \int_0^T \{1 - F(x)\}dx = \frac{\mu_T}{F(T)}.$$

The sequence of random variables for interoccurrence times of the bona fide failure $\{Y_i\}$ generates a renewal process whose mean rate of failures per unit time in the long run is $1/E[Y_1]$. This inference again relies on the elementary renewal theorem. Depending on F, the modified failure rate $1/E[Y_1]$ may possibly yield a lower failure rate than $1/\mu$, the rate when replacements are made only upon failure.

Let us suppose that each replacement, whether planned or not, costs $\$K$, and that each failure incurs an additional penalty of $\$c$. Multiplying these costs by the appropriate rates gives the long run mean cost per unit time as a function of the replacement age T:

$$C(T) = \frac{K}{\mu_T} + \frac{c}{E[Y_1]}$$

$$= \frac{K + cF(T)}{\displaystyle\int_0^T [1 - F(x)]dx},$$

In any particular situation a routine calculus exercise or recourse to numerical computation produces the value of T that minimizes the long run cost rate. For example, if $K = 1$, $c = 4$, and lifetimes are uniformly distributed on $[0, 1]$, then $F(x) = x$ for $0 \leq x \leq 1$ and

$$\int_0^T [1 - F(x)]dx = T\left(1 - \frac{1}{2}T\right),$$

and

$$C(T) = \frac{1 + 4T}{T(1 - T/2)}.$$

To obtain the cost minimizing T, we differentiate $C(T)$ with respect to T and equate to zero, thereby obtaining

$$\frac{dC(T)}{dT} = 0 = \frac{4T(1 - T/2) - (1 + 4T)(1 - T)}{[T(1 - T/2)]^2},$$

$$4T - 2T^2 - 1 + T - 4T + 4T^2 = 0,$$

$$2T^2 + T - 1 = 0,$$

$$T = \frac{-1 \pm \sqrt{1 + 8}}{4} = \frac{1}{2}, -1,$$

and the optimal choice is $T^\star = \frac{1}{2}$. Routine calculus will verify that this choice leads to a minimum cost, and not a maximum or inflection point.

7.4.2 The Renewal Theorem for Continuous Lifetimes

The elementary renewal theorem asserts that

$$\lim_{t \to \infty} \frac{M(t)}{t} = \frac{1}{\mu}.$$

It is tempting to conclude from this that $M(t)$ behaves like t/μ as t grows large, but the precise meaning of the phrase "behaves like" is rather subtle. For example, suppose that all of the lifetimes are deterministic, say $X_k = 1$ for $k = 1, 2, \ldots$. Then it is straightforward to calculate

$$M(t) = N(t) = 0 \quad \text{for} \quad 0 \leq t < 1$$

$$= 1 \quad \text{for} \quad 1 \leq t < 2$$

$$= k \quad \text{for} \quad k \leq t < k + 1.$$

That is, $M(t) = [t]$ where $[t]$ denotes the greatest integer not exceeding t. Since $\mu = 1$ in this example, then $M(t) - t/\mu = [t] - t$, a function that oscillates indefinitely between 0 and -1. While it remains true in this illustration that $M(t)/t = [t]/t \to 1 = 1/\mu$, it is not clear in what sense $M(t)$ "behaves like" t/μ. If we rule out the periodic behavior that is exemplified in the extreme by this deterministic example, then $M(t)$ behaves like t/μ in the sense described by the *renewal theorem,* which we now explain. Let $M(t, t + h] = M(t + h) - M(t)$ denote the mean number of renewals in the interval $(t, t + h]$. The renewal theorem asserts that, when periodic behavior is precluded, then

$$\lim_{t \to \infty} M(t, t + h] = h/\mu \qquad \text{for any fixed} \quad h > 0. \qquad (7.12)$$

In words, asymptotically, the mean number of renewals in an interval is proportional to the interval's length, with proportionality constant $1/\mu$.

A simple and prevalent situation in which the renewal theorem (7.12) is valid occurs when the lifetimes $X_1, X_2, \ldots$ are continuous random variables having the probability density function $f(x)$. In this circumstance, the renewal function is differentiable and

$$m(t) = \frac{dM(t)}{dt} = \sum_{n=1}^{\infty} f_n(t) \qquad (7.13)$$

where $f_n(t)$ is the probability density function for $W_n = X_1 + \ldots + X_n$. Now (7.12) may be written in the form

$$\frac{M(t + h) - M(t)}{h} \to \frac{1}{\mu} \qquad \text{as} \quad t \to \infty,$$

which, when h is small, suggests that

$$\lim_{t \to \infty} m(t) = \lim_{t \to \infty} \frac{dM(t)}{dt} = \frac{1}{\mu}, \qquad (7.14)$$

and, indeed, this is the case in all but the most pathological of circumstances when $X_1, X_2, \ldots$ are continuous random variables.

If in addition to being continuous, the lifetimes $X_1, X_2, \ldots$ have a finite mean μ and finite variance σ^2, then the renewal theorem can be refined to include a second term. Under the stated conditions we have

$$\lim_{t \to \infty} \left[M(t) - \frac{t}{\mu} \right] = \frac{\sigma^2 - \mu^2}{2\mu^2}. \qquad (7.15)$$

Example When the lifetimes $X_1, X_2, \ldots$ have the gamma density function

$$f(x) = xe^{-x} \qquad \text{for} \quad x > 0, \qquad (7.16)$$

then the waiting times $W_n = X_1 + \ldots + X_n$ have the gamma density

$$f_n(x) = \frac{x^{2n-1}}{(2n - 1)!} e^{-x} \qquad \text{for} \quad x > 0,$$

as may be verified by performing the appropriate convolutions. (See Section 1.2.5.) Substitution into (7.13) yields

$$m(x) = \sum_{n=1}^{\infty} f_n(x) = e^{-x} \sum_{n=1}^{\infty} \frac{x^{2n-1}}{(2n-1)!}$$

$$= e^{-x} \frac{e^x - e^{-x}}{2} = \frac{1}{2}(1 - e^{-2x}),$$

and

$$M(t) = \int_0^t m(x)dx = \frac{1}{2} t - \frac{1}{4} [1 - e^{-2t}].$$

Since the gamma density in (7.16) has moments $\mu = 2$ and $\sigma^2 = 2$, we verify that $m(t) \to 1/\mu$ as $t \to \infty$ and $M(t) - t/\mu \to -\frac{1}{4} = (\sigma^2 - \mu^2)/2\mu^2$, in agreement with (7.14) and (7.15).

7.4.3 The Asymptotic Distribution of N(t)

The elementary renewal theorem

$$\lim_{t \to \infty} \frac{E[N(t)]}{t} = \frac{1}{\mu} \tag{7.17}$$

implies that the asymptotic mean of $N(t)$ is approximately t/μ. When $\mu = E[X_k]$ and $\sigma^2 = \text{Var}[X_k] = E[(X_k - \mu)^2]$ are finite, then the asymptotic variance of $N(t)$ behaves according to

$$\lim_{t \to \infty} \frac{\text{Var}[N(t)]}{t} = \frac{\sigma^2}{\mu^3}. \tag{7.18}$$

That is, the asymptotic variance of $N(t)$ is approximately $t\sigma^2/\mu^3$. If we standardize $N(t)$ by subtracting its asymptotic mean and dividing by its asymptotic standard deviation, we get the following convergence to the normal distribution:

$$\lim_{t \to \infty} \Pr\left\{ \frac{N(t) - t/\mu}{\sqrt{t\sigma^2/\mu^3}} \geq x \right\} = \frac{1}{\sqrt{2\pi}} \int_{-\infty}^{x} e^{-y^2/2} \, dy.$$

In words, for large values of t the number of renewals $N(t)$ is approximately normally distributed with mean and variance given by (7.17) and (7.18), respectively.

7.4.4 The Limiting Distribution of Age and Excess Life

Again we assume that the lifetimes $X_1, X_2, \ldots$ are continuous random variables with finite mean μ. Let $\gamma_t = W_{N(t)+1} - t$ be the excess life at time t. The excess life has the limiting distribution

$$\lim_{t\to\infty} \Pr\{\gamma_t \le x\} = \frac{1}{\mu}\int_0^x [1 - F(y)]dy. \tag{7.19}$$

The reader should verify that the right side of (7.19) defines a valid distribution function which we denote by $H(x)$. The corresponding probability density function is $h(y) = \mu^{-1}[1 - F(y)]$. The mean of this limiting distribution is determined according to

$$\begin{aligned}
\int_0^\infty yh(y)dy &= \frac{1}{\mu}\int_0^\infty y[1 - F(y)]dy \\
&= \frac{1}{\mu}\int_0^\infty y\left\{\int_y^\infty f(t)dt\right\}dy \\
&= \frac{1}{\mu}\int_0^\infty f(t)\left\{\int_0^t ydy\right\}dt \\
&= \frac{1}{2\mu}\int_0^\infty t^2 f(t)dt \\
&= \frac{\sigma^2 + \mu^2}{2\mu},
\end{aligned}$$

where σ^2 is the common variance of the lifetimes X_1, X_2, ...

The limiting distribution for the current life or age $\delta_t = t - W_{N(t)}$ can be deduced from the corresponding result (7.19) for the excess life. With the aid of Figure 7.4, corroborate the equivalence

$$\{\gamma_t \ge x \text{ and } \delta_t \ge y\} \quad \text{if and only if} \quad \{\gamma_{t-y} \ge x + y\}. \tag{7.20}$$

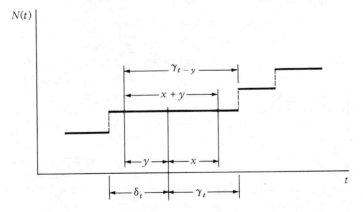

Figure 7.4 $\{\delta_t \ge y \text{ and } \gamma_t \ge x\}$ if and only if $\{\gamma_{t-y} \ge x + y\}$.

It follows that

$$\lim_{t \to \infty} \Pr\{\gamma_t \geq x, \, \delta_t \geq y\} = \lim_{t \to \infty} \Pr\{\gamma_{t-y} \geq x + y\}$$

$$= \mu^{-1} \int_{x+y}^{\infty} [1 - F(z)]dz,$$

exhibiting the joint limiting distribution of (γ_t, δ_t). In particular,

$$\lim_{t \to \infty} \Pr\{\delta_t \geq y\} = \lim_{t \to \infty} \Pr\{\gamma_t \geq 0, \, \delta_t \geq y\}$$

$$= \mu^{-1} \int_{y}^{\infty} [1 - F(z)]dz$$

$$= 1 - H(y).$$

Problems 7.4

1. Consider the triangular lifetime density function $f(x) = 2x$, for $0 < x < 1$. Determine the optimal replacement age in an age replacement model with replacement cost $K = 1$ and failure penalty $c = 4$ (cf. pp. 286–288).

2. Show that the optimal age replacement policy is to replace upon failure alone when lifetimes are exponentially distributed with parameter λ. Can you provide an intuitive explanation?

3. What is the limiting distribution of excess life when renewal lifetimes have the uniform density $f(x) = 1$, for $0 < x < 1$?

7.5 Generalizations and Variations on Renewal Processes

7.5.1 Delayed Renewal Processes

We continue to assume that $\{X_k\}$ are all independent positive random variables, but only $X_2, X_3, \ldots$ (from the second on) are identically distributed with distribution function F, while X_1 has possibly a different distribution function G. Such a process is called a *delayed renewal process*. We have all the ingredients for an ordinary renewal process except that the initial time to the first renewal has a distribution different from that of the other interoccurrence times.

   A delayed renewal process will arise when the component in operation at time $t = 0$ is not new, but all subsequent replacements are new. For example, suppose that the time origin is taken y time units after the start of an ordinary renewal process. Then the time to the first renewal after the origin in the delayed process will have the distribution of the excess life at time y of an ordinary renewal process.

As before, let $W_0 = 0$ and $W_n = X_1 + \ldots + X_n$, and let $N(t)$ count the number of renewals up to time t. But now it is essential to distinguish between the mean number of renewals in the delayed process

$$M_D(t) = E[N(t)], \tag{7.21}$$

and the renewal function associated with the distribution F,

$$M(t) = \sum_{k=1}^{\infty} F_k(t). \tag{7.22}$$

The elementary renewal theorem

$$\lim_{t \to \infty} \frac{M_D(t)}{t} = \frac{1}{\mu} \qquad \text{where} \quad \mu = E[X_2] \tag{7.23}$$

maintains, as does the renewal theorem

$$\lim_{t \to \infty} [M_D(t) - M_D(t - h)] = \frac{h}{\mu}$$

where $X_2, X_3, \ldots$ are continuous random variables.

7.5.2 Stationary Renewal Processes

A delayed renewal process for which the first life has the distribution function

$$G(x) = \mu^{-1} \int_0^x \{1 - F(y)\} dy$$

is called a stationary renewal process. We are attempting to model a renewal process that began indefinitely far in the past, so that the remaining life of the item in service at the origin has the limiting distribution of the excess life in an ordinary renewal process. We recognize G as this limiting distribution.

It is anticipated that such a process exhibits a number of stationary or time-invariant properties. For a stationary renewal process, then

$$M_D(t) = E[N(t)] = \frac{t}{\mu}, \tag{7.24}$$

and

$$\Pr\{\gamma_t^D \le x\} = G(x),$$

for all t. Thus, what is in general only an asymptotic renewal relation becomes an identity, holding for all t, in a stationary renewal process.

7.5.3 Cumulative and Related Processes

Suppose associated with the ith unit or lifetime interval is a second random variable Y_i ($\{Y_i\}$ identically distributed) in addition to the lifetime X_i. We allow X_i and Y_i to be dependent, but assume that the pairs (X_1, Y_1), $(X_2, Y_2), \ldots$ are independent. We use the notation $F(x) = \Pr\{X_i \leq x\}$, $G(y) = \Pr\{Y_i \leq y\}$, $\mu = E[X_i]$, and $\nu = E[Y_i]$.

A number of problems of practical and theoretical interest have a natural formulation in those terms.

Renewal Processes Involving Two Components to Each Renewal Interval

Suppose that Y_i represents a portion of the duration X_i. Figure 7.5 illustrates the model. There we have depicted the Y portion occurring at the beginning of the interval, but this assumption is not essential for the results that follow.

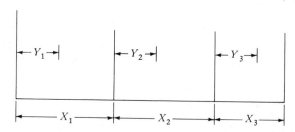

Figure 7.5 A renewal process in which an associated random variable Y_i represents a portion of the *i*th renewal interval.

Let $p(t)$ be the probability that t falls in a Y portion of some renewal interval. When $X_1, X_2, \ldots$ are continuous random variables, the renewal theorem implies the following important asymptotic evaluation:

$$\lim_{t \to \infty} p(t) = \frac{E[Y_1]}{E[X_1]}. \tag{7.25}$$

Here are some concrete examples.

A Replacement Model Consider a replacement model in which replacement is not instantaneous. Let Y_i be the operating time and Z_i the lag period preceding installment of the $(i + 1)$st operating unit. (The delay in replacement can be conceived as a period of repair of the service unit.) We assume that the sequence of times between successive replacements $X_k = Y_k + Z_k$, $k = 1, 2, \ldots$, constitutes a renewal process. Then $p(t)$, the probability that the system is in operation at time t, converges to $E[Y_1]/E[X_1]$.

A Queueing Model A queueing process is a process in which customers arrive at some designated place where a service of some kind is being rendered, for example, at the teller's window in a bank or beside the cashier at a supermarket. It is assumed that the time between arrivals, or interarrival time, and the time that is spent in providing service for a given customer are governed by probabilistic laws.

If arrivals to a queue follow a Poisson process of intensity λ, then the successive times X_k from the commencement of the kth busy period to the start of the next busy period form a renewal process. (A busy period is an uninterrupted duration when the queue is not empty.) Each X_k is composed of a busy portion Z_k and an idle portion Y_k. Then $p(t)$, the probability that the queue is empty at time t, converges to $E[Y_1]/E[X_1]$. This example is treated more fully in Chapter 9 on Queueing Systems.

The Peter Principle   The "Peter Principle" asserts that a worker will be promoted until first reaching a position in which he or she is incompetent. When this happens, the person stays in that job until retirement. Consider the following single job model of the "Peter Principle": A person is selected at random from the population and placed in the job. If the person is competent, he or she remains in the job for a random time having cumulative distribution function F and mean μ and is promoted. If incompetent, the person remains for a random time having cumulative distribution function G and mean $v > \mu$ and retires. Once the job is vacated, another person is selected at random and the process repeats. Assume that the infinite population contains the fraction p of competent people and $q = 1 - p$ incompetent ones.

In the long run, what fraction of time is the position held by an incompetent person?

A renewal occurs every time that the position is filled, and therefore the mean duration of a renewal cycle is

$$E[X_k] = p\mu + (1 - p)v.$$

To answer the question, we let $Y_k = X_k$ if the kth person is incompetent, and $Y_k = 0$ if the kth person is competent. Then the long run fraction of time that the position is held by an incompetent person is

$$\frac{E[Y_1]}{E[X_1]} = \frac{(1 - p)v}{p\mu + (1 - p)v}.$$

Suppose that $p = \frac{1}{2}$ of the people are competent, and that $v = 10$ while $\mu = 1$. Then

$$\frac{E[Y_1]}{E[X_1]} = \frac{(1/2)(10)}{(1/2)(10) + (1/2)(1)} = \frac{10}{11} = .91.$$

Thus, while half of the people in the population are competent, the job is filled by a competent person only 9 percent of the time!

Cumulative Processes

Interpret Y_i as a cost or value associated with the ith renewal cycle. A class of problems with natural setting in this general context of pairs (X_i, Y_i), where X_i generates a renewal process, will now be considered. Interest here focuses on the so-called cumulative process

$$W(t) = \sum_{k=1}^{N(t)+1} Y_k,$$

the accumulated costs or value up to time t (assuming transactions are made at the beginning of a renewal cycle).

The elementary renewal theorem asserts in this case that

$$\lim_{t \to \infty} \frac{1}{t} E[W(t)] = \frac{E[Y_1]}{\mu}. \tag{7.26}$$

This equation justifies the interpretation of $E[Y_1]/\mu$ as a long run mean cost or value per unit time, an interpretation that was used repeatedly in the examples of Section 7.2.

Here are some examples of cumulative processes.

Replacement Models   Suppose Y_i is the cost of the ith replacement. Let us suppose that under an age replacement strategy (see Section 7.3 and the example entitled "Age Replacement Policies" in Section 7.4) a planned replacement at age T costs c_1 dollars, while a failure replaced at time $x < T$ costs c_2 dollars. If Y_k is the cost incurred at the kth replacement cycle, then

$$Y_k = \begin{cases} c_1 & \text{with probability } 1 - F(T), \\ \\ c_2 & \text{with probability } F(T), \end{cases}$$

and $E[Y_k] = c_1[1 - F(T)] + c_2 F(T)$. Since the expected length of a replacement cycle is

$$E[\min\{X_k, T\}] = \int_0^T [1 - F(x)]dx,$$

we have that the long run cost per unit time is

$$\frac{c_1[1 - F(T)] + c_2 F(T)}{\int_0^T [1 - F(x)]dx},$$

and in any particular situation a routine calculus exercise or recourse to numerical computation produces the value of T that minimizes the long run cost per unit time.

Under a block replacement policy, there is one planned replacement every T units of time and, on the average, $M(T)$ failure replacements, so the expected cost is $E[Y_k] = c_1 + c_2 M(T)$, and the long run mean cost per unit time is $\{c_1 + c_2 M(T)\}/T$.

Risk Theory   Suppose claims arrive at an insurance company according to a renewal process with interoccurrence times X_1, X_2, ... Let Y_k be the magnitude of the kth claim. Then $W(t) = \sum_{k=1}^{N(t)+1} Y_k$ represents the cumulative amount claimed up to time t, and the long run mean claim rate is

$$\lim_{t\to\infty} \frac{1}{t} E[W(t)] = \frac{E[Y_1]}{E[X_1]}.$$

Maintaining Current Control of a Process   A production process produces items one by one. At any instance, the process is in one of two possible states, which we label *in-control* and *out-of-control*. These states are not directly observable. Production begins with the process in-control, and it remains in-control for a random and unobservable length of time before a breakdown occurs, after which the process is out-of-control. A control chart is to be used to help detect when the out-of-control state occurs so that corrective action may be taken.

To be more specific, we assume that the quality of an individual item is a normally distributed random variable having an unknown mean and a known variance σ^2. If the process is in-control, the mean equals a standard target or design value μ_0. Process breakdown takes the form of shift in mean away from standard to $\mu_1 = \mu_0 \pm \delta\sigma$, where δ is the amount of the shift in standard deviation units.

The Shewhart control chart method for maintaining process control calls for measuring the qualities of the items as they are produced and then plotting these qualities versus time on a chart that has lines drawn at the target value μ_0 and above and below this target value at $\mu_0 \pm k\sigma$, where k is a parameter of the control scheme being used. As long as the plotted qualities fall inside these so-called *action lines* at $\mu_0 \pm k\sigma$, the process is assumed to be operating in-control, but if ever a point falls outside these lines, the process is assumed to have left the in-control state, and investigation and repair are instituted. There are obviously two possible types of errors that can be made while thus controlling the process: (1) needless investigation and repair when the process is in-control yet an observed quality purely by chance falls outside the action lines and (2) continued operation with the process out-of-control because the observed qualities are falling inside the action lines, again by chance.

Our concern is the rational choice of the parameter k, that is, the rational spacing of the action lines, so as to balance, in some sense, these two possible errors.

The probability that a single quality will fall outside the action lines when the process is in-control is given by an appropriate area under the normal density curve. Denoting this probability by α we have

$$\alpha = \Phi(-k) + 1 - \Phi(k) = 2\Phi(-k)$$

where $\Phi(x) = (2\pi)^{-1/2} \int_{-\infty}^{x} \exp(-y^2/2)dy$ is the standard cumulative normal distribution function. Representative values are

k	α
1.645	.10
1.96	.05

Similarly, the probability that a single point will fall outside the action lines when the process is out-of-control, denoted p, is given by

$$p = \Phi(-\delta - k) + 1 - \Phi(-\delta + k).$$

Let S denote the number of items inspected before an out-of-control signal arises assuming the process is out-of-control. Then $\Pr\{S = 1\} = p$, $\Pr\{S = 2\} = (1 - p)p$ and, in general, $\Pr\{S = n\} = (1 - p)^{n-1}p$. Thus, S has a geometric distribution and

$$E[S] = \frac{1}{p}.$$

Let T be the number of items produced while the process is in-control. We suppose that the mean operating time in-control $E[T]$ is known from past records.

The sequence of durations between detected and repaired out-of-control conditions forms a renewal process because each such duration begins with a newly repaired process and is a probabilistic replica of all other such intervals. It follows from the general elementary renewal theorem that the long run fraction of time spent out-of-control (O.C.) is

$$\text{O.C.} = \frac{E[S]}{E[S] + E[T]} = \frac{1}{1 + pE[T]}.$$

The long run number of repairs per unit time is

$$R = \frac{1}{E[S] + E[T]} = \frac{p}{1 + pE[T]}.$$

Let N be the random number of "false alarms" while the process is in-control, that is during the time up to T, the first out-of-control. Then, conditioned on T, the random variable N has a binomial distribution with probability parameter α and thus $E[N|T] = \alpha T$ and $E[N] = \alpha E[T]$. Again, it follows from the general elementary renewal theorem that the long run false alarms per unit time (F.A.) is

$$\text{F.A.} = \frac{E[N]}{E[S] + E[T]} = \frac{\alpha p E[T]}{1 + p E[T]}.$$

If each false alarm costs c dollars, each repair costs K dollars, and the cost rate while operating out-of-control is C dollars, then we have the long run average cost per unit time of

$$\text{A.C.} = C(\text{O.C.}) + K(R) + c(\text{F.A.})$$
$$= \frac{C + kp + c\alpha p E[T]}{1 + p E[T]}.$$

By trial and error one may now choose k, which determines α and p, so as to minimize this average cost expression.

Problems 7.5

1. The weather in a certain locale consists of alternating wet and dry spells. Suppose that the number of days in each rainy spell is Poisson distributed with parameter 2, and that a dry spell follows a geometric distribution with a mean of 7 days. Assume that the successive durations of rainy and dry spells are statistically independent random variables. In the long run, what is the probability on a given day that it will be raining?

2. The random lifetime X of an item has a distribution function $F(x)$. What is the mean remaining life $E[X|X > x]$ of an item of age x?

3. At the beginning of each period, customers arrive at a taxi stand at times of a renewal process with distribution law $F(x)$. Assume an unlimited supply of cabs, such as might occur at an airport. Suppose that each customer pays a random fee at the stand following the distribution law $G(x)$, for $x > 0$. Write an expression for the sum $W(t)$ of money collected at the stand by time t, and then determine the limit expectation

$$\lim_{t \to \infty} \frac{E[W(t)]}{t}.$$

7.6 Discrete Renewal Theory

In this section we outline the renewal theory that pertains to nonnegative integer-valued lifetimes. We emphasize renewal equations, the renewal argument, and the renewal theorem (Theorem 7.1).

Consider a light bulb whose life, measured in discrete units, is a random variable X where $\Pr\{X = k\} = p_k$ for $k = 0, 1, \ldots$. If one starts with a fresh bulb and if each bulb when it burns out is replaced by a new one, then $M(n)$, the expected number of renewals (not including the initial bulb) up to time n, solves the equation

$$M(n) = F_X(n) + \sum_{k=0}^{n} p_k M(n - k), \qquad (7.27)$$

where $F_X(n) = p_0 + \ldots + p_n$ is the cumulative distribution function of the random variable X. A vector or functional equation of the form (7.27) in the unknowns $M(0)$, $M(1)$, . . . is termed a *renewal equation*. The equation is established by a *renewal argument*, a first step analysis that proceeds by conditioning on the life of the first bulb and then invoking the law of total probability. In the case of (7.27), for example, if the first bulb fails at time $k \leq n$, then we have its failure plus, on the average, $M(n - k)$ additional failures in the interval $[k, k + 1, \ldots, n]$. We weight this conditional mean by the probability $p_k = \Pr\{X_1 = k\}$ and sum according to the law of total probability to obtain

$$M(n) = \sum_{k=0}^{n} [1 + M(n - k)] p_k$$

$$= F_X(n) + \sum_{k=0}^{n} p_k M(n - k).$$

Equation (7.27) is only a particular instance of what is called a renewal equation. In general, a renewal equation is prescribed by a given bounded sequence $\{b_k\}$ and takes the form

$$v_n = b_n + \sum_{k=0}^{n} p_k v_{n-k} \qquad \text{for} \quad n = 0, 1, \ldots \qquad (7.28)$$

The unknown variables are v_0, v_1, . . . , and p_0, p_1, . . . is a probability distribution for which, to avoid trivialities, we always assume $p_0 < 1$.

Let us first note that there is one and only one sequence v_0, v_1, . . . satisfying a renewal equation because we may solve (7.28) successively to get

$$v_0 = \frac{b_0}{1 - p_0},$$

$$v_1 = \frac{b_1 + p_1 v_0}{1 - p_0}, \qquad (7.29)$$

and so on.

Let u_n be the mean number of renewals that take place exactly in period n. When $p_0 = 0$, so that the lifetimes are strictly positive and at most one renewal can occur in any period, then u_n is the probability that a single renewal occurs in period n. The sequence u_0, u_1, . . . satisfies a renewal equation that is of fundamental importance in the general theory. Let

$$\delta_n = \begin{cases} 1 & \text{for} \quad n = 0, \\ 0 & \text{for} \quad n > 0. \end{cases} \qquad (7.30)$$

Then $\{u_n\}$ satisfies the renewal equation

$$u_n = \delta_n + \sum_{k=0}^{n} p_k u_{n-k} \qquad \text{for} \quad n = 0, 1, \ldots \qquad (7.31)$$

Again, Equation (7.31) is established via a renewal argument. First, observe that δ_n counts the initial bulb, the renewal at time 0. Next, condition on the lifetime of this first bulb. If it fails in period $k \le n$, which occurs with probability p_k, then the process begins afresh and the conditional probability of a renewal in period n becomes u_{n-k}. Weighting the contingency represented by u_{n-k} by its respective probability p_k and summing according to the law of total probability then yields (7.31).

The next lemma shows how the solution $\{v_n\}$ to the general renewal equation (7.28) can be expressed in terms of the solution $\{u_n\}$ to the particular equation (7.31).

Lemma 7.1   *If $\{v_n\}$ satisfies (7.28) and $\{u_n\}$ satisfies (7.31), then*

$$v_n = \sum_{k=0}^{n} b_{n-k} u_k \qquad \text{for} \quad n = 0, 1, \ldots$$

Proof   In view of our remarks on the existence and uniqueness of solutions to Equation (7.28), we need only verify that $v_n = \sum_{k=0}^{n} b_{n-k} u_k$ satisfies (7.28). We have

$$\begin{aligned}
v_n &= \sum_{k=0}^{n} b_{n-k} u_k \\
&= \sum_{k=0}^{n} b_{n-k} \left\{ \delta_k + \sum_{l=0}^{k} p_{k-l} u_l \right\} \\
&= b_n + \sum_{k=0}^{n} \sum_{l=0}^{k} b_{n-k} p_{k-l} u_l \\
&= b_n + \sum_{l=0}^{n} \sum_{k=l}^{n} b_{n-k} p_{k-l} u_l \\
&= b_n + \sum_{l=0}^{n} \sum_{j=0}^{n-l} p_j b_{n-l-j} u_l \\
&= b_n + \sum_{j=0}^{n} \sum_{l=0}^{n-j} p_j b_{n-j-l} u_l \\
&= b_n + \sum_{j=0}^{n} p_j v_{n-j}. \qquad \square
\end{aligned}$$

Example   Let $X_1, X_2, \ldots$ be the successive lifetimes of the bulbs and let $W_0 = 0$ and $W_n = X_1 + \ldots + X_n$ be the replacement times. We assume that $p_0 = \Pr\{X_1 = 0\} = 0$. The number of replacements (not including the initial bulb) up to time n is given by

$$N(n) = k \qquad \text{for} \quad W_k \le n < W_{k+1}.$$

The $M(n) = E[N(n)]$ satisfies the renewal equation (7.27)

$$M(n) = p_0 + \cdots + p_n + \sum_{k=0}^{n} p_k M(n-k),$$

and elementary algebra shows that $m_n = E[N(n) + 1] = M(n) + 1$ satisfies

$$m_n = 1 + \sum_{k=0}^{n} p_k m_{n-k} \qquad \text{for} \quad n = 0, 1, \dots \qquad (7.32)$$

Then (7.32) is a renewal equation for which $b_n \equiv 1$ for all n. In view of Lemma 7.1 we conclude that

$$m_n = \sum_{k=0}^{n} 1 u_k = u_0 + \cdots + u_n.$$

Conversely, $u_n = m_n - m_{n-1} = M(n) - M(n-1)$.

To continue with the example, let $g_n = E[W_{N(n)+1}]$. The definition is illustrated in Figure 7.6. We will argue that g_n satisfies a certain renewal equation. As shown in Figure 7.1, $W_{N(n)+1}$ always includes the first renewal duration X_1. In addition, if $X_1 = k \le n$, which occurs with probability p_k, then the conditional mean of the added lives constituting $W_{N(n)+1}$ is g_{n-k}. Weighting these conditional means by their respective probabilities and summing according to the law of total probability then gives

$$g_n = E[X_1] + \sum_{k=0}^{n} g_{n-k} p_k.$$

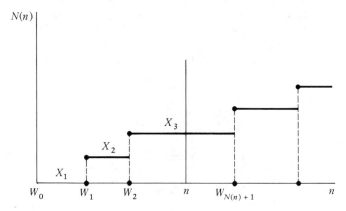

N(n)

Figure 7.6 $W_{N(n)+1}$ always contains X_1 and contains additional durations when $X_1 = k \le n$.

Hence, by Lemma 7.1,

$$g_n = \sum_{k=0}^{n} E[X_1] u_k = E[X_1] m_n.$$

We get the interesting formula

$$E[X_1 + \cdots + X_{N(n)+1}] = E[X_1] \times E[N(n) + 1]. \qquad (7.33)$$

Note that $N(n)$ is not independent of $\{X_k\}$, and yet (7.33) still prevails.

7.6.1 The Discrete Renewal Theorem

The renewal theorem provides conditions under which the solution $\{v_n\}$ to a renewal equation will converge as n grows large. Certain periodic behavior, such as failures occurring only at even ages, must be precluded, and the simplest assumption assuring this preclusion is that $p_1 > 0$.

Theorem 7.1 *Suppose that $0 < p_1 < 1$ and that $\{u_n\}$ and $\{v_n\}$ are the solutions to the renewal equations (7.31) and (7.28), respectively. Then (a) $\lim_{n\to\infty} u_n = 1/ \sum_{k=0}^{\infty} kp_k$; and (b) if $\sum_{k=0}^{\infty} |b_k| < \infty$, then $\lim_{n\to\infty} v_n = \{\sum_{k=0}^{\infty} b_k\}/\{\sum_{k=0}^{\infty} kp_k\}$.*

We recognize that $\sum_{k=0}^{\infty} kp_k = E[X_1]$ is the mean lifetime of a unit. Thus (a) in Theorem 7.1 asserts that, in the long run, the probability of a renewal occurring in a given interval is one divided by the mean life of a unit.

Remark 7.1 Theorem 7.1 holds in certain circumstances when $p_1 = 0$. It suffices to assume that the greatest common divisor of the integers k for which $p_k > 0$ is one.

Example Let $\gamma_n = W_{N(n)+1} - n$ be the excess life at time n. For a fixed integer m, let $f_n(m) = \Pr\{\gamma_n > m\}$. We will establish a renewal equation for $f_n(m)$ by conditioning on the first life X_1. For $m \le n$ then

$$\Pr\{\gamma_n > m | X_1 = k\} = \begin{cases} f_{n-k}(m) & \text{if } 0 \le k \le n, \\ 0 & \text{if } n < k \le n + m, \\ 1 & \text{if } n + m < k. \end{cases}$$

(The student is urged to diagram the alternatives arising in $\Pr\{\gamma_n > m | X_1 = k\}$.) Then, by the law of total probability

$$f_n(m) = \Pr\{\gamma_n > m\} = \sum_{k=0}^{\infty} \Pr\{\gamma_n > m | X_1 = k\} p_k$$

$$= \sum_{k>n+m} p_k + \sum_{k=0}^{n} f_{n-k}(m) p_k$$

or

$$f_n(m) = [1 - F_X(n + m)] + \sum_{k=0}^{n} f_{n-k}(m) p_k,$$

where $F_X(j) = p_0 + \ldots + p_j$. We apply Theorem 7.1 with $b_n = 1 - F(n + m)$ to conclude that

$$\lim_{n\to\infty} \Pr\{\gamma_n > m\} = \frac{\sum_{k=0}^{\infty} [1 - F_X(k + m)]}{\sum_{k=0}^{\infty} kp_k}.$$

The limiting probability mass function for the excess life γ_n is given by

$$\lim_{n\to\infty} \Pr\{\gamma_n = m\} = \lim_{n\to\infty} \Pr\{\gamma_n > m - 1\} - \lim_{n\to\infty} \Pr\{\gamma_n > m\}$$

$$= \frac{\sum_{k=0}^{\infty} [1 - F_X(k + m - 1)] - \sum_{k=0}^{\infty} [1 - F_X(k + m)]}{\sum_{k=0}^{\infty} kp_k}$$

$$= \frac{1 - F_X(m - 1)}{\sum_{k=0}^{\infty} kp_k}$$

$$= \frac{1 - F_X(m - 1)}{E[X_1]}, \qquad m = 1, 2, \ldots \qquad (7.34)$$

The limit is a bona fide probability mass function since its terms

$$\pi(m) = \frac{1 - F_X(m - 1)}{E[X_1]}, \qquad m = 1, 2, \ldots$$

are nonnegative and sum to one because

$$\sum_{m=1}^{\infty} \pi(m) = \frac{\sum_{m=1}^{\infty} [1 - F_X(m - 1)]}{E[X_1]}$$

$$= \frac{\sum_{m=1}^{\infty} \Pr\{X_1 \geq m\}}{E[X_1]} = \frac{E[X_1]}{E[X_1]} = 1.$$

Problems 7.6

1. Solve for v_n for $n = 0, 1, \ldots, 10$ in the renewal equation

$$v_n = b_n + \sum_{k=0}^{n} p_k v_{n-k} \qquad \text{for} \quad n = 0, 1, \ldots$$

where $b_0 = b_1 = \frac{1}{2}$, $b_2 = b_3 = \ldots = 0$ and $p_0 = \frac{1}{4}$, $p_1 = \frac{1}{2}$, and $p_2 = \frac{1}{4}$.

2. (Continuation)
   (a) Solve for u_n for $n = 0, 1, \ldots, 10$ in the renewal equation

$$u_n = \delta_n + \sum_{k=0}^{n} p_k u_{n-k} \qquad \text{for} \quad n = 0, 1, \ldots,$$

   where $\delta_0 = 1$, $\delta_1 = \delta_2 = \ldots = 0$ and $\{p_k\}$ is as defined in Problem 1.
   (b) Verify that the solution v_n in Problem 1 and u_n are related according to $v_n = \sum_{k=0}^{n} b_k u_{n-k}$.

3. Using the data of Problems 1 and 2, determine
   (a) $\lim_{n\to\infty} u_n$.
   (b) $\lim_{n\to\infty} v_n$.

4. Suppose the lifetimes $X_1, X_2, \ldots$ have the geometric distribution

$$\Pr\{X_1 = k\} = \alpha(1 - \alpha)^{k-1} \qquad \text{for} \quad k = 1, 2, \ldots$$

   where $0 < \alpha < 1$. Determine the limiting distribution of excess life γ_n.

Chapter 8 | Branching Processes and Population Growth

8.1 Branching Processes

Suppose an organism at the end of its lifetime produces a random number ξ of offspring with probability distribution

$$\Pr\{\xi = k\} = p_k \qquad \text{for} \quad k = 0, 1, 2, \ldots, \qquad (8.1)$$

where, as usual $p_k \geq 0$ and $\sum_{k=0}^{\infty} p_k = 1$. We assume that all offspring act independently of each other and at the end of their lifetime (for simplicity, the lifespans of all organisms are assumed to be the same) individually have progeny in accordance with the probability distribution (8.1), thus propagating their species. The process $\{X_n\}$, where X_n is the population size at the nth generation, is a Markov chain of special structure, called a *branching process*.

The Markov property may be reasoned simply as follows. In the nth generation the X_n individuals independently give rise to numbers of offspring $\xi_1^{(n)}, \xi_2^{(n)}, \ldots, \xi_{X_n}^{(n)}$, and hence the cumulative number produced for the $(n + 1)$st generation is

$$X_{n+1} = \xi_1^{(n)} + \xi_2^{(n)} + \cdots + \xi_{X_n}^{(n)}. \qquad (8.2)$$

8.1.1 Examples of Branching Processes

There are numerous examples of Markov branching processes that arise naturally in various scientific disciplines. We list some of the more prominent cases.

Electron Multipliers

An electron multiplier is a device that amplifies a weak current of electrons. A series of plates are set up in the path of electrons emitted by a source. Each electron, as it strikes the first plate, generates a random number of new electrons, which in turn strike the next plate and produce more electrons, and so forth. Let X_0 be the number of electrons initially emitted, X_1 the number of electrons produced on the first plate by the impact due to the X_0 initial electrons; in general let X_n be the number of electrons emitted from the nth plate due to electrons emanating from the $(n - 1)$st plate. The sequence of random variables X_0, X_1, X_2, . . . , X_n, . . . constitutes a branching process.

Neutron Chain Reaction

A nucleus is split by a chance collision with a neutron. The resulting fission yields a random number of new neutrons. Each of these secondary neutrons may hit some other nucleus producing a random number of additional neutrons, and so forth. In this case the initial number of neutrons is $X_0 = 1$. The first generation of neutrons comprises all those produced from the fission caused by the initial neutron. The size of the first generation is a random variable X_1. In general, the population X_n at the nth generation is produced by the chance hits of the X_{n-1} individual neutrons of the $(n - 1)$st generation.

Survival of Family Names

The family name is inherited by sons only. Suppose that each individual has probability p_k of having k male offspring. Then from one individual there result the 1st, 2nd, . . . , nth, . . . generations of descendants. We may investigate the distribution of such random variables as the number of descendants in the nth generation, or the probability that the family name will eventually become extinct. Such questions will be dealt with in the general analysis of branching processes of this chapter.

Survival of Mutant Genes

Each individual gene has a chance to give birth to k offspring, $k = 1$, 2, . . . , which are genes of the same kind. Any individual, however, has a chance to transform into a different type of mutant gene. This gene may become the first in a sequence of generations of a particular mutant gene. We may inquire about the chances of survival of the mutant gene within the population of the original genes. In this example, the number of offspring is often assumed to follow a Poisson distribution.

The rationale behind this choice of distribution is as follows. In many populations a large number of zygotes (fertilized eggs) are produced, only a small number of which grow to maturity. The events of fertilization and maturation of different zygotes obey the law of independent binomial trials.

The number of trials (i.e., number of zygotes) is large. The Law of Rare Events (see p. 23) then implies that the number of progeny that mature will approximately follow the Poisson distribution. The Poisson assumption seems quite appropriate in the model of population growth of a rare mutant gene. If the mutant gene carries a biological advantage (or disadvantage), then the probability distribution is taken to be the Poisson distribution with mean $\lambda > 1$ or (< 1).

All of the preceding examples possess the following structure. Let X_0 denote the size of the initial population. Each individual gives birth to k new individuals with probability p_k *independently of the others*. The totality of all the direct descendants of the initial population constitutes the first generation whose size we denote by X_1. Each individual of the first generation independently bears a progeny set whose size is governed by the probability distribution (8.1). The descendants produced constitute the second generation of size X_2. In general the nth generation is composed of descendants of the $(n - 1)$st generation each of whose members independently produces k progeny with probability p_k, $k = 0, 1, 2, \ldots$. The population size of the nth generation is denoted by X_n. The X_n form a sequence of integer-valued random variables that generate a Markov chain in the manner described by (8.2).

8.1.2 The Mean and Variance of a Branching Process

Equation (8.2) characterizes the evolution of the branching process as successive random sums of random variables. Random sums were studied in Section 2.3, and we can use the moment formulas developed there to compute the mean and variance of the population size X_n. First some notation. Let $\mu = E[\xi]$ and $\sigma^2 = \text{Var}[\xi]$ be the mean and variance, respectively, of the offspring distribution (8.1). Let $M(n)$ and $V(n)$ be the mean and variance of X_n under the initial condition $X_0 = 1$. Then direct application of Equations (2.30), with respect to the random sum (8.2), gives the recursions

$$M(n + 1) = \mu M(n), \tag{8.3}$$

and

$$V(n + 1) = \sigma^2 M(n) + \mu^2 V(n). \tag{8.4}$$

The initial condition $X_0 = 1$ starts the recursions (8.3) and (8.4) at $M(0) = 1$ and $V(0) = 0$. Then, from (8.3) we obtain $M(1) = \mu 1 = \mu$, $M(2) = \mu M(1) = \mu^2$ and, in general,

$$M(n) = \mu^n \quad \text{for} \quad n = 0, 1, \ldots \tag{8.5}$$

Thus, the mean population size increases geometrically when $\mu > 1$, decreases geometrically when $\mu < 1$, and remains constant when $\mu = 1$.

Next, substitution of $M(n) = \mu^n$ into (8.4) gives $V(n + 1) = \sigma^2\mu^n + \mu^2 V(n)$, which with $V(0) = 0$ yields

$$V(1) = \sigma^2$$

$$V(2) = \sigma^2\mu + \mu^2 V(1) = \sigma^2\mu + \sigma^2\mu^2$$

$$V(3) = \sigma^2\mu^2 + \mu^2 V(2)$$
$$= \sigma^2\mu^2 + \sigma^2\mu^3 + \sigma^2\mu^4,$$

and, in general,

$$V(n) = \sigma^2[\mu^{n-1} + \mu^n + \cdots + \mu^{2n-2}]$$
$$= \sigma^2\mu^{n-1}[1 + \mu + \cdots + \mu^{n-1}]$$

$$= \sigma^2\mu^{n-1} \times \begin{cases} n & \text{if } \mu = 1, \\[2ex] \dfrac{1 - \mu^n}{1 - \mu} & \text{if } \mu \neq 1. \end{cases} \tag{8.6}$$

Thus the variance of the population size increases geometrically if $\mu > 1$, increases linearly if $\mu = 1$, and decreases geometrically if $\mu < 1$.

8.1.3 Extinction Probabilities

Population extinction occurs when and if the population size is reduced to zero. The random time of extinction N is thus the first time n for which $X_n = 0$, and then, obviously, $X_k = 0$ for all $k \geq N$. In Markov chain terminology 0 is an absorbing state and we may calculate the probability of extinction by invoking a first step analysis. Let

$$u_n = \Pr\{N \leq n\} = \Pr\{X_n = 0\} \tag{8.7}$$

be the probability of extinction at or prior to the nth generation, beginning with a single parent $X_0 = 1$. Suppose that the single parent represented by $X_0 = 1$ gives rise to $\xi_1^{(0)} = k$ offspring. In turn, each of these offspring will generate a population of its own descendants, and if the original population is to die out in n generations, then each of these k lines of descent must die out in $n - 1$ generations. The analysis is depicted in Figure 8.1.

Now the k subpopulations generated by the distinct offspring of the original parent are independent, and they have the same statistical properties as the original population. Therefore the probability that any particular one of them dies out in $n - 1$ generations is u_{n-1}, by definition, and the probability that all k subpopulations die out in $n - 1$ generations is the kth power $(u_{n-1})^k$ because they are independent. Upon weighting this factor by the

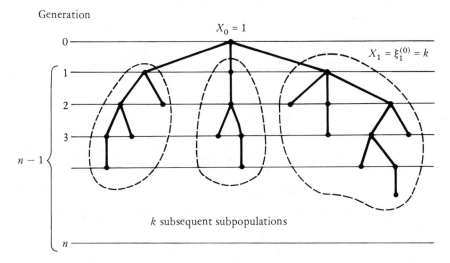

Figure 8.1   The diagram illustrates that if the original population is to die out by generation n, then the subpopulations generated by distinct initial offspring must all die out in $n - 1$ generations.

probability of k offspring and summing according to the law of total probability, we obtain

$$u_n = \sum_{k=0}^{\infty} p_k (u_{n-1})^k, \qquad n = 1, 2, \ldots \qquad (8.8)$$

Of course $u_0 = 0$, and $u_1 = p_0$, the probability that the original parent had no offspring.

Example   Suppose a parent has no offspring with probability $\frac{1}{4}$ and two offspring with probability $\frac{3}{4}$. Then the recursion (8.8) specializes to

$$u_n = \frac{1}{4} + \frac{3}{4}(u_{n-1})^2 = \frac{1 + 3(u_{n-1})^2}{4}.$$

Beginning with $u_0 = 0$ we successively compute

$$
\begin{aligned}
u_1 &= .2500 & u_6 &= .3313 \\
u_2 &= .2969 & u_7 &= .3323 \\
u_3 &= .3161 & u_8 &= .3328 \\
u_4 &= .3249 & u_9 &= .3331 \\
u_5 &= .3292 & u_{10} &= .3332.
\end{aligned}
$$

Problems 8.1

1. Let $Z = \sum_{n=0}^{\infty} X_n$ be the total family size in a branching process whose offspring distribution has a mean $\mu = E[\xi] < 1$. Assuming that $X_0 = 1$, show that $E[Z] = 1/(1 - \mu)$.

8.2 Branching Processes and Generating Functions

Consider a nonnegative integer-valued random variable ξ whose probability distribution is given by

$$\Pr\{\xi = k\} = p_k \qquad \text{for} \quad k = 0, 1, \ldots . \tag{8.9}$$

The *generating function* $\phi(s)$ associated with the random variable ξ (or equivalently, with the distribution $\{p_k\}$) is defined by

$$\phi(s) = E[s^\xi] = \sum_{k=0}^{\infty} p_k s^k \qquad \text{for} \quad 0 \leq s \leq 1. \tag{8.10}$$

Much of the importance of generating functions derives from the following three results.

First, the relation between probability mass functions (8.9) and generating functions (8.10) is one-to-one. Thus knowing the generating function is equivalent, in some sense, to knowing the distribution. The relation that expresses the probability mass function $\{p_k\}$ in terms of the generating function $\phi(s)$ is

$$p_k = \frac{1}{k!} \frac{d^k \phi(s)}{ds^k} \bigg|_{s=0} \tag{8.11}$$

For example

$$\phi(s) = p_0 + p_1 s + p_2 s^2 + \cdots$$

whence

$$p_0 = \phi(0),$$

and

$$\frac{d\phi(s)}{ds} = p_1 + 2p_2 s + 3p_3 s^2 + \cdots$$

whence

$$p_1 = \frac{d\phi(s)}{ds} \bigg|_{s=0}$$

Second, if $\xi_1, \ldots, \xi_n$ are independent random variables having generating functions $\phi_1(s), \ldots, \phi_n(s)$, respectively, then the generating function of their sum $X = \xi_1 + \ldots + \xi_n$ is simply the product

$$\phi_X(s) = \phi_1(s)\ \phi_2(s) \cdots \phi_n(s). \tag{8.12}$$

This simple result makes generating functions extremely helpful in dealing with problems involving sums of independent random variables. It is to be expected, then, that generating functions might provide a major tool in the analysis of branching processes.

Third, the moments of a nonnegative integer-valued random variable may be found by differentiating the generating function. For example, the first derivative is

$$\frac{d\phi(s)}{ds} = p_1 + 2p_2s + 3p_3s^2 + \cdots$$

whence

$$\left.\frac{d\phi(s)}{ds}\right|_{s=1} = p_1 + 2p_2 + 3p_3 + \cdots = E[\xi], \tag{8.13}$$

and the second derivative is

$$\frac{d^2\phi(s)}{ds^2} = 2p_2 + 3(2)p_3s + 4(3)p_4s^2 + \cdots$$

whence

$$\left.\frac{d^2\phi(s)}{ds^2}\right|_{s=1} = 2p_2 + 3(2)p_3 + 4(3)p_4 + \cdots$$

$$= \sum_{k=2}^{\infty} k(k-1)p_k = E[\xi(\xi-1)]$$

$$= E[\xi^2] - E[\xi].$$

Thus

$$E[\xi^2] = \left.\frac{d^2\phi(s)}{ds^2}\right|_{s=1} + E[\xi]$$

$$= \left.\frac{d^2\phi(s)}{ds^2}\right|_{s=1} + \left.\frac{d\phi(s)}{ds}\right|_{s=1}$$

and

$$\mathrm{Var}[\xi] = E[\xi^2] - \{E[\xi]\}^2$$

$$= \left.\frac{d^2\phi(s)}{ds^2}\right|_{s=1} + \left.\frac{d\phi(s)}{ds}\right|_{s=1} - \left\{\left.\frac{d\phi(s)}{ds}\right|_{s=1}\right\}^2. \tag{8.14}$$

Example If ξ has a Poisson distribution with mean λ for which

$$p_k = \mathrm{Pr}\{\xi = k\} = \frac{\lambda^k e^{-\lambda}}{k!} \qquad \text{for} \quad k = 0, 1, \ldots$$

then

$$\phi(s) = E[s^\xi] = \sum_{k=0}^{\infty} s^k \frac{\lambda^k e^{-\lambda}}{k!}$$

$$= e^{-\lambda} \sum_{k=0}^{\infty} \frac{(\lambda s)^k}{k!}$$

$$= e^{-\lambda} e^{\lambda s} = e^{-\lambda(1-s)} \qquad \text{for} \quad |s| < 1.$$

Then

$$\frac{d\phi(s)}{ds} = \lambda e^{-\lambda(1-s)}; \qquad \frac{d\phi(s)}{ds}\bigg|_{s\,=\,1} = \lambda;$$

$$\frac{d^2\phi(s)}{ds^2} = \lambda^2 e^{-\lambda(1-s)}; \qquad \frac{d^2\phi(s)}{ds^2}\bigg|_{s\,=\,1} = \lambda^2.$$

From (8.13) and (8.14) we verify that

$$E[\xi] = \lambda$$

$$\text{Var}[\xi] = \lambda^2 + \lambda - (\lambda)^2 = \lambda.$$

8.2.1 Generating Functions and Extinction Probabilities

Consider a branching process whose population size at stage n is denoted by X_n. Assume that the offspring distribution $p_k = \Pr\{\xi = k\}$ has the generating function $\phi(s) = E[s^\xi] = \sum_k s^k p_k$. If $u_n = \Pr\{X_n = 0\}$ is the probability of extinction by stage n, then the recursion (8.8) in terms of generating functions becomes

$$u_n = \sum_{k=0}^{\infty} p_k (u_{n-1})^k = \phi(u_{n-1}).$$

That is, knowing the generating function $\phi(s)$, we may successively compute the extinction probabilities u_n beginning with $u_0 = 0$ and then $u_1 = \phi(u_0)$, $u_2 = \phi(u_1)$, and so on.

Example The extinction probabilities when there are no offspring with probability $p_0 = \frac{1}{4}$ and two offspring with probability $p_2 = \frac{3}{4}$ were computed on page 309. We now reexamine this example using the offspring generating function $\phi(s) = \frac{1}{4} + \frac{3}{4}s^2$. This generating function is plotted as Figure 8.2. From the picture it is clear that the extinction probabilities converge upward to the smallest solution of the equation $u = \phi(u)$. This, in fact, occurs in the most general case. If u_∞ denotes this smallest solution to $u = \phi(u)$, then u_∞ gives the probability that the population eventually becomes extinct at some indefinite, but finite time. The alternative is that the population grows infinitely large, and this occurs with probability $1 - u_\infty$.

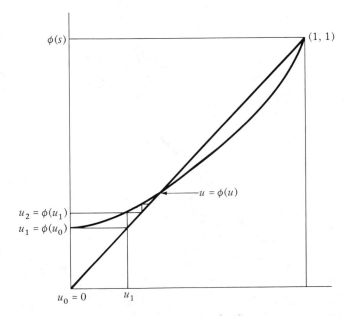

Figure 8.2   The generating function corresponding to the off-
spring distribution $p_0 = \frac{1}{4}$ and $p_2 = \frac{3}{4}$. Here $u_k = \Pr\{X_k = 0\}$ is
the probability of extinction by generation k.

For the example at hand, $\phi(s) = \frac{1}{4} + \frac{3}{4}s^2$ and the equation $u = \phi(u)$ is
the simple quadratic $u = \frac{1}{4} + \frac{3}{4}u^2$ which solves to give

$$u = \frac{4 \pm \sqrt{16 - 12}}{6} = 1, \frac{1}{3}.$$

The smaller solution is $u_\infty = \frac{1}{3}$ which is to be compared with the apparent
limit of the sequence u_n computed in the Example on page 309.

It may happen that $u_\infty = 1$, that is, the population is sure to die out at
some time. An example is depicted in Figure 8.3; the offspring distribution
is $p_0 = \frac{3}{4}$ and $p_2 = \frac{1}{4}$. We solve $u = \phi(u) = \frac{3}{4} + \frac{1}{4}u^2$ to obtain

$$u = \frac{4 \pm \sqrt{16 - 12}}{2} = 1, 3.$$

The smaller solution is $u_\infty = 1$, the probability of eventual extinction.

In general, the key is whether or not the generating function $\phi(s)$
crosses the 45° line $s = s$, and this, in turn, can be determined from the slope

$$\phi'(1) = \left.\frac{d\phi(s)}{ds}\right|_{s = 1}$$

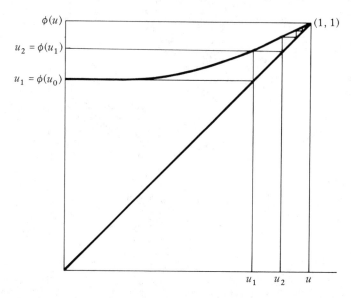

Figure 8.3   The generating function corresponding to the off-spring distribution $p_0 = \frac{3}{4}$ and $p_2 = \frac{1}{4}$.

of the generating function at $s = 1$. If this slope is less than or equal to one, then no crossing takes place, and the probability of eventual extinction is $u_\infty = 1$. On the other hand, if the slope $\phi'(1)$ exceeds one, then the equation $u = \phi(u)$ has a smaller solution that is less than one, and extinction is not a certain event.

But the slope $\phi'(1)$ of a generating function at $s = 1$ is the mean $E[\xi]$ of the corresponding distribution. We have thus arrived at the following important conclusion: If the mean offspring size $E[\xi] \leq 1$, then $u_\infty = 1$ and extinction is certain. If $E[\xi] > 1$, then $u_\infty < 1$ and the population may grow unboundedly with positive probability.

The borderline case $E[\xi] = 1$ merits some special attention. Here $E[X_n | X_0 = 1] = 1$ for all n, so the mean population size is constant. Yet the population is sure to die out eventually! This is a simple example in which the mean population size alone does not adequately describe the population behavior.

8.2.2 Probability Generating Functions and Sums of Independent Random Variables

Let ξ and η be independent nonnegative integer-valued random variables having the probability generating functions (p.g.f.'s)

$$\phi(s) = E[s^\xi] \quad \text{and} \quad \psi(s) = E[s^\eta] \quad \text{for} \quad |s| < 1.$$

The probability generating function of the sum $\xi + \eta$ is simply the product $\phi(s)\psi(s)$ because

$$E[s^{\xi+\eta}] = E[s^\xi s^\eta]$$

$$= E[s^\xi]\,E[s^\eta] \quad \text{(because } \xi \text{ and } \eta \text{ are independent)} \quad (8.15)$$

$$= \phi(s)\psi(s).$$

The converse is also true. Specifically, if the product of the p.g.f.'s of two independent random variables is a p.g.f. of a third random variable, then the third random variable equals (in distribution) the sum of the other two.

Let $\xi_1, \xi_2, \ldots$ be independent and identically distributed nonnegative integer-valued random variables with p.g.f. $\phi(s) = E[s^\xi]$. Direct induction of (8.15) implies that the sum $\xi_1 + \ldots + \xi_m$ has p.g.f.

$$E[s^{\xi_1 + \cdots + \xi_m}] = [\phi(s)]^m. \tag{8.16}$$

We extend this result to determine the p.g.f. of a sum of a random number of independent summands. Accordingly, let N be a nonnegative integer-valued random variable, independent of $\xi_1, \xi_2, \ldots$, with p.g.f. $g_N(s) = E[s^N]$, and consider the random sum (see Section 2.3)

$$X = \xi_1 + \cdots + \xi_N.$$

Let $h_X(s) = E[s^X]$ be the p.g.f. of X. We claim that $h_X(s)$ takes the simple form

$$h_X(s) = g_N[\phi(s)]. \tag{8.17}$$

To establish (8.17), consider

$$h_X(s) = \sum_{k=0}^{\infty} \Pr\{X = k\}\,s^k$$

$$= \sum_{k=0}^{\infty}\left(\sum_{n=0}^{\infty}\Pr\{X = k|N = n\}\Pr\{N = n\}\right)s^k$$

$$= \sum_{k=0}^{\infty}\left(\sum_{n=0}^{\infty}\Pr\{\xi_1 + \cdots + \xi_n = k|N = n\}\Pr\{N = n\}\right)s^k$$

$$= \sum_{k=0}^{\infty}\sum_{n=0}^{\infty}\Pr\{\xi_1 + \cdots + \xi_n = k\}\Pr\{N = n\}s^k$$

$$\text{(because } N \text{ is independent of } \xi_1, \xi_2, \ldots\text{)}$$

$$= \sum_{n=0}^{\infty}\left(\sum_{k=0}^{\infty}\Pr\{\xi_1 + \cdots + \xi_n = k\}s^k\right)\Pr\{N = n\}$$

$$= \sum_{n=0}^{\infty}\phi(s)^n\,\Pr\{N = n\} \quad \text{[using (8.16)]}$$

$$= g_N[\phi(s)] \quad \text{[by the definition of } g_N(s)\text{].}$$

With the aid of (8.16), the basic branching process equation

$$X_{n+1} = \xi_1^{(n)} + \cdots + \xi_{X_n}^{(n)} \tag{8.18}$$

can be expressed equivalently and succinctly by means of generating functions. To this end, let $\phi_n(s) = E[s^{X_n}]$ be the p.g.f. of the population size X_n at generation n, assuming that $X_0 = 1$. Then, easily, $\phi_0(s) = E[s^1] = s$, and $\phi_1(s) = \phi(s) = E[s^\xi]$. To obtain the general expression, we apply (8.17) to (8.18) to yield

$$\phi_{n+1}(s) = \phi_n[\phi(s)]. \tag{8.19}$$

This expression may be iterated in the manner

$$
\begin{aligned}
\phi_{n+1}(s) &= \phi_{n-1}\{\phi[\phi(s)]\} \\
&= \underbrace{\phi\{\cdots\phi[\phi(s)]\}}_{(n+1)\text{ iterations}} \\
&= \phi[\phi_n(s)]. \tag{8.20}
\end{aligned}
$$

That is, we obtain the generating function for the population size X_n at generation n, given that $X_0 = 1$, by repeated substitution in the probability generating function of the offspring distribution.

For general initial population sizes $X_0 = k$, the p.g.f. is

$$\sum_{j=0}^{\infty} \Pr\{X_n = j | X_0 = k\} s^j = [\phi_n(s)]^k, \tag{8.21}$$

exactly that of a sum of k independent lines of descents. From this perspective, the branching process evolves as the sum of k independent branching processes, one for each initial parent.

Example   Let $\phi(s) = q + ps$, where $0 < p < 1$ and $p + q = 1$. The associated branching process is a pure death process. In each period each individual dies with probability q and survives with probability p. The iterates $\phi_n(s)$ in this case are readily determined e.g., $\phi_2(s) = q + p(q + ps) = 1 - p^2 + p^2 s$, and generally, $\phi_n(s) = 1 - p^n + p^n s$. If we follow (8.21), the nth generation p.g.f. starting from an initial population size of k is $[\phi_n(s)]^k = [1 - p^n + p^n s]^k$.

The probability distribution of the time T to extinction may be determined from the p.g.f. as follows:

$$
\begin{aligned}
\Pr\{T = n | X(0) = k\} &= \Pr\{X_n = 0 | X_0 = k\} - \Pr\{X_{n-1} = 0 | X_0 = k\} \\
&= [\phi_n(0)]^k - [\phi_{n-1}(0)]^k \\
&= (1 - p^n)^k - (1 - p^{n-1})^k.
\end{aligned}
$$

Problems 8.2

1. Suppose that the offspring distribution is Poisson with mean $\lambda = 1.1$. Compute the extinction probabilities $u_n = \Pr\{X_n = 0 | X_0 = 1\}$ for $n = 0, 1, \ldots, 5$. What is u_∞, the probability of ultimate extinction?

2. Determine the probability generating function for the offspring distribution in which an individual either dies, with probability p_0, or is replaced by two progeny, with probability p_2, where $p_0 + p_2 = 1$.

3. Determine the probability generating function corresponding to the offspring distribution in which each individual produces 0 or N direct descendants, with probabilities p or q, respectively.

4. Consider a large region consisting of many subareas. Each subarea contains a branching process that is characterized by a Poisson distribution with parameter λ. Assume, furthermore, that the value of λ varies with the subarea, and its distribution over the whole region is that of a gamma distribution. Formally, suppose that the offspring distribution is given by

$$\pi(k|\lambda) = \frac{e^{-\lambda}\lambda^k}{k!} \qquad \text{for} \quad k = 0, 1, \ldots,$$

where λ itself is a random variable having the density function

$$f(\lambda) = \frac{\theta^\alpha \lambda^{\alpha-1} e^{-\theta\lambda}}{\Gamma(\alpha)} \qquad \text{for} \quad \lambda > 0,$$

where θ and α are positive constants. Determine the marginal offspring distribution $p_k = \int \pi(k|\lambda) f(\lambda) \, d\lambda$.
Hint: Refer to the example on page 64.

5. Let $\phi(s) = 1 - p(1 - s)^\beta$, where p and β are constants with $0 < p, \beta < 1$. Prove that $\phi(s)$ is a probability generating function and that its iterates are

$$\phi_n(s) = 1 - p^{1+\beta+\cdots+\beta^{n-1}}(1 - s)^{\beta^n} \qquad \text{for} \quad n = 1, 2, \ldots$$

6. At time 0, a blood culture starts with one red cell. At the end of one minute, the red cell dies and is replaced by one of the following combinations with the probabilities as indicated:

2 red cells	$\frac{1}{4}$
1 red, 1 white	$\frac{1}{2}$
2 white	$\frac{1}{4}$

Each red cell lives for one minute and gives birth to offspring in the same way as the parent cell. Each white cell lives for one minute and dies without reproducing. Assume that individual cells behave independently.

(a) At time $n + \frac{1}{2}$ minutes after the culture begins, what is the probability that no white cells have yet appeared?

(b) What is the probability that the entire culture eventually dies out entirely?

7. Let $\phi(s) = as^2 + bs + c$, where a, b, c are positive and $\phi(1) = 1$. Assume that the probability of extinction is u_∞, where $0 < u_\infty < 1$. Prove that $u_\infty = c/a$.

8.3 Geometrically Distributed Offspring

Consider a branching process whose offspring follow the modified geometric distribution defined by

$$p_k = bc^{k-1} \qquad \text{for} \quad k = 1, 2, \ldots, \qquad (8.22)$$

and

$$p_0 = 1 - \sum_{k=1}^{\infty} p_k,$$

where b, $c > 0$ and $b + c \le 1$. When $b = c(1 - c)$, this is the usual geometric distribution $p_k = (1 - c)c^k$, for $k = 0, 1, \ldots$. When $b = (1 - c)$, then this is the geometric distribution with $p_0 = 0$ and $p_k = (1 - c)c^{k-1}$, for $k = 1, 2, \ldots$.

For the generalized geometric distribution of (8.22), a number of quantities relevant to a branching process may be determined explicitly. First,

$$p_0 = 1 - b\sum_{k=1}^{\infty} c^{k-1} = 1 - \frac{b}{1 - c} = \frac{1 - b - c}{1 - c}.$$

The corresponding probability generating function is

$$\phi(s) = 1 - \frac{b}{1 - c} + bs\sum_{k=1}^{\infty} (cs)^{k-1} = \frac{1 - (b + c)}{1 - c} + \frac{bs}{1 - cs}. \qquad (8.23)$$

The mean is $m = \phi'(1) = b/(1 - c)^2$.

Notice that $\phi(s)$ has the form of a linear fractional transformation

$$f(s) = \frac{\alpha + \beta s}{\gamma + \delta s}, \qquad \alpha\delta - \beta\gamma \ne 0. \qquad (8.24)$$

We now record several elementary properties of linear fractional transformations:

(i) Iterates of linear fractional transformations are again linear fractional transformations, for if $f(s)$ is defined by (8.24), simple algebra gives

$$f(f(s)) = \frac{\alpha(\gamma + \beta) + (\alpha\delta + \beta^2)s}{\alpha\delta + \gamma^2 + \delta(\gamma + \beta)s}. \tag{8.25}$$

(ii) There always exist two finite (possibly identical) solutions to the equation $f(s) = s$. The solutions are called fixed points of $f(\cdot)$. If $f(s)$ is a probability-generating function, then $s_1 = 1$ is one of the fixed points and we shall see that the other fixed point s_0 is less than one, equal to one, or greater than one, according to whether $f'(1)$ is greater than, equal to, or less than one.

For the generating function given by (8.24), one can verify by straightforward algebra that the second fixed point, for $c > 0$, and $b + c < 1$, is

$$s_0 = \frac{1 - b - c}{c(1 - c)}. \tag{8.26}$$

(iii) For any two points s_i, $i = 0, 1$, it is easily seen that

$$\frac{f(s) - f(s_i)}{s - s_i} = \frac{\gamma\beta - \alpha\delta}{(\gamma + \delta s)(\gamma + \delta s_i)}.$$

Hence

$$\frac{f(s) - f(s_0)}{f(s) - f(s_1)} = \left(\frac{\gamma + \delta s_1}{\gamma + \delta s_0}\right)\left(\frac{s - s_0}{s - s_1}\right). \tag{8.27}$$

If we now let s_0 and s_1 be the two (nonidentical) fixed points of $f(\cdot)$ and write $w = f(s)$, (8.27) becomes

$$\frac{w - s_0}{w - s_1} = \kappa\left(\frac{s - s_0}{s - s_1}\right), \tag{8.28}$$

where κ can be calculated from (8.27) or more simply from (8.26) by setting $s = 0$.

Using (8.28) we easily obtain the iterates $f_n(s) = w_n$ of $f(s)$:

$$\frac{w_2 - s_0}{w_2 - s_1} = \kappa\frac{w_1 - s_0}{w_1 - s_1} = \kappa\left(\kappa\frac{s - s_0}{s - s_1}\right),$$

and in general

$$\frac{w_n - s_0}{w_n - s_1} = \kappa^n\left(\frac{s - s_0}{s - s_1}\right). \tag{8.29}$$

For the generating function of the geometric distribution given by (8.25), noting that the fixed points are $s_0 = (1 - b - c)/c(1 - c)$ and $s_1 = 1$, we obtain

$$\kappa = \frac{(1 - c)^2}{b} = \frac{1}{m},$$

where m is the mean of the geometric distribution. For $m \neq 1$ the two fixed points s_0 and 1 are different; hence, solving for w_n in (8.29) gives

$$w_n = \frac{s_0 - (1/m^n)[(s - s_0)/(s - 1)]}{1 - (1/m^n)[(s - s_0)/(s - 1)]}, \qquad m \neq 1, \qquad (8.30)$$

which may be written in the form

$$\phi_n(s) = 1 - m^n\left(\frac{1 - s_0}{m^n - s_0}\right) + \frac{m^n[(1 - s_0)/(m^n - s_0)]^2 s}{1 - [(m^n - 1)/(m^n - s_0)]s} \qquad (8.31)$$

Then the probabilities of extinction at the nth generation are

$$\Pr\{X_n = 0\} = \phi_n(0) = 1 - m^n\left(\frac{1 - s_0}{m^n - s_0}\right).$$

Note that this expression converges to s_0 as $n \to \infty$ if $m > 1$ and to 1 if $m < 1$. The probability of a given population size in the nth generation, $\Pr\{X_n = k\}$, $k = 1, 2, \ldots$, can be computed by simply expanding (8.31) as a power series in s. If we define the time to extinction T as the smallest subscript n such that $X_n = 0$, i.e., the first passage time into state 0, then

$$\Pr\{T \leq n\} = \Pr\{X_n = 0\} = \phi_n(0)$$

and

$$\Pr\{T = n\} = \Pr\{T \leq n\} - \Pr\{T \leq n - 1\} = \phi_n(0) - \phi_{n-1}(0).$$

In the case $m \neq 1$, we have

$$\Pr\{T = n\} = 1 - m^n\left(\frac{1 - s_0}{m^n - s_0}\right) - 1 + m^{n-1}\left(\frac{1 - s_0}{m^{n-1} - s_0}\right)$$

$$= m^{n-1}s_0 \frac{(m - 1)(1 - s_0)}{(m^n - s_0)(m^{n-1} - s_0)} \qquad \text{for} \quad n = 1, 2, \ldots.$$

If $m = 1$, then $b = (1 - c)^2$ and the equation $\phi(s) = s$ has the double root $s = 1$ and no other root. In fact,

$$\phi(s) = c + \frac{(1 - c)^2 s}{1 - cs} = \frac{c - (2c - 1)s}{1 - cs}.$$

Then

$$\phi_2(s) = \phi(\phi(s)) = \frac{c - (2c - 1)[(c - (2c - 1)s)/(1 - cs)]}{1 - c[(c - (2c - 1)s)/(1 - cs)]}$$

$$= \frac{2c - (3c - 1)s}{1 + c - 2cs}$$

and by induction

$$\phi_n(s) = \frac{nc - [(n + 1)c - 1]s}{1 + (n - 1)c - ncs}. \qquad (8.32)$$

In the case $m = 1$ we have the extinction probabilities

$$\Pr\{X_n = 0\} = \phi_n(0) = \frac{nc}{1 + (n - 1)c} \qquad \text{for} \quad n = 1, 2, \ldots$$

Further, the time to extinction T has the distribution

$$\Pr\{T = n\} = \phi_n(0) - \phi_{n-1}(0) = \frac{nc}{1 + (n - 1)c} - \frac{(n - 1)c}{1 + (n - 2)c}$$

$$= \frac{c(1 - c)}{[1 + (n - 1)c][1 + (n - 2)c]}.$$

8.3.1 Some Conditional Limit Theorems

Assuming the offspring distribution given in (8.22), we will determine the long term probabilities conditioned on nonextinction; that is, we find

$$\lim_{n \to \infty} \Pr\{X_n = k \mid X_n > 0\}.$$

For this purpose, we form the corresponding conditional p.g.f. (Note that the $k = 0$ term is absent.)

$$\sum_{k=1}^{\infty} \Pr\{X_n = k \mid X_n > 0\} s^k = \sum_{k=1}^{\infty} \frac{\Pr\{X_n = k, X_n > 0\}}{\Pr\{X_n > 0\}} s^k$$

$$= \sum_{k=1}^{\infty} \frac{\Pr\{X_n = k\}}{\Pr\{X_n > 0\}} s^k$$

$$= \frac{1}{\Pr\{X_n > 0\}} \sum_{k=1}^{\infty} \Pr\{X_n = k\} s^k$$

$$= \frac{\phi_n(s) - \phi_n(0)}{1 - \phi_n(0)}.$$

From the expression (8.31) for the particular branching process at hand, we now have $(m = \phi'(1))$

$$\sum_{k=1}^{\infty} \Pr\{X_n = k \mid X_n > 0\} s^k$$

$$= \left(\frac{1}{m^n[(1 - s_0)/(m^n - s_0)]} \right) \left(\frac{m^n[(1 - s_0)/(m^n - s_0)]^2 s}{1 - [(m^n - 1)/(m^n - s_0)]s} \right)$$

$$= \frac{(1 - s_0)}{m^n - s_0} \times \frac{s}{1 - [(m^n - 1)/(m^n - s_0)s]}$$

and expanding into a geometric series

$$= \sum_{k=0}^{\infty} \left(\frac{1 - s_0}{m^n - s_0} \right) \left(\frac{m^n - 1}{m^n - s_0} \right)^k s^{k+1}$$

or

$$\sum_{k=1}^{\infty} \Pr\{X_n = k | X_n > 0\} s^k = \sum_{k=1}^{\infty} \left(\frac{1 - s_0}{m^n - s_0} \right) \left(\frac{m^n - 1}{m^n - s_0} \right)^{k-1} s^k.$$

Equating coefficients of s^k gives

$$\Pr\{X_n = k | X_n > 0\} = \left(\frac{1 - s_0}{m^n - s_0} \right) \left(\frac{m^n - 1}{m^n - s_0} \right)^{k-1}. \tag{8.33}$$

When $(1 - b - c)/[c(1 - c)] > 1$ we necessarily have $m < 1$ and therefore $m^n \to 0$ as $n \to \infty$. Taking $n \to \infty$ in (8.33) gives

$$\lim_{n \to \infty} \Pr\{X_n = k | X_n > 0\} = \frac{1 - s_0}{-s_0} \left(\frac{1}{s_0} \right)^{k-1}$$

$$= \left(1 - \frac{1}{s_0} \right) \left(\frac{1}{s_0} \right)^{k-1}, \qquad k = 1, 2, \ldots$$

Problems 8.3

1. Consider a branching process whose offspring follow the geometric distribution $p_k = (1 - c)c^k$ for $k = 0, 1, \ldots$ where $0 < c < 1$. Determine the probability of eventual extinction.

8.4 Variations on Branching Processes

8.4.1 Multiple Branching Processes

Population growth processes often involve several life history phases (e.g., juvenile, reproductive adult, senescence) with different viability and behavioral patterns. We consider a number of examples of branching processes that take account of this characteristic.

   For the first example, suppose that a mature individual produces offspring according to the p.g.f. $\phi(s)$. Consider a population of immature individuals each of which grows to maturity with probability p and then reproduces independently of the status of the remaining members of the population. With probability $1 - p$ an immature individual will not attain maturity and thus will leave no descendants. With probability p an individual will reach maturity and reproduce an offspring number determined according to the p.g.f. $\phi(s)$. Therefore the progeny size distribution (or equivalently the p.g.f.) of a typical immature individual taking account of both contingencies is

$$(1 - p) + p\phi(s). \tag{8.34}$$

If individuals are censused at the adult (mature) stage, the aggregate number of mature individuals contributed by a mature individual will now have p.g.f.

$$\phi(1 - p + ps). \tag{8.35}$$

(The student should verify this finding.)

It is worth emphasis that the p.g.f.'s (8.34) and (8.35) have the same mean $p\phi'(1)$ but generally not the same variance, the first being

$$p[\phi''(1) + \phi'(1) - (\phi'(1))^2]$$

as compared to

$$p^2\phi''(1) + p\phi'(1) - p^2(\phi'(1))^2.$$

A second example leading to (8.35) as opposed to (8.34) concerns the different forms of mortality that affect a population. We appraise the strength (*stability*) of a population as the probability of indefinite survivorship = 1 − probability of eventual extinction.

In the absence of mortality the offspring number X of a single individual has the p.g.f. $\phi(s)$. Assume, consistent with the postulates of a branching process, that all offspring in the population behave independently governed by the same probability laws. Assume also an adult population of size $X = k$. We consider three types of mortality.

Mortality of Individuals   Let p be the probability of an offspring surviving to reproduce independent of what happens to others. Thus the contribution of each litter (family) to the adult population of the next generation has a binomial distribution with parameters (N, p) where N is the progeny size of the parent with p.g.f. $\phi(s)$. The p.g.f. of the adult numbers contributed by a single grandparent is therefore $\phi(q + ps)$, $q = 1 - p$, and for the population as a whole is

$$\psi_1(s) = [\phi(q + ps)]^k. \tag{8.36}$$

This type of mortality might reflect predation on adults.

Mortality on Litters   Independent of what happens to other litters each litter survives with probability p and is wiped out with probability $q = 1 - p$. That is, given an actual litter size N, the effective litter size is N with probability p and 0 with probability q. The p.g.f. of adults in the following generation is accordingly

$$\psi_2(s) = [q + p\phi(s)]^k. \tag{8.37}$$

This type of mortality might reflect predation on juveniles or on nests and eggs in the case of birds.

Mortality on Generations   An entire generation survives with probability p and is wiped out with probability q. This type of mortality might represent environmental catastrophes (e.g., forest fire, flood). The p.g.f. of population size in the next generation in this case is

$$\psi_3(s) = q + p[\phi(s)]^k. \tag{8.38}$$

All the p.g.f.'s (8.36) through (8.38) have the same mean but usually different variances.

It is interesting to assess the relative stability of these three models. That is, we need to compare the smallest positive roots of $\psi_i(s) = s$, $i = 1$, 2, 3, which we will denote by $s_i^{\star}$, $i = 1, 2, 3$, respectively.

We will show by convexity analysis that

$$\psi_1(s) \le \psi_2(s) \le \psi_3(s).$$

A function $f(x)$ is convex in x if for every x_1 and x_2 and $0 < \lambda < 1$, then $f(\lambda x_1 + (1 - \lambda)x_2) \le \lambda f(x_1) + (1 - \lambda)f(x_2)$. In particular, the function $\phi(s) = \Sigma_{k=0}^{\infty} p_k s^k$ for $0 < s < 1$ is convex in s since for each positive integer k, $[(\lambda s_1) + (1 - \lambda)s_2]^k \le \lambda s_1^k + (1 - \lambda)s_2^k$ for $0 < \lambda$, $s_1, s_2 < 1$. Now $\psi_1(s) = [\phi(q + ps)]^k < [q\phi(1) + p\phi(s)]^k = [q + p\phi(s)]^k = \psi_2(s)$ and then $s_1^{\star} < s_2^{\star}$. Thus the first model is more stable than the second.

Observe further that due to the convexity of $f(x) = x^k$, $x > 0$, that $\psi_2(s) = [p\phi(s) + q]^k < p[\phi(s)]^k + q \times 1^k = \psi_3(s)$, and thus $s_2^{\star} < s_3^{\star}$, implying the second model is more stable than the third model. In conjunction we get the ordering $s_1^{\star} < s_2^{\star} < s_3^{\star}$.

8.4.2 Branching Processes with Immigration

A Markov chain $\{Y_n\}$ with transition probabilities $\{P_{ij}\}$ whose one-step p.g.f.'s have the form

$$f_i(s) = \sum_{j=0}^{\infty} P_{ij} s^j = [\phi(s)]^i \psi(s) \tag{8.39}$$

can be regarded as a branching process with immigration for the following reasons: $\phi(s)$ can be interpreted as the p.g.f. of the number of progeny of each individual in the branching process. The p.g.f. $\psi(s)$ refers to the number of immigrants into the population during a single generation. These events are independent since the p.g.f.'s are multiplied in (8.39) (see Section 8.2).

In a branching process with immigration assume that $\phi'(1) = m < 1$ (that is, on the average each parent produces less than one offspring). Ordinarily the population would go extinct. But immigration may maintain a positive number despite this regressive tendency. We would expect therefore the ultimate attainment of a stationary measure. Let π_i be the stationary probability of state i (population size i) and let its p.g.f. be denoted by

$$\pi(s) = \sum_{i=0}^{\infty} \pi_i s^i.$$

Since

$$\pi_k = \sum_{k=0}^{\infty} \pi_i P_{ik}, \qquad i = 0, 1, 2, \ldots, \tag{8.40}$$

we multiply (8.40) by s^k and sum to get

$$\sum_{i=0}^{\infty} \pi_k s^k = \sum_{i=0}^{\infty} \pi_i \sum_{k=0}^{\infty} P_{ik} s^k$$

yielding [substituting from (8.39)] the functional equation

$$\pi(s) = \psi(s) \sum_{i=0}^{\infty} \pi_i [\phi(s)]^i = \psi(s) \pi(\phi(s)). \tag{8.41}$$

For the special case $\phi(s) = q + ps$ (a pure death branching process) and $\psi(s) = e^{s-1}$ (a Poisson distributed number of immigrants) the functional equation (8.41) becomes

$$\pi(s) = e^{s-1} \pi(q + ps).$$

We try a solution of the form $\pi(s) = e^{\lambda(s-1)}$ which gives

$$e^{-\lambda} e^{\lambda s} = e^{(\lambda p + 1)s} e^{-1 + \lambda q - \lambda}$$

and $\lambda = 1/q$ works in the above identity yielding $\pi(s) = e^{(s-1)/q}$. We recognize this as the p.g.f. of a Poisson distribution whose mean is $1/q$.

8.4.3 Branching Processes with Killing

We consider a population of individuals that reproduces in the following way. Each individual alive at a particular time produces, independently of the others alive at that time, a random number of offspring, each with the distribution of a random variable Z satisfying

$$\Pr\{Z = k\} = p_k, \qquad k \geq 0; \qquad p_0 > 0, \qquad p_0 + p_1 < 1.$$

An offspring born to a particular individual has a probability $1 - \alpha$ of being found to be defective in some way. We assume in this simplest case that detection of defectives is independent over all individuals in a family and over families. To avoid trivialities, we will assume that $0 < \alpha < 1$. A family of size k survives to reproduce if and only if all k individuals are normal. This has probability $p_k \alpha^k$, $k \geq 0$. It follows that if $f(s) = \sum_{k=0}^{\infty} p_k s^k$ is the p.g.f. of Z, then the (improper) p.g.f. of the number of offspring born to an individual with no defective offspring is

$$g(s) = f(\alpha s) \tag{8.42}$$

and $1 - g(1) = 1 - f(\alpha)$ is the probability that a family contains at least one defective individual. The population now evolves as follows. Let X_n be the number of individuals alive at time n. The population continues to the next generation only if no defective individuals are born. Otherwise, we say the process has ended by a killing (or detection) event. Under the simple detection scheme, it is in principle straightforward to analyze the process. We take $X_0 = 1$ and define the iterates of $g(\cdot)$ by

$$g_0(s) = s, \qquad g_n(s) = g(g_{n-1}(s)), \qquad n \geq 1. \tag{8.43}$$

Intuitively, it is clear that the process ends either in extinction or in detection. Let q be the probability that extinction prevails. It is simple to show that if $0 < \alpha < 1$, then q is the unique root satisfying $0 < q < 1$ of the functional equation $g(s) = f(\alpha s) = s$; if $X_0 = i$, then the extinction probability is q^i. For the special case where $f(s) = \frac{1}{2} + \frac{1}{2} s^2$ the functional equation becomes $q = f(\alpha q) = \frac{1}{2} + \frac{1}{2}\alpha^2 q^2$ and therefore $q = (1 \pm \sqrt{1 - \alpha^2})/\alpha^2$ but since $0 \leq q \leq 1$, necessarily $q = (1 - \sqrt{1 - \alpha^2})/\alpha^2$.

The probability that the detection time T_D is greater than n is given by

$$\Pr\{T_D > n\} = \Pr\{X_n \geq 0\} = g_n(1).$$

Since $g_n(s)$ is decreasing in n for s in (q, s_1), where s_1 is the larger root of $f(\alpha s) = s$ satisfying $s_1 > 1$, we conclude that $g_n(1) \to q$ as $n \to \infty$ (why?). This conclusion implies that the process terminates either by detection or extinction. The probability that detection prevails is then $1 - q^i$ if $X_0 = i$.

Two relevant distributions in the study of detection times are the (conditional) detection time, T_D, and $T = \min(T_0, T_D)$, the time to extinction or detection. We have

$$\Pr[T_D > n \mid T_D < \infty] = \frac{g_n(1) - q}{1 - q}, \qquad n \geq 0 \tag{8.44}$$

and

$$\Pr[T > n] = g_n(1) - g_n(0), \qquad n \geq 0.$$

Consider the foregoing model with

$$f(s) = \frac{r + s(1 - r - p)}{1 - ps}, \qquad 0 < r < 1, \quad 0 < p < 1, \quad p + r \leq 1,$$

whose iterates by (8.25) to (8.28) are accessible (Section 8.3). The probabilities are

$$p_k = \begin{cases} r, & k = 0; \\ (1 - r)(1 - p)p^{k-1}, & k \geq 1. \end{cases}$$

In this latter case, we have

$$g(s) = \frac{r + s\alpha(1 - r - p)}{1 - p\alpha s}.$$

Let $0 < s_0 < 1 < s_1$ be the roots of the equation $g(s) = s$, and define

$$K = \frac{1 - p\alpha s_1}{1 - p\alpha s_0}. \tag{8.45}$$

It is readily checked that $0 < K < 1$, and using a standard method for iterating a linear fractional function (see Section 8.3), we obtain

$$g_n(s) = \frac{s_0 s_1 (K^n - 1) + s(s_0 - K^n s_1)}{(K^n s_0 - s_1) - s(K^n - 1)}, \qquad 0 < s < s_1. \tag{8.46}$$

It follows immediately from (8.46) that $P(T_0 \le n) = g_n(0) \to s_0$ as $n \to \infty$, so that $q = P(T_0 < \infty) = s_0$. Further, from (8.44)

$$P[T_D > n \mid T_D < \infty] = \frac{K^n(s_1 - s_0)}{K^n(1 - s_0) + (s_1 - 1)}.$$

Another pertinent conditioning focuses on those paths that become extinct, rather than those that end in detection. The process $\{X_n, n \ge 0\}$ that arises by conditioning on extinction is again a branching process, with offspring p.g.f. given by

$$\tilde{f}(s) = \frac{g(sq)}{q} = \frac{p\alpha s_1 + s(1 - p\alpha(s_1 + s_0))}{1 - sp\alpha s_0}, \qquad 0 \le s < 1.$$

The process has offspring mean $\tilde{f}'(1) = K < 1$, and this conditional process, of course, goes to extinction.

Let T_0 be the time to reach $\{0\}$, the extinction time. It is not difficult to show that if $\tilde{X}_0 = 1$,

$$\tilde{a}_j = \lim_{n \to \infty} P[\tilde{X}_n = j \mid \tilde{T}_0 > n] = \left(1 - \frac{s_0}{s_1}\right)\left(\frac{s_0}{s_1}\right)^{j-1}, \qquad j \ge 1.$$

This distribution has mean $1/[1 - (s_0/s_1)]$.

Problems 8.4

1. Suppose that in a branching process the number of offspring of an initial particle has a distribution whose generating function is $f(s)$. Each member of the first generation has a number of offspring whose distribution has generating function $g(s)$. The next generation has generating function f, the next has g, and the distributions continue to alternate in this way from generation to generation.
   (a) Determine the extinction probability of the process in terms of $f(s)$ and $g(s)$.

(b) Determine the mean population size at generation n.

(c) Would any of these quantities change if the process started with the $g(s)$ process and then continued to alternate?

8.5 Some Stochastic Models of Plasmid Reproduction and Plasmid Copy Number Partition

Plasmids are small circular DNA segments found mainly in bacterial cells that replicate autonomously. They play a role in resistance to antibiotics, in conjugation (sex in bacteria), in exchange (or joining) of genetic material between bacterial colonies under special circumstances, and in other functions of bacterial relevance.

We consider a cell line initiated by a single cell at time $n = 0$, and growing by binary splitting at times $n = 1, 2, \ldots$. The initial cell contains a number of plasmids, which replicate in the cell and are then distributed to the two daughter cells at cell division. We will analyze two ingredients of this process. The first is the *replication* mechanism, the second the *partition* mechanism. The replication process determines how plasmids within a given cell replicate before cell division, whereas the partition mechanism describes how the plasmids, after replication in a cell, are divided among the two daughter cells.

The models proposed describe the behavior of the number of plasmids contained in a cell in any specific generation of a randomly chosen line of descent through the cell population. Interest centers on two facets:

(i) stability of plasmid copy number which includes the mean and variance of the number of plasmids in the line of descent at particular times, and

(ii) the cure rate. Along a line of descent the cells must eventually contain no plasmids (be cured). At what rate does such curing occur?

8.5.1 The Replication Process

There are three such mechanisms that will be described, the multiplicative model, the additive model, and the equilibrium model.

The Multiplicative (or Doubling) Model (M)   Suppose that a cell contains i (> 0) plasmids just before replication. We assume that each plasmid replicates once independently of all others, giving rise to either one offspring plasmid (with probability $1 - p$) or two offspring (with probability p), where $0 < p \le 1$. We define

$$R_{il} = \Pr\{l \text{ plasmids in cell before partition} | i \text{ before replication}\}. \tag{8.47}$$

In the present case in order to produce a total of l plasmids on replication there are i trials, and if k produced singly while $(i - k)$ doubled, then the total would be $k + 2(i - k) = l$ or $k = 2i - l$. The probability of this outcome is

$$R_{il} = \binom{i}{i - k} p^{i-k}(1 - p)^k = \binom{i}{l - i} p^{l-i}(1 - p)^{2i-l},$$

$$l = i, i + 1, \ldots, 2i \quad (8.48)$$

When $p = 1$, the model reduces to one in which each plasmid produces exactly two offspring, the "doubling" case. The parameter p takes account of random variation in this reproduction scheme.

The Additive Model (A)   Consider a situation in which a cell contains i (> 0) plasmids. In this case, there is a fixed number N of replication events, each of which is successful with probability p, independently for each such event. The number of plasmids in the cell prior to partition and after replication is then equal to $i + Y$, where Y is a binomial random variable with parameters N and p. This model can be viewed as providing a regulation mechanism to control the replication process. It is easily seen that for $i > 0$,

$$R_{il} = \Pr\{Y = l\} = \binom{N}{l - i} p^{l-i}(1 - p)^{N-l+i},$$

$$l = i, i + 1, \ldots, N + i \quad (8.49)$$

The Equilibrium Model (E)   This process also seeks to describe a replication control mechanism. Given i (> 0) plasmids in a cell, and a replication event success probability of $p = 1$, the number of plasmids in the cell before partition is $2N$, regardless of the value of i. This model reflects the inability of a given cell to contain more than a given number of plasmids. When $p < 1$, we view the process as leaving $i + Y$ offspring plasmids, where Y is a binomial random variable with parameters $2N - i$ and p. It follows that

$$R_{il} = \binom{2N - i}{l - i} p^{l-i}(1 - p)^{2N-l}, \qquad l = i, \ldots, 2N \quad (8.50)$$

In all three models, it is assumed that p and the form of the replication probabilities R_{il} remain constant through time.

We are supposing that a cured cell (one containing no plasmids) leaves cured daughter cells, and therefore, $R_{00} = 1$.

8.5.2 Random Lines of Descent

A random line of descent through the cell line is defined as follows: We choose with probability $\frac{1}{2}$ one of the two daughter cells at time $n = 1$, with probability $\frac{1}{2}$ one of the daughter cells of that chosen at time $n = 2$, and so on. Now let X_n be the number of plasmids in the cell (before replication) at generation time n, $n = 0, 1, 2, \ldots$. For the three models described, it is clear that $\{X_n, n \geq 0\}$ is a Markov chain, and its one-step transition probabilities $\|P_{ij}\|$ are given by

$$P_{ij} = \Pr\{X_{n+1} = j | X_n = i\} = \sum_l R_{il} Q_{lj}, \tag{8.51}$$

where, as in (8.47),

$$R_{il} = \Pr\{l \text{ plasmids in cell before partition} | i \text{ before replication}\},$$

and

$$Q_{lj} = \Pr\{\text{randomly chosen daughter cell inherits } j \text{ plasmids} | \text{mother cell}$$

$$\text{had } l \text{ before partition}\}.$$

So as not to obscure the effects of the replication process, we will in all cases assume a *random* partition process. This means that any plasmid in a cell before partition is assigned independently and at random to either of the two daughter cells. Accordingly

$$Q_{lj} = \binom{l}{j} 2^{-l}, \qquad j = 0, 1, \ldots, l. \tag{8.52}$$

The type of Markov chain represented by the transition probabilities (8.51) is most easily ascertained by looking at the probability generating functions $f_i(s)$, $i > 0$ given by

$$f_i(s) = \sum_j P_{ij} s^j, \qquad 0 \leq s \leq 1. \tag{8.53}$$

Using (8.52) and the replication probabilities (8.48) through (8.50), these p.g.f.'s are easily evaluated. For example, in the multiplicative model

$$\sum_j P_{ij} s^j = \sum_j \left(\sum_l R_{il} Q_{lj} \right) s^j$$

$$= \sum_l R_{il} \sum_j Q_{lj} s^j$$

$$= \sum_l R_{il} \sum_j \binom{l}{j} \frac{1}{2^l} s^j = \sum_l R_{il} \left(\frac{1}{2} + \frac{s}{2} \right)^l$$

$$\text{[substituting from (8.52)]}$$

$$= \sum_{l=i}^{2i} \binom{i}{l-i} p^{l-i} (1 - p)^{2i-l} \left(\frac{1}{2} + \frac{s}{2} \right)^l$$

(making the change of summation variable $l - i = m$, i.e., equivalently rearranging the order of summation)

$$= \sum_{m=0}^{i} \binom{i}{m} p^m (1 - p)^{i-m} \left(\frac{1}{2} + \frac{s}{2} \right)^m \left(\frac{1}{2} + \frac{s}{2} \right)^i$$

$$= \left[1 - p + \left(\frac{1}{2} + \frac{s}{2} \right) p \right]^i \left(\frac{1}{2} + \frac{s}{2} \right)^i$$

$$= \left(\frac{1 + s}{2}\right)^i \left(1 - \frac{p}{2} + \frac{p}{2}s\right)^i.$$

The results for all three models are given in Table 8.1.

Table 8.1

Model	p.g.f. $f_i(s)$, $i > 0$
Multiplicative	$\left(\frac{1}{2} + \frac{1}{2}s\right)^i \left(1 - \frac{p}{2} + \frac{p}{2}s\right)^i$
Additive	$\left(\frac{1}{2} + \frac{1}{2}s\right)^i \left(1 - \frac{p}{2} + \frac{p}{2}s\right)^N$
Equilibrium	$\left(\frac{1}{2} + \frac{1}{2}s\right)^i \left(1 - \frac{p}{2} + \frac{p}{2}s\right)^{2N-i}$

From Table 8.1, it is immediate that for the multiplicative model, X_n is a branching process with offspring p.g.f. $\phi(s)$ (see Section 8.2) given by

$$\phi(s) = \left(\frac{1}{2} + \frac{1}{2}s\right)\left(1 - \frac{p}{2} + \frac{p}{2}s\right).$$

Similarly, the additive model may be viewed as a branching process with immigration (see Section 8.4.2), however, stopped wherever 0 is reached. The offspring p.g.f. $\phi(s)$ and immigration p.g.f. $\psi(s)$ are given by

$$\phi(s) = \frac{1}{2} + \frac{1}{2}s, \qquad \psi(s) = \left(1 - \frac{p}{2} + \frac{p}{2}s\right)^N.$$

The equilibrium model has no such convenient interpretation.

All three processes have an absorbing state at 0, which corresponds to curing of the line of descent. We will write the transition matrix $\mathbf{P} = \|P_{i,j}\|$ in the form

$$\mathbf{P} = \begin{pmatrix} 1 & \mathbf{0}' \\ \mathbf{q} & \mathbf{T} \end{pmatrix},$$

where $\mathbf{T}$ corresponds to transitions among the transient states $\{1, 2, \ldots\}$.

In the multiplicative model the expected progeny size is $f'(1) = (1 + p)/2$, and therefore the expected plasmid number in a line of descent at generation t is $i[(1 + p)/2]^t$ where i is the plasmid number in the initial cell.

From general theory the rate of approach to state 0 is governed by the largest eigenvalue $\rho^\star$ of the matrix $\mathbf{T}$. This is difficult to obtain explicitly but can be approximated by numerical means. In the additive model the following bound can be obtained

$$\rho^\star \leq 1 - \left(\frac{1}{2}\right)^{2N}.$$

For the equilibrium model which is a finite Markov chain of size $2N + 1$ numerical work suggests that $\frac{1}{2} < \rho^\star < 1$.

8.6 Population Growth Processes with Interacting Types

A problem of importance in the study of certain stochastic population growth processes is the ascertainment of conditions delineating cases of certain or noncertain extinction of a population. In this section we discuss this problem for certain classes of growth processes allowing various kinds of interaction phenomena among the types. The trend associated with changes of expected sizes generally overwhelms any effects of statistical fluctuation so that we would expect the conditions for extinction to reduce to natural conditions on expected values. This statement is true but its proof is beyond the scope of this book. Effectively, for branching processes involving multi-types with interactions when the expected growth rate exceeds one, extinction does not happen with certainty, and for a realization where extinction does not occur, the population actually grows at an exponential rate.

Multiple Mating

Consider a population consisting of two types of individuals, males and females. We assume that all females can produce offspring governed by the p.g.f. $\phi(s)$ provided the population contains at least one male. The biological motivation perhaps comes from the situation of cows and bulls. Each cow bears progeny provided there is at least one bull (stud) available for mating.

Let the probability that an offspring is female be α, $0 < \alpha < 1$. Following the reasoning of Equation (8.35), interpreting a male offspring as a mutant type we find that the p.g.f. of the number of female offspring is $\phi(\alpha s + (1 - \alpha))$.

The probability of producing at least one male is clearly $1 - \phi(\alpha)$. The probability that exclusively female progeny occur in a litter of k is $p_k \alpha^k$. The p.g.f. summarizing these realizations is $\Sigma p_k \alpha^k s^k = \phi(\alpha s)$.

Combining these possibilities we deduce that the p.g.f. for the number of females produced in the next generation *conditioned* that at least one male is produced as well is

$$\frac{\phi(\alpha s + 1 - \alpha) - \phi(\alpha s)}{1 - \phi(\alpha)}. \tag{8.54}$$

The expected number of females under this conditioning is then

$$\gamma = \alpha\left(\frac{\phi'(1) - \phi'(\alpha)}{1 - \phi(\alpha)}\right). \tag{8.55}$$

We would expect that if the quantity γ exceeds 1 then with positive probability extinction does not occur and the population size grows indefinitely provided the initial population makeup involves at least one male and female. This is true and can be stated more formally as follows: If each female contributes to the next generation on an average more than one offspring and provided each litter has positive chance of containing a male offspring, then extinction is not a certain event and in this case the population grows at an exponential rate. Note that $\phi'(1) = m$ is the unrestricted expected progeny size and αm the corresponding expected female number. It is nontrivial but correct that if $\alpha m > 1$, then $\gamma > 1$.

Permanent Pairing

Consider a population composed of females and males whose numbers are denoted by (X_n, X_n) in the nth generation. Suppose in each generation permanent pairings of a male with a female take place. Thus,

$$V_n = \min(X_n, Y_n) \tag{8.56}$$

pairs are formed. For each couple let $f(s_1, s_2)$ be the joint p.g.f. of the numbers of female and male progeny produced. We assume that couples behave independently of each other with the same p.g.f. $f(s_1, s_2)$. Let

$$m_1 = \text{expected number of female offspring per couple,}$$

$$m_2 = \text{expected number of male offspring per couple.}$$

The condition for extinction of this V_n process is the content of the next result.

Consider the V_n process as defined in (8.56). Then

$$V_n \to 0 \qquad \text{if and only if} \quad \min(m_1, m_2) \le 1.$$

Thus, the natural requirements on expected progeny sizes determine completely whether certain or noncertain extinction of the population occurs.

A more general version can be developed where each female acquires as a family p males (or vice versa).

8.7 Deterministic Population Growth with Age Distribution

In this section we will discuss a simple deterministic model of population growth that takes into account the age structure of the population. Both the renewal theorem (Section 7.6) and generating functions (Section 8.2) will play a role in the analysis. As the language will suggest, the deterministic model that we treat may be viewed as describing the mean population size in a more elaborate stochastic model that is beyond our scope to develop fully.

8.7.1 A Simple Growth Model

We consider a single species evolving in discrete time $t = 0, 1, 2, \ldots$, and we let N_t be the population size at time t. We assume that each individual present in the population at time t gives rise to a constant number λ of offspring that form the population at time $t + 1$. (If death does not occur in the model, then we include the parent as one of the offspring, and then necessarily $\lambda \geq 1$.) If N_0 is the initial population size, and each individual gives rise to λ offspring, then

$$N_1 = \lambda N_0,$$
$$N_2 = \lambda N_1 = \lambda^2 N_0,$$

and, in general,

$$N_t = \lambda^t N_0. \tag{8.57}$$

If $\lambda > 1$, then the population grows indefinitely in time, if $\lambda < 1$, then the population dies out, while if $\lambda = 1$, then the population size remains constant at $N_t = N_0$ for all $t = 0, 1, \ldots$.

8.7.2 The Model with Age Structure

We shall now introduce an age structure in the population. We need the following notation:

$n_{u,t}$ = the number of individuals of age u in the population at time t;

$N_t = \sum_{u=0}^{\infty} n_{u,t}$ = the total number of individuals in the population at time t;

b_t = the number of new individuals created in the population at time t, the number of births;

β_u = the expected number of progeny of a single individual of age u in one time period;

l_u = the probability that an individual will survive, from birth, at least to age u.

The conditional probability that an individual survives at least to age u, given that it has survived to age $u - 1$ is simply the ratio l_u/l_{u-1}. The *net maternity function* is the product

$$m_u = l_u \beta_u$$

and is the birth rate adjusted for the death of some fraction of the population. That is, m_u is the expected number of offspring at age u of an individual now of age 0.

Let us derive the total progeny of a single individual during its lifespan. An individual survives at least to age u with probability l_u, and then during the next unit of time gives rise to β_u offspring. Summing $l_u\beta_u = m_u$ over all ages u then gives the total progeny of a single individual:

$$M = \sum_{u=0}^{\infty} l_u\beta_u = \sum_{u=0}^{\infty} m_u. \tag{8.58}$$

If $M > 1$, then we would expect the population to increase over time; if $M < 1$, then we would expect the population to decrease; while if $M = 1$, then the population size should neither increase nor decrease in the long run. This is indeed the case, but the exact description of the population evolution is more complex, as we will now determine.

In considering the effect of age structure on a growing population, our interest will center on b_t, the number of new individuals created in the population at time t. We regard β_u, l_u, and $n_{u,0}$ as known, and the problem is to determine b_t for $t \geq 0$. Once b_t is known, then $n_{u,t}$ and N_t may be determined according to, for example,

$$n_{0,1} = b_1, \tag{8.59}$$

$$n_{u,1} = n_{u-1,0}\left[\frac{l_u}{l_{u-1}}\right] \quad \text{for} \quad u \geq 1, \tag{8.60}$$

and

$$N_1 = \sum_{u=0}^{\infty} n_{u,1}. \tag{8.61}$$

In the first of these simple relations, $n_{0,1}$ is the number in the population at time 1 of age 0, which obviously is the same as b_1, those born in the population at time 1. For the second equation, $n_{u,1}$ is the number in the population at time 1 of age u. These individuals must have survived from the $n_{u-1,0}$ individuals in the population at time 0 of age $u - 1$; the conditional probability of survivorship is $[l_u/l_{u-1}]$, which explains the second equation. The last relation simply asserts that the total population size results by summing the numbers of individuals of all ages. The generalizations of (8.59) through (8.61) are

$$n_{0,t} = b_t, \tag{8.62}$$

$$n_{u,t} = n_{u-1,t-1}\left[\frac{l_u}{l_{u-1}}\right] \quad \text{for} \quad u \geq 1, \tag{8.63}$$

and

$$N_t = \sum_{u=0}^{\infty} n_{u,t} \quad \text{for} \quad t \geq 1. \tag{8.64}$$

Having explained how $n_{u,t}$ and N_t are found once b_t is known, we turn to determining b_t. The number of individuals created at time t has two components. One component, a_t, say, counts the offspring of those individuals in the population at time t who already existed at time 0. There were $n_{u,0}$ individuals of age u at time 0. The probability that an individual of age u at time 0 will survive to time t (at which time he will be of age $t + u$) is l_{t+u}/l_u. Hence the number of individuals of age u at time 0 that survive to time t is $n_{u,0}(l_{t+u}/l_u)$ and each of these individuals, now of age $t + u$, will produce β_{t+u} new offspring. Adding over all ages we obtain

$$a_t = \sum_{u=0}^{\infty} \beta_{t+u} n_{u,0} \frac{l_{t+u}}{l_u}$$

$$= \sum_{u=0}^{\infty} \frac{m_{t+u} n_{u,0}}{l_u}. \tag{8.65}$$

The second component of b_t counts those individuals created at time t whose parents were not initially in the population but were born after time 0. Now the number of individuals created at time τ is b_τ. The probability that one of these individuals survives to time t, at which time he will be of age $t - \tau$, is $l_{t-\tau}$. The rate of births for individuals of age $t - \tau$ is $\beta_{t-\tau}$. The second component results from summing over τ and gives

$$b_t = a_t + \sum_{\tau=0}^{t} \beta_{t-\tau} l_{t-\tau} b_\tau$$

$$= a_t + \sum_{\tau=0}^{t} m_{t-\tau} b_\tau. \tag{8.66}$$

Example   Consider an organism that produces two offspring at age 1, and two more at age 2, and then dies. The population begins with a single organism of age 0 at time 0. We have the data

$$n_{0,0} = 1, \quad n_{u,0} = 0 \quad \text{for} \quad u \geq 1,$$

$$b_1 = b_2 = 2,$$

$$l_0 = l_1 = l_2 = 1, \quad \text{and} \quad l_u = 0 \quad \text{for} \quad u > 2.$$

We calculate from (8.65) that

$$a_0 = 0, \quad a_1 = 2, \quad a_2 = 2, \quad \text{and} \quad a_t = 0 \quad \text{for} \quad t > 2.$$

Finally, (8.66) is solved recursively, as

$$b_0 = 0$$

$$b_1 = a_1 + m_0 b_1 + m_1 b_0$$

$$= 2 + \quad 0 \quad + \quad 0 \quad = 2$$

$$b_2 = a_2 + m_0 b_2 + m_1 b_1 + m_2 b_0$$

$$= 2 + \quad 0 \quad + (2)(2) + \quad 0 \quad = 6$$

$$b_3 = a_3 + m_0 b_3 + m_1 b_2 + m_2 b_1 + m_3 b_0$$

$$= 0 + \quad 0 \quad + (2)(6) + (2)(2) + \quad 0 \quad = 16.$$

Thus, for example, an individual of age 0 at time 0 gives rise to 16 new individuals entering the population at time 3.

A second approach to (8.66) results from introducing the generating functions

$$b\star(s) = \sum_{t=0}^{\infty} b_t s^t,$$

$$a\star(s) = \sum_{t=0}^{\infty} a_t s^t,$$

and

$$m\star(s) = \sum_{t=0}^{\infty} m_t s^t \qquad \text{for } |s| < 1.$$

We multiply (8.66) by s^t and sum, thereby obtaining

$$\sum_{t=0}^{\infty} b_t s^t = \sum_{t=0}^{\infty} a_t s^t + \sum_{t=0}^{\infty} s^t \sum_{\tau=0}^{t} m_{t-\tau} b_\tau.$$

Then, after we write $s^t = (s^{t-\tau})(s^\tau)$ and interchange the order of summation, the equation becomes,

$$b\star(s) = a\star(s) + \sum_{t=0}^{\infty} \sum_{\tau=0}^{t} (s^{t-\tau} m_{t-\tau})(s^\tau b_\tau)$$

$$= a\star(s) + \sum_{\tau=0}^{\infty} \sum_{t=\tau}^{\infty} (s^{t-\tau} m_{t-\tau})(s^\tau b_\tau)$$

$$= a\star(s) + \sum_{\tau=0}^{\infty} \sum_{v=0}^{\infty} (s^v m_v)(s^\tau b_\tau)$$

$$= a\star(s) + m\star(s) \sum_{\tau=0}^{\infty} (s^\tau b_\tau)$$

$$= a\star(s) + m\star(s) b\star(s).$$

That is, in terms of generating functions we may solve explicitly to get

$$b\star(s) = \frac{a\star(s)}{1 - m\star(s)}. \tag{8.67}$$

When the population begins with a single individual of age 0 at time 0, then $a_t = m_t$ for all t, $a\star(s) = m\star(s)$, and (8.67) reduces to

$$b\star(s) = \frac{m\star(s)}{1 - m\star(s)} = \frac{1}{1 - m\star(s)} - 1. \tag{8.68}$$

Let us investigate (8.68) in the circumstances of the previous example: two offspring at year 1 and two at year 2. Then $m\star(s) = 2s + 2s^2$. Now b_t is the coefficient of s^t in $b\star(s)$, which from (8.68) is

$$b\star(s) = \frac{1}{1 - (2s + 2s^2)} - 1$$

$$= 2s + 6s^2 + 16s^3 + 44s^4 + 120s^5 + 328s^6 + \cdots$$

showing that the original parent gives rise to 328 new offspring in the sixth year.

The Long Run Behavior

Somewhat surprisingly, since no "renewals" are readily apparent, the discrete renewal theorem (Theorem 7.1) will be invoked to deduce the long run behavior of this age-structured population model. Observe that (8.66)

$$b_t = a_t + \sum_{\tau=0}^{t} m_{t-\tau} b_\tau$$

$$= a_t + \sum_{v=0}^{t} m_v b_{t-v}$$

(8.69)

has the form of a renewal equation except that $\{m_v\}$ is not necessarily a bona fide probability distribution in that, typically, $\{m_v\}$ will not sum to one. Fortunately, there is a trick that overcomes this difficulty. We introduce a variable s, whose value will be chosen later, and let

$$m_v^\# = m_v s^v, \quad b_v^\# = b_v s^v, \quad \text{and} \quad a_v^\# = a_v s^v.$$

Now multiply (8.69) by s^t and observe that $s^t m_v b_{t-v} = (m_v s^v)(b_{t-v} s^{t-v}) = m_v^\# b_{t-v}^\#$ to get

$$b_t^\# = a_t^\# + \sum_{v=0}^{t} m_v^\# b_{t-v}^\#.$$

(8.70)

This renewal equation holds no matter what value we choose for s. We therefore choose s so that $\{m_v^\#\}$ is a bona fide probability distribution. That is we fix the value of s such that

$$\sum_{v=0}^{\infty} m_v^\# = \sum_{v=0}^{\infty} m_v s^v = 1.$$

There is always a unique such s whenever $1 < \sum_{v=0}^{\infty} m_v < \infty$. We may now apply the renewal theorem to (8.70), provided its hypothesis concerning nonperiodic behavior is satisfied. For this it suffices, for example, that $m_1 > 0$. Then we conclude that

$$\lim_{t \to \infty} b_t^\# = \lim_{t \to \infty} b_t s^t = \frac{\sum_{v=0}^{\infty} a_v^\#}{\sum_{v=0}^{\infty} v m_v^\#}.$$

(8.71)

We set $\lambda = 1/s$ and $K = \sum_{v=0}^{\infty} a_v^\# / \sum_{v=0}^{\infty} v m_v^\#$ to write (8.71) in the form

$$b_t \sim K\lambda^t \quad \text{for } t \text{ large.}$$

In words, asymptotically the population grows at rate λ where $\lambda = 1/s$ is the solution to

$$\sum_{v=0}^{\infty} m_v \lambda^{-v} = 1.$$

When t is large ($t > u$) then (8.63) may be iterated in the manner

$$n_{u,t} = n_{u-1,t-1}\left[\frac{l_u}{l_{u-1}}\right]$$

$$= n_{u-2,t-2}\left[\frac{l_{u-1}}{l_{u-2}}\right]\left[\frac{l_u}{l_{u-1}}\right]$$

$$= n_{u-2,t-2}\left[\frac{l_u}{l_{u-2}}\right]$$

.

.

.

$$= n_{0,t-u}\left[\frac{l_u}{l_0}\right] = b_{t-u}l_u.$$

This simply expresses that those of age u at time t were born $t - u$ time units ago and survived. Since for large t we have $b_{t-u} \sim K\lambda^{t-u}$, then

$$n_{u,t} \sim Kl_u\lambda^{t-u} = K(l_u\lambda^{-u})\lambda^t,$$
$$N_t = \sum_{u=0}^{\infty} n_{u,t} \sim K\sum_{u=0}^{\infty} (l_u\lambda^{-u})\lambda^t,$$

and

$$\lim_{t \to \infty} \frac{n_{u,t}}{N_t} = \frac{l_u\lambda^{-u}}{\sum_{v=0}^{\infty} l_v\lambda^{-v}}.$$

This last expression furnishes the asymptotic or *stable age distribution* in the population.

Example Continuing the example in which $m_1 = m_2 = 2$ and $m_k = 0$, otherwise, then we have

$$\sum_{v=0}^{\infty} m_v s^v = 2s + 2s^2 = 1$$

which solves to give

$$s = \frac{-2 \pm \sqrt{4 + 8}}{4} = \frac{-1 \pm \sqrt{3}}{2}$$

$$= 0.366, \ -1.366.$$

The relevant root is $s = 0.366$ whence $\lambda = 1/s = 2.732$. Thus asymptotically the population grows geometrically at rate $\lambda = 2.732 \ldots$ and the stable age distribution is

Age	Fraction of Population
0	$1/(1 + s + s^2) = .6667$
1	$s/(1 + s + s^2) = .2440$
2	$s^2/(1 + s + s^2) = .0893$

Problems 8.7

1. Determine the long run population growth rate for a population whose individual net maternity function is $m_2 = m_3 = 2$, and $m_k = 0$, otherwise. Why does delaying the age at which offspring are first produced cause a reduction in the population growth rate? (The population growth rate when $m_1 = m_2 = 2$, and $m_k = 0$, otherwise, was determined on page 339.)

2. Determine the long run population growth rate for a population whose individual net maternity function is $m_0 = m_1 = 0$ and $m_2 = m_3 = \ldots = a > 0$. Compare this with the population growth rate when $m_2 = a$, and $m_k = 0$ for $k \neq 2$.

Chapter 9 | Queueing Systems

9.1 Queueing Processes

A queueing system consists of "customers" arriving at random times to some facility where they receive service of some kind and depart. We use "customer" as a generic term. It may refer, for example, to bona fide customers demanding service at a counter, to ships entering a port, to batches of data flowing into a computer subsystem, to broken machines awaiting repair, and so on. Queueing systems are classified according to

(1) *the input process,* the probability distribution of the pattern of arrivals of customers in time;
(2) *the service distribution,* the probability distribution of the random time to serve a customer (or group of customers in the case of batch service); and
(3) *the queue discipline,* the number of servers and the order of customer service.

While a variety of input processes may arise in practice, two simple and frequently occurring types are mathematically tractable and give insights into more complex cases. First is the scheduled input where customers arrive at fixed times $T, 2T, 3T, \ldots$. The second most common model is the "completely random" arrival process where the times of customer arrivals form a Poisson process. Understanding the axiomatic development of the Poisson process in Chapter 5 may help one to evaluate the validity of the Poisson assumption in any given application. Many theoretical results are available when the times of customer arrivals form a renewal process.

Exponentially distributed interarrival times then correspond to a Poisson process of arrivals as a special case.

We will always assume that the durations of service for individual customers are independent and identically distributed nonnegative random variables and are independent of the arrival process. The situation in which all service times are the same fixed duration D is, then, a special case.

The most common queue discipline is *first come, first served* where customers are served in the same order in which they arrive. All of the models that we consider in this chapter are of this type.

Queueing models aid the design process by predicting system performance. For example, a queueing model might be used to evaluate the costs and benefits of adding a server to an existing system. The models enable us to calculate system performance measures in terms of more basic quantities. Some important measures of system behavior are

> (1) *The probability distribution of the number of customers in the system.* Not only do customers in the system often incur costs, but in many systems, physical space for waiting customers must be planned for and provided. Large numbers of waiting customers can also adversely affect the input process by turning potential new customers away. (See Section 9.4.1 on queueing with balking.)
> (2) *The utilization of the server(s).* Idle servers may incur costs without contributing to system performance.
> (3) *System throughput.* The long run number of customers passing through the system is a direct measure of system performance.
> (4) *Customer waiting time.* Long waits for service are annoying in the simplest queueing situations and directly associated with major costs in many large systems such as those describing ships waiting to unload at a port facility or patients awaiting emergency care at a hospital.

The Queueing Formula $L = \lambda W$

Consider a queueing system that has been operating sufficiently long to have reached an approximate steady state, or a position of statistical equilibrium. Let

> L = the average number of customers in the system;
> λ = the rate of arrival of customers to the system; and
> W = the average time spent by a customer in the system.

The equation $L = \lambda W$ is valid under great generality for such systems and is of basic importance in the theory of queues since it directly relates two of our most important measures of system performance, the mean queue size and the mean customer waiting time in the steady state, that is, mean queue size and mean customer waiting time evaluated with respect to a limiting or stationary distribution for the process.

The validity of $L = \lambda W$ does not rest on the details of any particular model, but depends only upon long run mass flow balance relations. To sketch this reasoning, consider a time T sufficiently long so that statistical fluctuations have averaged out. Then the total number of customers to have entered the system is λT, the total number to have departed is $\lambda(T - W)$, and the net number remaining in the system L must be the difference

$$L = \lambda T - [\lambda(T - W)] = \lambda W.$$

Figure 9.1 depicts the relation $L = \lambda W$.

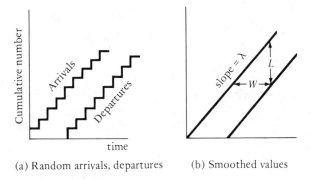

(a) Random arrivals, departures   (b) Smoothed values

Figure 9.1   The cumulative number of arrivals and departures in a queueing system. The smoothed values in (b) are meant to symbolize long run averages. The rate of arrivals per unit time is λ, the mean number in the system is L and the mean time a customer spends in the system is W.

Of course what we have done is by no means a proof, and, indeed, we shall give no proof. We shall, however, provide several sample verifications of $L = \lambda W$ where L is the mean of the stationary distribution of customers in the system, W is the mean customer time in the system determined from the stationary distribution, and λ is the arrival rate in a Poisson arrival process.

Let L_0 be the average number of customers waiting in the system who are not yet being served, and let W_0 be the average waiting time in the system excluding service time. In parallel to $L = \lambda W$, we have the formula

$$L_0 = \lambda W_0. \tag{9.1}$$

The total waiting time in the system is the sum of the waiting time before service, plus the service time. In terms of means, we have

$$W = W_0 + \text{Mean Service Time}. \tag{9.2}$$

In the remainder of this chapter we will study a variety of queueing systems. A standard shorthand is used in much of the queueing literature for identifying simple queueing models. The shorthand assumes that the arrival times form a renewal process, and the format $A/B/c$ uses A to describe the interarrival distribution, B to specify the individual customer service time distribution, and c to indicate the number of servers. The common cases for the first two positions are $G = GI$ for a general or arbitrary distribution, M (memoryless) for the exponential distribution, E_k (Erlang) for the gamma distribution of order k, and D for a deterministic distribution, a schedule of arrivals or fixed service times.

Some examples discussed in the sequel are

The $M/M/1$ queue   Arrivals follow a Poisson process; service times are exponentially distributed; and there is a single server. The number $X(t)$ of customers in the system at time t forms a birth and death process. (See Section 9.2).

The $M/M/\infty$ queue   There are Poisson arrivals and exponentially distributed service times. Any number of customers are processed simultaneously and independently. Often self-service situations may be described by this model. In the older literature this was called the "telephone trunking problem."

The $M/GI/1$ queue   In this model there are Poisson arrivals but arbitrarily distributed service times. The analysis proceeds with the help of an embedded Markov chain.

More elaborate variations will also be set forth. *Balking* is the refusal of new customers to enter the system if the waiting line is too long. More generally, in a queueing system with balking, an arriving customer enters the system with a probability that depends on the size of the queue. Here it is important to distinguish between the *arrival process* and the *input process* as shown in Figure 9.2. A special case is a queue with *overflow* in which an arriving customer enters the queue if and only if there is at least one server free to begin service immediately.

In a *priority queue,* customers are allowed to be of different types. Both the service discipline and the service time distribution may vary with the customer type.

A *queueing network* is a collection of service facilities where the departure from some stations form the arrivals of others. The network is *closed* if the total number of customers is fixed, these customers continuously circulating through the system. The machine repair model (see the example entitled "Repairman Models" in Section 6.4) is an example of a closed queueing network. In an open queueing network, customers may arrive from, and depart to, places outside the network, as well as move from station to station. Queueing network models have found much recent application in the design of complex information processing systems.

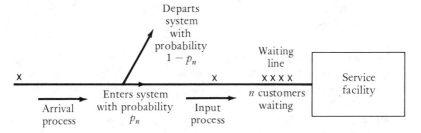

Figure 9.2 If n customers are waiting in a queueing system with balking, an arriving customer enters the system with probability p_n, and does not enter with probability $1 - p_n$.

Problems 9.1

1. What design questions might be answered by modeling the following queueing systems?

	The Customer	The Server
(a)	Arriving airplanes	The runway
(b)	Cars	A parking lot
(c)	Broken TV's	Repairman
(d)	Patients	Doctor
(e)	Fires	Fire engine company

What might be reasonable assumptions concerning the arrival process, service distribution, and priority in these instances?

2. Consider a system, such as a barber shop, where the service required is essentially identical for each customer. Then actual service times would tend to cluster near the mean service time. Argue that the exponential distribution would not be appropriate in this case. For what types of service situations might the exponential distribution be quite plausible?

3. Two dump trucks cycle between a gravel loader and a gravel unloader. Suppose that the travel times are insignificant relative to the load and unload times, which are exponentially distributed with parameters μ and λ, respectively. Model the system as a closed queueing network. Determine the long run gravel loads moved per unit time. *Hint:* Refer to the example entitled "Repairman Models" in Section 6.4.

9.2 Poisson Arrivals and Exponentially Distributed Service Times

The simplest and most extensively studied queueing models are those having a Poisson arrival process and exponentially distributed service times. In this case the queue size forms a birth and death process (see Sections 6.3 and 6.4), and the corresponding stationary distribution is readily found.

We let λ denote the intensity or rate of the Poisson arrival process and assume that the service time distribution is exponential with parameter μ. The corresponding density function is

$$g(x) = \mu e^{-\mu x} \quad \text{for} \quad x > 0. \tag{9.3}$$

For the Poisson arrival process we have

$$\Pr\{\text{An arrival in } [t, t + h)\} = \lambda h + o(h) \tag{9.4}$$

and

$$\Pr\{\text{No arrivals in } [t, t + h)\} = 1 - \lambda h + o(h). \tag{9.5}$$

Similarly, the memoryless property of the exponential distribution as expressed by its constant hazard rate (see Section 1.4.2) implies that

$$\Pr\{\text{A service is completed in } [t, t + h)|\text{Service in progress at time } t\}$$

$$= \mu h + o(h), \tag{9.6}$$

and

$$\Pr\{\text{Service not completed in } [t, t + h)|\text{Service in progress at time } t\}$$

$$= 1 - \mu h + o(h). \tag{9.7}$$

The service rate μ applies to a particular server. If k servers are simultaneously operating, the probability that one of them completes service in a time interval of duration h is $(k\mu)h + o(h)$ so that the system service rate is $k\mu$. The principle used here is the same as that used in deriving the infinitesimal parameters of the Yule process (Section 6.1).

We let $X(t)$ denote the number of customers in the system at time t, counting the customers undergoing service as well as those awaiting service. The independence of arrivals in disjoint time intervals together with the memoryless property of the exponential service time distribution implies that $X(t)$ is a time homogeneous Markov chain, in particular, a birth and death process. (See Sections 6.3 and 6.4).

The *M/M/*1 System

We consider first the case of a single server and let $X(t)$ denote the number of customers in the system at time t. An increase in $X(t)$ by one unit corresponds to a customer arrival, and in view of (9.4) and (9.7) and the postulated independence of service times and the arrival process we have

$$\Pr\{X(t + h) = k + 1 | X(t) = k\} = [\lambda h + o(h)] \times [1 - \mu h + o(h)]$$

$$= \lambda h + o(h) \quad \text{for} \quad k = 0, 1, \ldots.$$

Similarly, a decrease in $X(t)$ by one unit corresponds to a completion of service, whence

$$\Pr\{X(t + h) = k - 1 | X(t) = k\} = \mu h + o(h) \quad \text{for} \quad k = 1, 2, \ldots.$$

Then $X(t)$ is a birth and death process with birth parameters

$$\lambda_k = \lambda \quad \text{for} \quad k = 0, 1, 2, \ldots$$

and death parameters

$$\mu_k = \mu \quad \text{for} \quad k = 1, 2, \ldots.$$

Of course no completion of service is possible when the queue is empty. We thus specify $\mu_0 = 0$.

Let

$$\pi_k = \lim_{t \to \infty} \Pr\{X(t) = k\} \quad \text{for} \quad k = 0, 1, \ldots$$

be the limiting or equilibrium distribution of queue length. Section 6.4 describes a straightforward procedure for determining the limiting distribution π_k from the birth and death parameters λ_k and μ_k. The technique is to first obtain intermediate quantities θ_j defined by

$$\theta_0 = 1 \quad \text{and} \quad \theta_j = \frac{\lambda_0 \lambda_1 \cdots \lambda_{j-1}}{\mu_1 \mu_2 \cdots \mu_j} \quad \text{for} \quad j \geq 1, \tag{9.8}$$

and then

$$\pi_0 = \frac{1}{\sum\limits_{j=0}^{\infty} \theta_j} \quad \text{and} \quad \pi_k = \theta_k \pi_0 = \frac{\theta_k}{\sum\limits_{j=0}^{\infty} \theta_j} \quad \text{for} \quad k \geq 1. \tag{9.9}$$

When $\sum_{j=0}^{\infty} \theta_j = \infty$, then $\lim_{t \to \infty} \Pr\{X(t) = k\} = 0$ for all k and the queue length grows unboundedly in time.

For the $M/M/1$ queue at hand we readily compute $\theta_0 = 1$ and $\theta_j = (\lambda/\mu)^j$ for $j = 1, 2, \ldots.$ Then

$$\sum_{j=0}^{\infty} \pi_j = \sum_{j=0}^{\infty} \left(\frac{\lambda}{\mu}\right)^j = \frac{1}{(1 - \lambda/\mu)} \quad \text{if} \quad \lambda < \mu,$$

$$= \infty \quad \text{if} \quad \lambda \geq \mu.$$

Thus, no equilibrium distribution exists when the arrival rate λ is equal to or greater than the service rate μ. In this case the queue length grows without bound.

When $\lambda < \mu$ a *bona fide* limiting distribution exists given by

$$\pi_0 = \frac{1}{\sum \theta_j} = 1 - \frac{\lambda}{\mu} \tag{9.10}$$

and

$$\pi_k = \pi_0 \theta_k = \left(1 - \frac{\lambda}{\mu}\right)\left(\frac{\lambda}{\mu}\right)^k \qquad \text{for} \quad k = 0, 1, \ldots \qquad (9.11)$$

The equilibrium distribution (9.11) gives us the answer to many questions involving the limiting behavior of the system. We recognize the form of (9.11) as that of a geometric distribution, and then reference to Section 1.3.3 gives us the mean queue length in equilibrium to be

$$L = \frac{\lambda}{\mu - \lambda}. \qquad (9.12)$$

The ratio $\rho = \lambda/\mu$ is called the *traffic intensity*,

$$\rho = \frac{\text{Arrival rate}}{\text{System service rate}} = \frac{\lambda}{\mu}. \qquad (9.13)$$

As the traffic intensity approaches one, the mean queue length $L = \rho/(1 - \rho)$ becomes infinite. Again using (9.8), the probability of being served immediately upon arrival is

$$\pi_0 = 1 - \frac{\lambda}{\mu},$$

the probability, in the long run, of finding the server idle. The server utilization, or long run fraction of time that the server is busy, is $1 - \pi_0 = \lambda/\mu$.

We can also calculate the distribution of waiting time in the stationary case when $\lambda < \mu$. If an arriving customer finds n people in front of him, his total waiting time T, including his own service time, is the sum of the service times of himself and those ahead, all distributed exponentially with parameter μ, and since the service times are independent of the queue size, T has a gamma distribution of order $n + 1$ with scale parameter μ,

$$\Pr\{T \le t | n \text{ ahead}\} = \int_0^t \frac{\mu^{n+1} \tau^n e^{-\mu\tau}}{\Gamma(n + 1)} \, d\tau. \qquad (9.14)$$

By the law of total probabilities, we have

$$\Pr\{T \le t\} = \sum_{n=0}^{\infty} \Pr\{T \le t | n \text{ ahead}\} \times \left(\frac{\lambda}{\mu}\right)^n\left(1 - \frac{\lambda}{\mu}\right),$$

since $(\lambda/\mu)^n(1 - \lambda/\mu)$ is the probability that in the stationary case a customer on arrival will find n ahead in line. Now, substituting from (9.14), we obtain

$$\Pr\{T \le t\} = \sum_{n=0}^{\infty} \int_0^t \frac{\mu^{n+1} \tau^n e^{-\mu\tau}}{\Gamma(n + 1)} \left(\frac{\lambda}{\mu}\right)^n\left(1 - \frac{\lambda}{\mu}\right) d\tau$$

$$= \int_0^t \mu e^{-\mu\tau}\left(1 - \frac{\lambda}{\mu}\right)\sum_{n=0}^{\infty} \frac{\tau^n \lambda^n}{\Gamma(n + 1)} \, d\tau$$

$$= \int_0^t \left(1 - \frac{\lambda}{\mu}\right)\mu \, \exp\left\{-\tau\mu\left(1 - \frac{\lambda}{\mu}\right)\right\}d\tau = 1 - \exp[-t(\mu - \lambda)],$$

which is also an exponential distribution.

The mean of this exponential waiting time distribution is the reciprocal of the exponential parameter, or

$$W = \frac{1}{\mu - \lambda}. \tag{9.15}$$

Reference to (9.12) and (9.15) verifies the fundamental queueing formula $L = \lambda W$.

A queueing system alternates between durations when the servers are busy and durations when the system is empty and the servers are idle. An *idle period* begins the instant the last customer leaves and endures until the arrival of the next customer. When the arrival process is Poisson of rate λ, then an idle period is exponentially distributed with mean

$$E[I_1] = \frac{1}{\lambda}.$$

A busy period is an uninterrupted duration in which the system is not empty. When arrivals to a queue follow a Poisson process, then the successive durations X_k from the commencement of the kth busy period to the start of the next busy period form a renewal process (see Figure 9.3). Each X_k is comprised of a busy portion B_k and an idle portion I_k. Then the renewal theorem (see "A Queueing Model," p. 294) applies to tell us that $p_0(t)$, the probability that the system is empty at time t, converges to

$$\lim_{t \to \infty} p_0(t) = \pi_0 = \frac{E[I_1]}{E[I_1] + E[B_1]}.$$

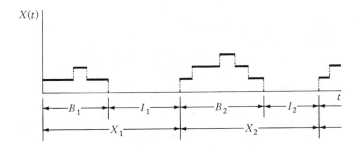

Figure 9.3 The busy periods B_k and idle periods I_k of a queueing system. When arrivals form a Poisson process, then $X_k = B_k + I_k$, $k = 1, 2, \ldots$ are independent identically distributed non-negative random variables, and thus form a renewal process.

We substitute the known quantities $\pi_0 = 1 - \lambda/\mu$ and $E[I_1] = 1/\lambda$ to obtain

$$1 - \frac{\lambda}{\mu} = \frac{1/\lambda}{1/\lambda + E[B_1]}$$

which solves to give

$$E[B_1] = \frac{1}{\mu - \lambda}$$

for the mean length of a busy period.

In Section 9.3 in studying the $M/G/1$ system we will reverse this reasoning, calculate the mean busy period directly, and then use renewal theory to determine the server idle fraction π_0.

The $M/M/\infty$ System

When an unlimited number of servers are always available, then all customers in the system at any instant are simultaneously being served. The departure rate of a single customer being μ, the departure rate of k customers is $k\mu$, and we obtain the birth and death parameters

$$\lambda_k = \lambda \quad \text{and} \quad \mu_k = k\mu \quad \text{for} \quad k = 0, 1, \ldots .$$

The auxiliary quantities of (9.8) are

$$\theta_k = \frac{\lambda_0\lambda_1 \cdots \lambda_{k-1}}{\mu_1\mu_2 \cdots \mu_k} = \frac{1}{k!}\left(\frac{\lambda}{\mu}\right)^k \quad \text{for} \quad k = 0, 1, \ldots$$

which sum to

$$\sum_{k=0}^{\infty} \theta_k = \sum_{k=0}^{\infty} \frac{1}{k!}\left(\frac{\lambda}{\mu}\right)^k = e^{\lambda/\mu}$$

whence

$$\pi_0 = \frac{1}{\displaystyle\sum_{k=0}^{\infty} \theta_k} = e^{-\lambda/\mu}$$

and

$$\pi_k = \theta_k\pi_0 = \frac{(\lambda/\mu)^k e^{-\lambda/\mu}}{k!} \quad \text{for} \quad k = 0, 1, \ldots, \quad (9.16)$$

a Poisson distribution with mean queue length

$$L = \frac{\lambda}{\mu}.$$

Since a customer in this system begins service immediately upon arrival, customer waiting time consists only of the exponentially distributed service time, and the mean waiting time is $W = 1/\mu$. Again, the basic queueing formula $L = \lambda W$ is verified.

The $M/G/\infty$ queue will be developed extensively in the next section.

The $M/M/s$ System

When a fixed number s of servers are available and the assumption is made that a server is never idle if customers are waiting, then the appropriate birth and death parameters are

$$\lambda_k = \lambda \quad \text{for} \quad k = 1, 2, \ldots$$

$$\mu_k = \begin{cases} k\mu & \text{for} \quad k = 0, 1, \ldots, s \\ s\mu & \text{for} \quad k > s. \end{cases}$$

If $X(t)$ is the number of customers in the system at time t, then the number undergoing service is $\min\{X(t), s\}$ and the number waiting for service is $\max\{X(t) - s, 0\}$. The system is depicted in Figure 9.4.

The auxiliary quantities are given by

$$\theta_k = \frac{\lambda_0 \lambda_1 \cdots \lambda_{k-1}}{\mu_1 \mu_2 \cdots \mu_k} = \begin{cases} \dfrac{1}{k!}\left(\dfrac{\lambda}{\mu}\right)^k & \text{for} \quad k = 0, 1, \ldots, s \\[2ex] \dfrac{1}{s!}\left(\dfrac{\lambda}{\mu}\right)^s\left(\dfrac{\lambda}{s\mu}\right)^{k-s} & \text{for} \quad k \geq s, \end{cases}$$

and, when $\lambda < s\mu$, then

$$\sum_{j=0}^{\infty} \theta_j = \sum_{j=0}^{s-1} \frac{1}{j!}\left(\frac{\lambda}{\mu}\right)^j + \sum_{j=s}^{\infty} \frac{1}{s!}\left(\frac{\lambda}{\mu}\right)^s\left(\frac{\lambda}{s\mu}\right)^{j-s}$$

$$= \sum_{j=0}^{s-1} \frac{1}{j!}\left(\frac{\lambda}{\mu}\right)^j + \frac{(\lambda/\mu)^s}{s!(1 - \lambda/s\mu)} \quad \text{for} \quad \lambda < s\mu. \tag{9.17}$$

The traffic intensity in an $M/M/s$ system is $\rho = \lambda/s\mu$. Again as the

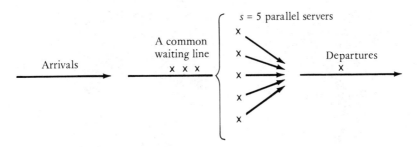

Figure 9.4 A queueing system with s servers

traffic intensity approaches one, the mean queue length becomes unbounded. When $\lambda < s\mu$, then from (9.9) and (9.17),

$$\pi_0 = \left\{ \sum_{j=0}^{s-1} \frac{1}{j!} \left(\frac{\lambda}{\mu} \right)^j + \frac{(\lambda/\mu)^s}{s!(1 - \lambda/s\mu)} \right\}^{-1},$$

and

$$\pi_k = \begin{cases} \dfrac{1}{k!} \left(\dfrac{\lambda}{\mu} \right)^k \pi_0 & \text{for} \quad k = 0, 1, \ldots, s \\[3mm] \dfrac{1}{s!} \left(\dfrac{\lambda}{\mu} \right)^s \left(\dfrac{\lambda}{s\mu} \right)^{k-s} \pi_0 & \text{for} \quad k \geq s. \end{cases} \tag{9.18}$$

We evaluate L_0, the mean number of customers in the system waiting for,
and not undergoing, service. Then

$$\begin{aligned} L_0 &= \sum_{j=s}^{\infty} (j - s)\pi_j = \sum_{k=0}^{\infty} k\pi_{s+k} \\ &= \pi_0 \sum_{k=0}^{\infty} k \frac{1}{s!} \left(\frac{\lambda}{\mu} \right)^s \left(\frac{\lambda}{s\mu} \right)^k \\ &= \frac{\pi_0}{s!} \left(\frac{\lambda}{\mu} \right)^s \sum_{k=0}^{\infty} k \left(\frac{\lambda}{s\mu} \right)^k \\ &= \frac{\pi_0}{s!} \left(\frac{\lambda}{\mu} \right)^s \frac{(\lambda/s\mu)}{(1 - \lambda/s\mu)^2}. \end{aligned} \tag{9.19}$$

Then

$$W_0 = \frac{L_0}{\lambda},$$

$$W = W_0 + \frac{1}{\mu}$$

and

$$L = \lambda W = \lambda \left(W_0 + \frac{1}{\mu} \right) = L_0 + \frac{\lambda}{\mu}.$$

Problems 9.2

1. On a single graph, plot the server utilization $1 - \pi_0 = \rho$ and the mean
   queue length $L = \rho/(1 - \rho)$ for the $M/M/1$ queue as a function of the
   traffic intensity $\rho = \lambda/\mu$ for $0 < \rho < 1$.

2. Determine explicit expressions for π_0 and L for the $M/M/s$ queue when
   $s = 2$. Plot $1 - \pi_0$ and L as a function of the traffic intensity $\rho = \lambda/2\mu$.

3. Determine the mean waiting time W for an $M/M/2$ system when $\lambda = 2$
   and $\mu = 1.2$. Compare this with the mean waiting time in an $M/M/1$

system whose arrival rate is $\lambda = 1$ and service rate is $\mu = 1.2$. Why is there a difference when the arrival rate per server is the same in both cases?

4. The problem is to model a queueing system having finite capacity. We assume arrivals according to a Poisson process of rate λ, independent exponentially distributed service times having mean $1/\mu$, a single server, and a finite system capacity N. By this we mean that if an arriving customer finds that there are already N customers in the system, then that customer does not enter the system and is lost.

 Let $X(t)$ be the number of customers in the system at time t. Suppose that $N = 3$ (2 waiting, 1 being served).
 (a) Specify the birth and death parameters for $X(t)$.
 (b) In the long run, what fraction of time is the system idle?
 (c) In the long run, what fraction of customers are lost?

5. Customers arrive at a service facility according to a Poisson process having rate λ. There is a single server whose service times are exponentially distributed with parameter μ. Let $N(t)$ be the number of people in the system at time t. Then $N(t)$ is a birth and death process with parameters $\lambda_n = \lambda$ for $n \geq 0$ and $\mu_n = \mu$ for $n \geq 1$. Assume $\lambda < \mu$. Then $\pi_k = (1 - \lambda/\mu)(\lambda/\mu)^k$, $k \geq 0$, is a stationary distribution for $N(t)$, cf. Equation (9.11).

 Suppose the process begins according to the stationary distribution. That is, suppose $\Pr\{N(t) = k\} = \pi_k$ for $k = 0, 1, \ldots$. Let $D(t)$ be the number of people completing service up to time t. Show that $D(t)$ has a Poisson distribution with mean λt.
 Hint: Let $P_{kj}(t) = \Pr\{D(t) = j | X(0) = k\}$ and $P_j(t) = \Sigma \pi_k P_{kj}(t) = \Pr\{D(t) = j\}$. Use a first step analysis to show that $P_{0\,j}(t + \Delta t) = (\lambda t)P_{1\,j}(t) + [1 - \lambda(\Delta t)]P_{0\,j}(t) + o(\Delta t)$, and for $k = 1, 2, \ldots$

$$P_{kj}(t + \Delta t) = \mu(\Delta t)P_{k-1,\,j-1}(t) + \lambda(\Delta t)P_{k+1,\,j}(t)$$
$$+ [1 - (\lambda + \mu)(\Delta t)]P_{k\,j}(t) + o(t).$$

 Then use $P_j(t) = \Sigma_k \pi_k P_{k\,j}(t)$ to establish a differential equation. Use the explicit form of π_k given in the problem.

9.3 The M/G/I and M/G/∞ Systems

We continue to assume that the arrivals follow a Poisson process of rate λ. The successive customer service times $Y_1, Y_2, \ldots$, however, are now allowed to follow an arbitrary distribution $G(y) = \Pr\{Y_k \leq y\}$ having a finite mean service time $v = E[Y_k]$. The long run service rate is $\mu = 1/v$. Deterministic service times of an equal fixed duration are an important special case.

The *M/G/1* System

If arrivals to a queue follow a Poisson process, then the successive durations X_k from the commencement of the kth busy period to the start of the next busy period form a renewal process. (A busy period is an uninterrupted duration when the queue is not empty. See Figure 9.3.) Each X_k is comprised of a busy portion B_k and an idle portion I_k. Then $p_0(t)$, the probability that the system is empty at time t, converges to

$$\lim_{t \to \infty} p_0(t) = \pi_0 = \frac{E[I_1]}{E[X_1]}$$

$$= \frac{E[I_1]}{E[I_1] + E[B_1]}$$

(9.20)

by the renewal theorem (see "A Queueing Model," p. 294).

The idle time is the duration from the completion of a service that empties the queue to the instant of the next arrival. Because of the memoryless property that characterizes the interarrival times in a Poisson process, each idle time is exponentially distributed with mean $E[I_1] = 1/\lambda$.

The busy period is comprised of the first service time Y_1, plus busy periods generated by all customers who arrive during this first service time. Let A denote this random number of new arrivals. We will evaluate the conditional mean busy period given that $A = n$ and $Y_1 = y$. First

$$E[B_1|A = 0, Y_1 = y] = y$$

because when no customers arrive, the busy period is comprised of the first customer's service time alone. Next consider the case in which $A = 1$ and let B' be the duration from the beginning of this customer's service to the next instant that the queue is empty. Then

$$E[B_1|A = 1, Y_1 = y] = y + E[B']$$

$$= y + E[B_1],$$

because upon the completion of service for the initial customer, the single arrival begins a busy period B' that is statistically identical to the first so that $E[B'] = E[B_1]$. Continuing in this manner we deduce that

$$E[B_1|A = n, Y_1 = y] = y + nE[B_1]$$

and then, using the law of total probability, that

$$E[B_1|Y_1 = y] = \sum_{n=0}^{\infty} E[B_1|A = n, Y_1 = y]\Pr\{A = n|Y_1 = y\}$$

$$= \sum_{n=0}^{\infty} \{y + nE[B_1]\} \frac{(\lambda y)^n e^{-\lambda y}}{n!}$$

$$= y + \lambda y E[B_1].$$

Finally

$$E[B_1] = \int_0^\infty E[B_1|Y_1 = \gamma]dG(\gamma)$$

$$= \int_0^\infty \{\gamma + \lambda\gamma E[B_1]\}dG(\gamma) \tag{9.21}$$

$$= \nu\{1 + \lambda E[B_1]\}.$$

Since $E[B_1]$ appears on both sides of (9.21) we may solve to obtain

$$E[B_1] = \frac{\nu}{1 - \lambda\nu} \quad \text{provided} \quad \lambda\nu < 1. \tag{9.22}$$

To compute the long run fraction of idle time, we use (9.20) and

$$\pi_0 = \frac{E[I_1]}{E[I_1] + E[B_1]}$$

$$= \frac{1/\lambda}{1/\lambda + \nu/(1 - \lambda\nu)} \tag{9.23}$$

$$= 1 - \lambda\nu \quad \text{if} \quad \lambda\nu < 1.$$

Note that (9.23) agrees, as it must, with the corresponding expression (9.10) obtained for the $M/M/1$ queue where $\nu = 1/\mu$. For example, if arrivals occur at the rate of $\lambda = 2$ per hour and the mean service time is 20 minutes or $\nu = \frac{1}{3}$ hours, then in the long run the server is idle $1 - 2(\frac{1}{3}) = \frac{1}{3}$ of the time.

The Embedded Markov Chain

The number $X(t)$ of customers in the system at time t is not a Markov process for a general $M/G/1$ system because, if one is to predict the future behavior of the system, one must know, in additon, the time expended in service for the customer currently in service. (It is the memoryless property of the exponential service time distribution that makes this additional information unnecessary in the $M/M/1$ case.)

Let X_n, however, denote the number of customers in the system immediately after the departure of the nth customer. Then $\{X_n\}$ is a Markov chain. Indeed, we can write

$$X_n = \begin{cases} X_{n-1} - 1 + A_n & \text{if} \quad X_{n-1} > 0, \\ \\ A_n & \text{if} \quad X_{n-1} = 0, \end{cases} \tag{9.24}$$

$$= (X_{n-1} - 1)^+ + A_n,$$

where A_n is the number of customers that arrive during the service of the nth customer and where $x^+ = \max\{x, 0\}$. Since the arrival process is Poisson, the number of customers A_n that arrive during the service of the nth customer is independent of earlier arrivals, and the Markov property follows instantly. We calculate

$$\alpha_k = \Pr\{A_n = k\} = \int_0^\infty \Pr\{A_n = k | Y_n = y\} dG(y)$$

$$= \int_0^\infty \frac{(\lambda y)^k e^{-\lambda y}}{k!} \, dG(y) \tag{9.25}$$

and then, for $j = 0, 1, \ldots$,

$$P_{ij} = \Pr\{X_n = j | X_{n-1} = i\} = \Pr\{A_n = j - (i-1)^+\}$$

$$= \begin{cases} \alpha_{j-i+1} & \text{for } i \geq 1, \quad j \geq i+1, \\ \\ \alpha_j & \text{for } i = 0. \end{cases} \tag{9.26}$$

The Mean Queue Length in Equilibrium L

The embedded Markov chain is of special interest in the $M/G/1$ queue because in this particular instance, the stationary distribution $\{\pi_j\}$ for the Markov chain $\{X_n\}$ equals the limiting distribution for the queue length process $\{X(t)\}$. That is, $\lim_{t \to \infty} \Pr\{X(t) = j\} = \lim_{n \to \infty} \Pr\{X_n = j\}$. We will use this helpful fact to evaluate the mean queue length L.

The equivalence between the stationary distribution for the Markov chain $\{X_n\}$ and that for the non-Markov process $\{X(t)\}$ is rather subtle. It is not the consequence of a general principle and should not be assumed to hold in other circumstances without careful justification. The equivalence in the case at hand is sketched in an appendix to this section.

We will calculate the expected queue length in equilibrium $L = \lim_{t \to \infty} E[X(t)]$ by calculating the corresponding quantity in the embedded Markov chain, $L = \lim_{n \to \infty} E[X_n]$. If $X = X_\infty$ is the number of customers in the system after a customer departs and X' is the number after the next departure, then in accordance with (9.24),

$$X' = X - \delta + N \tag{9.27}$$

where N is the number of arrivals during the service period and

$$\delta = \begin{cases} 1 & \text{if } X > 0 \\ \\ 0 & \text{if } X = 0. \end{cases}$$

In equilibrium, X has the same distribution as does X' and, in particular,

$$L = E[X] = E[X'], \tag{9.28}$$

and taking expectation in (9.27) gives

$$E[X'] = E[X] - E[\delta] + E[N],$$

and, by (9.28) and (9.23), then

$$E[N] = E[\delta] = 1 - \pi_0 = \lambda v. \tag{9.29}$$

Squaring (9.27) gives

$$(X')^2 = X^2 + \delta^2 + N^2 - 2\delta X + 2N(X - \delta)$$

and, since $\delta^2 = \delta$ and $X\delta = X$, then

$$(X')^2 = X^2 + \delta + N^2 - 2X + 2N(X - \delta). \tag{9.30}$$

Now N, the number of customers that arrive during a service period, is independent of X, and hence, of δ so that

$$E[N(X - \delta)] = E[N]E[X - \delta] \tag{9.31}$$

and because X and X' have the same distribution, then

$$E[(X')^2] = E[X^2]. \tag{9.32}$$

Taking expectations in (9.30) we deduce that

$$E[(X')^2] = E[X^2] + E[\delta] + E[N^2] - 2E[X] + 2E[N]E[X - \delta]$$

and then substituting from (9.29) and (9.32), we obtain

$$0 = \lambda v + E[N^2] - 2L + 2\lambda v\{L - \lambda v\}$$

or

$$L = \frac{\lambda v + E[N^2] - 2(\lambda v)^2}{2(1 - \lambda v)}. \tag{9.33}$$

It remains to evaluate $E[N^2]$ where N is the number of arrivals during a service time Y. Conditioned on $Y = y$, the random variable N has a Poisson distribution with a mean (and variance) equal to λy [see (9.25)], whence $E[N^2|Y = y] = \lambda y + (\lambda y)^2$. Using the law of total probability then gives

$$E[N^2] = \int_0^\infty E[N^2|Y = y]dG(y)$$

$$= \lambda \int_0^\infty y\,dG(y) + \lambda^2 \int_0^\infty y^2 dG(y) \tag{9.34}$$

$$= \lambda v + \lambda^2(\tau^2 + v^2)$$

where τ^2 is the variance of the service time distribution $G(y)$. Substituting (9.34) into (9.33) gives

$$L = \frac{2\lambda v + \lambda^2\tau^2 - (\lambda v)^2}{2(1 - \lambda v)}$$

$$= \rho + \frac{\lambda^2\tau^2 + \rho^2}{2(1 - \rho)} \tag{9.35}$$

where $\rho = \lambda v$ is the traffic intensity.

Finally, $W = L/\lambda$, which simplifies to

$$W = \nu + \frac{\lambda(\tau^2 + \nu^2)}{2(1 - \rho)}. \tag{9.36}$$

The results (9.35) and (9.36) express somewhat surprising facts. They say that for a given average arrival rate λ and mean service time ν, we can decrease the expected queue size L and waiting time W by decreasing the variance of service time. Clearly the best possible case in this respect corresponds to constant service times for which $\tau^2 = 0$.

The $M/G/\infty$ System

Complete results are available when each customer begins service immediately upon arrival independently of other customers in the system. Such situations may arise when modeling customer self-service systems. Let W_1, W_2, . . . be the successive arrival times of customers, and let V_1, V_2, . . . be the corresponding service times. In this notation, the kth customer is in the system at time t if and only if $W_k \le t$ (the customer arrived prior to t) and $W_k + V_k > t$ (the service extends beyond t).

The sequence of pairs (W_1, V_1), (W_2, V_2), . . . forms a *marked Poisson process* (see Section 5.6.2), and we may use the corresponding theory to quickly obtain results in this model. Figure 9.5 illustrates the marked Poisson process. Then $X(t)$, the number of customers in the system at time t, is also the number of points (W_k, V_k) for which $W_k \le t$ and $W_k + V_k > t$.

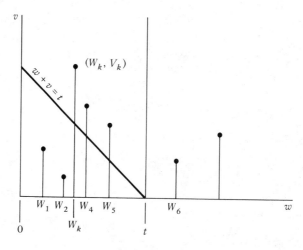

Figure 9.5   For the $M/G/\infty$ queue the number of customers in the system at time t corresponds to the number of pairs (W_k, V_k) for which $W_k \le t$ and $W_k + V_k > t$. In the sample illustrated here, the number of customers in the system at time t is 3.

That is, it is the number of points (W_k, V_k) in the unbounded trapezoid described by

$$A_t = \{(w, v): 0 \le w \le t \text{ and } v > t - w\}.$$

According to Theorem 5.6, the number of points in A_t follows a Poisson distribution with mean

$$
\begin{aligned}
\mu(A_t) &= \iint_{A_t} \lambda(dw)dG(v) \\
&= \lambda \int_0^t \left\{ \int_{t-w}^{\infty} dG(v) \right\} dw \\
&= \lambda \int_0^t [1 - G(t - w)]dw \\
&= \lambda \int_0^t [1 - G(x)]dx.
\end{aligned}
\tag{9.37}
$$

In summary,

$$
\begin{aligned}
p_k(t) &= \Pr\{X(t) = k\} \\
&= \frac{\mu(A_t)^k e^{-\mu(A_t)}}{k!} \qquad \text{for} \quad k = 0, 1, \dots
\end{aligned}
$$

where $\mu(A_t)$ is given by (9.37). As $t \to \infty$ then

$$\lim_{t \to \infty} \mu(A_t) = \lambda \int_0^{\infty} [1 - G(x)]dx = \lambda v$$

where v is the mean service time. Thus we obtain the limiting distribution

$$\pi_k = \frac{(\lambda v)^k e^{-\lambda v}}{k!} \qquad \text{for} \quad k = 0, 1, \dots .$$

Appendix

We sketch a proof of the equivalence between the limiting queue size distribution and the limiting distribution for the embedded Markov chain in an $M/G/1$ model. First, beginning at $t = 0$ let η_n denote those instants when the queue size $X(t)$ increases by one (an arrival), and let ξ_n denote those instants when $X(t)$ decreases by one (a departure). Let $Y_n = X(\eta_n-)$ denote the queue length immediately prior to an arrival and let $X_n = X(\xi_n+)$ denote the queue length immediately after a departure. For any queue length i and any time t the number of visits of Y_n to i up to time t differs from the number of visits of X_n to i by at most one unit. Therefore, in the long run the average visits per unit time of Y_n to i must equal the average visits of X_n to i, which is π_i, the stationary distribution of the Markov chain $\{X_n\}$. Thus we need only show that the limiting distribution of $\{X(t)\}$ is the same as that of $\{Y_n\}$, which is $X(t)$ just prior to an arrival. But because the arrivals are Poisson, and arrivals in disjoint time intervals are independent, it must be that $X(t)$ is independent of an arrival that occurs at time t. It follows that $\{X(t)\}$ and $\{Y_n\}$ have the same limiting distribution, and therefore $\{X(t)\}$ and the embedded Markov chain $\{X_n\}$ have the same limiting distribution.

Problems 9.3

1. Suppose that the service distribution in a single server queue is exponential with rate μ, i.e., $G(v) = 1 - e^{-\mu v}$ for $v \geq 0$. Substitute the mean and variance of this distribution into (9.35) and verify that the result agrees with that derived for the $M/M/1$ system in (9.12).

2. Consider a single server queueing system having Poisson arrivals at rate λ. Suppose that the service times have the gamma density

$$g(y) = \frac{\mu^\alpha \, y^{\alpha-1} \, e^{-\mu y}}{\Gamma(\alpha)} \qquad \text{for} \quad y \geq 0,$$

where $\alpha > 0$ and $\mu > 0$ are fixed parameters. The mean service time is α/μ and the variance is α/μ^2. Determine the equilibrium mean queue length L.

9.4 Variations and Extensions

In this section we consider a few variations on the simple queueing models studied so far. These examples do not exhaust the possibilities but serve only to suggest the richness of the area.

Throughout we restrict ourselves to Poisson arrivals and exponentially distributed service times.

9.4.1 Systems with Balking

Suppose that a customer who arrives when there are n customers in the systems enters with probability p_n and departs with probability $q_n = 1 - p_n$. If long queues discourage customers, then p_n would be a decreasing function of n. As a special case, if there is a finite waiting room of capacity C, we might suppose that

$$p_n = \begin{cases} 1 & \text{for} \quad n < C \\ 0 & \text{for} \quad n \geq C, \end{cases}$$

indicating that once the waiting room is filled, no more customers can enter the system.

Let $X(t)$ be the number of customers in the system at time t. If the arrival process is Poisson at rate λ and a customer who arrives when there are n customers in the system enters with probability p_n, then the appropriate birth parameters are

$$\lambda_n = \lambda p_n \quad \text{for} \quad n = 0, 1, \ldots$$

In the case of a single server, then $\mu_n = \mu$ for $n = 1, 2, \ldots$, and we may evaluate the stationary distribution π_k of queue length by the usual means.

In systems with balking, not all arriving customers enter the system, and some are lost. The *input rate* is the rate at which customers actually enter the system in the stationary state and is given by

$$\lambda_I = \lambda \sum_{n=0}^{\infty} \pi_n p_n.$$

The rate at which customers are lost is $\lambda \sum_{n=0}^{\infty} \pi_n q_n$, and the fraction of customers lost in the long run is

$$\text{Fraction Lost} = \sum_{n=0}^{\infty} \pi_n q_n.$$

Let us examine in detail the case of an $M/M/s$ system in which an arriving customer enters the system if and only if a server is free. Then

$$\lambda_k = \begin{cases} \lambda & \text{for} \quad k = 0, 1, \ldots, s-1 \\ 0 & \text{for} \quad k = s, \end{cases}$$

and,

$$\mu_k = k\mu \quad \text{for} \quad k = 0, 1, \ldots, s.$$

To determine the limiting distribution, we have

$$\theta_k = \frac{1}{k!} \left(\frac{\lambda}{\mu} \right)^k \quad \text{for} \quad k = 0, 1, \ldots, s$$

and then

$$\pi_k = \frac{\dfrac{1}{k!} \left(\dfrac{\lambda}{\mu} \right)^k}{\displaystyle\sum_{j=0}^{s} \dfrac{1}{j!} \left(\dfrac{\lambda}{\mu} \right)^j} \quad \text{for} \quad k = 0, 1, \ldots, s. \tag{9.38}$$

The long run fraction of customers lost is $\pi_s q_s = \pi_s$ since $q_s = 1$ in this case.

9.4.2 Variable Service Rates

In a similar vein, one can consider a system whose service rate depends on the number of customers in the system. For example, a second server might be added to a single server system whenever the queue length exceeds a critical point ξ. If arrivals are Poisson and service rates are memoryless, then the appropriate birth and death parameters are

$$\lambda_k = \lambda \quad \text{for} \quad k = 0, 1, \ldots \quad \text{and} \quad \mu_k = \begin{cases} \mu & \text{for} \quad k \leq \xi, \\ 2\mu & \text{for} \quad k > \xi. \end{cases}$$

More generally, let us consider Poisson arrivals $\lambda_k = \lambda$ for $k = 0$, $1, \ldots$, and arbitrary service rates μ_k for $k = 1, 2, \ldots$. The stationary distribution in this case is given by

$$\pi_k = \frac{\pi_0 \lambda^k}{\mu_1 \mu_2 \cdots \mu_k} \quad \text{for} \quad k \geq 1, \tag{9.39}$$

where

$$\pi_0 = \left\{ 1 + \sum_{k=1}^{\infty} \frac{\lambda^k}{\mu_1 \mu_2 \cdots \mu_k} \right\}^{-1}. \tag{9.40}$$

9.4.3 A System with Feedback

Consider a single server system with Poisson arrivals and exponentially distributed service times, but suppose that some customers, upon leaving the server, return to the end of the queue for additional service. In particular, suppose that a customer leaving the server departs from the system with probability q and returns to the queue for additional service with probability $p = 1 - q$. Suppose that all such decisions are statistically independent, and that a returning customer's demands for service are statistically the same as those of a customer arriving from outside the system. Let the arrival rate be λ and the service rate be μ. The queue system is depicted in Figure 9.6.

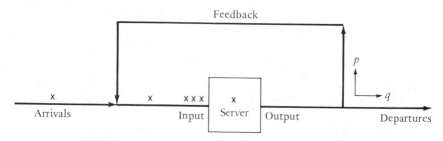

Figure 9.6   A queue with feedback

Let $X(t)$ denote the number of customers in the system at time t. Then $X(t)$ is a birth and death process with parameters $\lambda_n = \lambda$ for $n = 0, 1, \ldots$.

and $\mu_n = q\mu$ for $n = 1, 2, \ldots$ It is easily deduced that the stationary distribution in the case that $\lambda < q\mu$ is

$$\pi_k = \left(1 - \frac{\lambda}{q\mu}\right)\left(\frac{\lambda}{q\mu}\right)^k \qquad \text{for} \quad k = 0, 1, \ldots \qquad (9.41)$$

9.4.4 A Two Server Overflow Queue

Consider a two server system where server i has rate μ_i for $i = 1, 2$. Arrivals to the system follow a Poisson process of rate λ. A customer arriving when the system is empty goes to the first server. A customer arriving when the first server is occupied goes to the second server. If both servers are occupied, the customer is lost. The flow is depicted in Figure 9.7.

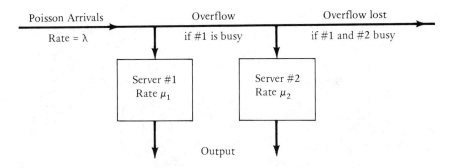

Figure 9.7 A two server overflow model

The system state is described by the pair $(X(t), Y(t))$ where

$$X(t) = \begin{cases} 1 & \text{if Server \#1 is busy,} \\ 0 & \text{if Server \#1 is idle;} \end{cases}$$

and

$$Y(t) = \begin{cases} 1 & \text{if Server \#2 is busy,} \\ 0 & \text{if Server \#2 is idle.} \end{cases}$$

The four states of the system are $\{(0, 0), (1, 0), (0, 1), (1, 1)\}$ and transitions among these states occur at the rates given in the following table:

From State	To State	Transition Rate	Description
(0, 0)	(1, 0)	λ	Arrival when system empty
(1, 0)	(0, 0)	μ_1	Service completion by #1 when #2 is free
(1, 0)	(1, 1)	λ	Arrival when #1 is busy
(1, 1)	(1, 0)	μ_2	Service completion by #2 when #1 is busy
(1, 1)	(0, 1)	μ_1	Service completion by #1 when #2 is busy
(0, 1)	(1, 1)	λ	Arrival when #2 is busy and #1 is free
(0, 1)	(0, 0)	μ_2	Service completion by #2 when #1 is free

The process $(X(t), Y(t))$ is a finite state, continuous time Markov chain (see Section 6.6) and the transition rates in the table furnish the infinitesimal matrix of the Markov chain:

$$
\mathbf{A} = \begin{array}{c} \\ (0, 0) \\ (0, 1) \\ (1, 0) \\ (1, 1) \end{array} \left\| \begin{array}{cccc} (0,0) & (0,1) & (1,0) & (1,1) \\ -\lambda & 0 & \lambda & 0 \\ \mu_2 & -(\lambda + \mu_2) & 0 & \lambda \\ \mu_1 & 0 & -(\lambda + \mu_1) & \lambda \\ 0 & \mu_1 & \mu_2 & -(\mu_1 + \mu_2) \end{array} \right\| .
$$

From (6.68) and (6.69), we find the stationary distribution $\boldsymbol{\pi} = (\pi_{(0,0)}, \pi_{(0,1)}, \pi_{(1,0)}, \pi_{(1,1)})$ by solving $\boldsymbol{\pi}\mathbf{A} = 0$, or

$$
\begin{aligned}
-\lambda\pi_{(0,0)} + \mu_2\pi_{(0,1)} + \mu_1\pi_{(1,0)} &= 0 \\
-(\lambda + \mu_2)\pi_{(0,1)} + \mu_1\pi_{(1,1)} &= 0 \\
\lambda\pi_{(0,0)} - (\lambda + \mu_1)\pi_{(1,0)} + \mu_2\pi_{(1,1)} &= 0 \\
\lambda\pi_{(0,1)} + \lambda\pi_{(1,0)} - (\mu_1 + \mu_2)\pi_{(1,1)} &= 0
\end{aligned}
$$

together with

$$
\pi_{(0,0)} + \pi_{(0,1)} + \pi_{(1,0)} + \pi_{(1,1)} = 1.
$$

Tedious but elementary algebra yields the solution:

$$
\pi_{(0,0)} = \frac{\mu_1\mu_2(2\lambda + \mu_1 + \mu_2)}{D}
$$

$$
\pi_{(0,1)} = \frac{\lambda^2\mu_1}{D}
$$

$$
\pi_{(1,0)} = \frac{\lambda\mu_2(\lambda + \mu_1 + \mu_2)}{D}
$$

(9.42)

$$
\pi_{(1,1)} = \frac{\lambda^2(\lambda + \mu_2)}{D}
$$

where

$$D = \mu_1\mu_2(2\lambda + \mu_1 + \mu_2) + \lambda^2\mu_1 + \lambda\mu_2(\lambda + \mu_1 + \mu_2) + \lambda^2(\lambda + \mu_2).$$

The fraction of customers that are lost, in the long run, is the same as the fraction of time that both servers are busy, $\pi_{(1,1)} = \lambda^2(\lambda + \mu_2)/D$.

9.4.5 Preemptive Priority Queues

Consider a single server queueing process that has two classes of customers, *priority* and *nonpriority,* forming independent Poisson arrival processes of rates α and β, respectively. The customer service times are independent and exponentially distributed with parameters γ and δ, respectively. Within classes there is a first come, first served discipline and the service of priority customers is never interrupted. If a priority customer arrives during the service of a nonpriority customer, then the latter's service is immediately stopped in favor of the priority customer. The interrupted customer's service is resumed when there are no priority customers present.

Let us introduce some convenient notation. The system arrival rate is $\lambda = \alpha + \beta$, the fraction $p = \alpha/\lambda$ of which are priority customers, and $q = \beta/\lambda$ of which are nonpriority customers. The system mean service time is given by the appropriately weighted means $1/\gamma$ and $1/\delta$ of the priority and nonpriority customers, respectively, or

$$\frac{1}{\mu} = p\left(\frac{1}{\gamma}\right) + q\left(\frac{1}{\delta}\right) = \frac{1}{\lambda}\left(\frac{\alpha}{\gamma} + \frac{\beta}{\delta}\right), \tag{9.43}$$

where μ is the system service rate. Finally, we introduce the traffic intensities $\rho = \lambda/\mu$ for the system, and $\sigma = \alpha/\gamma$ and $\tau = \beta/\delta$ for the priority and nonpriority customers, respectively. From (9.43) we see that $\rho = \sigma + \tau$.

The state of the system is described by the pair $(X(t), Y(t))$ where $X(t)$ is the number of priority customers in the system and $Y(t)$ is the number of nonpriority customers. Observe that the priority customers view the system as simply an $M/M/1$ queue. Accordingly, we have the limiting distribution from (9.11) to be

$$\lim_{t\to\infty} \Pr\{X(t) = m\} = (1 - \sigma)\sigma^m \quad \text{for} \quad m = 0, 1, \ldots \tag{9.44}$$

provided $\sigma = \alpha/\gamma < 1$.

Reference to (9.12) and (9.15), gives us, respectively, the mean queue length for priority customers

$$L_p = \frac{\alpha}{\gamma - \alpha} = \frac{\sigma}{1 - \sigma} \tag{9.45}$$

and the mean wait for priority customers

$$W_p = \frac{1}{\gamma - \alpha}. \tag{9.46}$$

To obtain information about the nonpriority customers is not as easy since these arrivals are strongly affected by the priority customers. Nevertheless, $(X(t), Y(t))$ is a discrete state, continuous time Markov chain, and the techniques of Section 6.6 enable us to describe the limiting distribution, when it exists. The transition rates of the $(X(t), Y(t))$ Markov chain are described in the following table:

From State	To State	Transition Rate	Description
(m, n)	$(m + 1, n)$	α	Arrival of priority customer
(m, n)	$(m, n + 1)$	β	Arrival of nonpriority customer
$(0, n)$ $n \geq 1$	$(0, n - 1)$	δ	Completion of nonpriority service
(m, n) $m \geq 1$	$(m - 1, n)$	γ	Completion of priority service

Let

$$\pi_{m,n} = \lim_{t \to \infty} \Pr\{X(t) = m, Y(t) = n\}$$

be the limiting distribution of the process. Reasoning analogous to that of (6.68) and (6.69) of Chapter 6 (where the theory was derived for a finite state Markov chain) leads to the following equations for the stationary distribution:

$$(\alpha + \beta)\pi_{0,0} = \gamma\pi_{1,0} \quad + \delta\pi_{0,1} \tag{9.47}$$

$$(\alpha + \beta + \gamma)\pi_{m,0} = \gamma\pi_{m+1,0} \quad + \quad \alpha\pi_{m-1,0}, \quad m \geq 1 \tag{9.48}$$

$$(\alpha + \beta + \delta)\pi_{0,n} = \gamma\pi_{1,n} \quad + \delta\pi_{0,n+1} + \beta\pi_{0,n-1} \quad , \quad n \geq 1 \tag{9.49}$$

$$(\alpha + \beta + \gamma)\pi_{m,n} = \gamma\pi_{m+1,n} \quad + \beta\pi_{m,n-1} + \alpha\pi_{m-1,n}, \quad m, n \geq 1 \tag{9.50}$$

The transition rates leading to Equation (9.50) are shown in Figure 9.8.

In principle, these equations, augmented with the condition $\Sigma_m\Sigma_n \pi_{m,n} = 1$, may be solved for the stationary distribution, when it exists. We will content ourselves with determining the mean number L_n of nonpriority customers in the system in steady state, given by

$$L_n = \sum_{m=0}^{\infty}\sum_{n=0}^{\infty} n\pi_{m,n}. \tag{9.51}$$

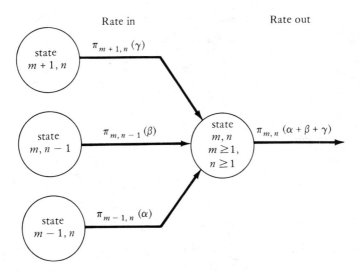

Rate in   Rate out

Figure 9.8 In equilibrium, the rate of flow into any state must equal the rate of flow out. Illustrated here is the state (m, n) when $m \geq 1$ and $n \geq 1$, leading to Equation (9.50).

We introduce the notation

$$M_m = \sum_{n=0}^{\infty} n\pi_{m,n} = \sum_{n=1}^{\infty} n\pi_{m,n} \tag{9.52}$$

so that

$$L_n = M_0 + M_1 + \cdots. \tag{9.53}$$

Using (9.44), let

$$p_m = \Pr\{X(t) = m\} = \sum_{n=0}^{\infty} \pi_{m,n} = (1 - \sigma)\sigma^m \tag{9.54}$$

and

$$\pi_n = \Pr\{Y(t) = n\} = \sum_{m=0}^{\infty} \pi_{m,n}. \tag{9.55}$$

We begin by summing both sides of (9.47) and (9.48) for $m = 0, 1, \ldots$ to obtain

$$(\alpha + \beta)\pi_0 + \gamma\sum_{m=1}^{\infty} \pi_{m,0} = \gamma\sum_{m=1}^{\infty} \pi_{m,0} + \delta\pi_{0,1} + \alpha\pi_0$$

which simplifies to give

$$\beta\pi_0 = \delta\pi_{0,1}. \tag{9.56}$$

Next, we sum (9.49) and (9.50) over $m = 0, 1, \ldots$ to obtain

$$(\alpha + \beta)\pi_n + \delta\pi_{0,n} + \gamma\sum_{m=1}^{\infty} \pi_{m,n} = \gamma\sum_{m=1}^{\infty} \pi_{m,n} + \delta\pi_{0,n+1} + \beta\pi_{n-1} + \alpha\pi_n$$

which simplifies to

$$\beta\pi_n + \delta\pi_{0,n} = \beta\pi_{n-1} + \delta\pi_{0,n+1}$$

and, inductively with (9.56), we obtain

$$\beta\pi_n = \delta\pi_{0,n+1} \quad \text{for} \quad n = 0, 1, \ldots \qquad (9.57)$$

Summing (9.57) over $n = 0, 1, \ldots$ and using $\Sigma \pi_n = 1$ we get

$$\beta = \delta\sum_{n=0}^{\infty}\pi_{0,n+1} = \delta \Pr\{X(t) = 0, Y(t) > 0\},$$

or

$$\Pr\{X(t) = 0, Y(t) > 0\} = \sum_{n=1}^{\infty}\pi_{0,n} = \frac{\beta}{\delta} = \tau. \qquad (9.58)$$

Since (9.54) asserts that $\Pr\{X(t) = 0\} = 1 - \dfrac{\alpha}{\gamma} = 1 - \sigma$, we have

$$\pi_{0,0} = \Pr\{X(t) = 0, Y(t) = 0\} = \Pr\{X(t) = 0\} - \Pr\{X(t) = 0, Y(t) > 0\}$$

$$= 1 - \frac{\alpha}{\gamma} - \frac{\beta}{\delta} = 1 - \sigma - \tau$$

$$\text{when} \quad \sigma + \tau < 1. \quad (9.59)$$

With these preliminary results in hand, we turn to determining $M_m = \Sigma_{n=1}^{\infty} n\pi_{m,n}$. Multiplying (9.49) by n and summing, we derive

$$(\alpha + \beta + \delta)M_0 = \gamma M_1 + \delta\sum_{n=1}^{\infty}n\pi_{0,n+1} + \beta\sum_{n=1}^{\infty}n\pi_{0,n-1}$$

$$= \gamma M_1 + \delta M_0 - \delta\sum_{n=0}^{\infty}\pi_{0,n+1} + \beta M_0 + \beta\sum_{n=1}^{\infty}\pi_{0,n-1}$$

$$= \gamma M_1 + \delta M_0 - \delta\left(\frac{\beta}{\delta}\right) + \beta M_0 + \beta(1 - \sigma),$$

where the last line results from (9.54) and (9.58). After simplification and rearrangement, the result is

$$M_1 = \sigma M_0 + \frac{\beta}{\gamma}\sigma. \qquad (9.60)$$

We next multiply (9.50) by n and sum to obtain

$$(\alpha + \beta + \gamma)M_m = \gamma M_{m+1} + \beta\sum_{n=1}^{\infty}n\pi_{m,n-1} + \alpha M_{m-1}$$

$$= \gamma M_{m+1} + \beta M_m + \beta\sum_{n=1}^{\infty}\pi_{m,n-1} + \alpha M_{m-1}.$$

Again, referring to (9.54) and simplifying, we see that

$$(\alpha + \gamma)M_m = \gamma M_{m+1} + \alpha M_{m-1} + \beta(1 - \sigma)\sigma^m$$

$$\text{for} \quad m = 1, 2, \ldots . \quad (9.61)$$

Equations (9.60) and (9.61) can be solved inductively to give

$$M_m = M_0 \sigma^m + \frac{\beta}{\gamma} m \sigma^m \qquad \text{for} \quad m = 0, 1, \ldots$$

which we sum to obtain

$$L_n = \sum_{m=0}^{\infty} M_m = \frac{1}{1 - \sigma} \left[M_0 + \frac{\beta}{\gamma} \frac{\sigma}{(1 - \sigma)} \right]. \tag{9.62}$$

This determines L_n in terms of M_0. To obtain a second relation, we multiply (9.57) by n and sum to obtain

$$\beta L_n = \delta \sum_{n=0}^{\infty} n \pi_{0,n+1} = \delta M_0 - \delta \sum_{n=0}^{\infty} \pi_{0,n+1}$$

$$= \delta M_0 - \delta \left(\frac{\beta}{\delta} \right) \qquad \text{[see (9.58)]}$$

or

$$M_0 = \frac{\beta}{\delta} (L_n + 1) = \tau(L_n + 1). \tag{9.63}$$

We substitute (9.63) into (9.62) and simplify, yielding

$$L_n = \frac{1}{1 - \sigma} \left[\tau(L_n + 1) + \frac{\beta}{\gamma} \frac{\sigma}{1 - \sigma} \right],$$

$$\left(1 - \frac{\tau}{1 - \sigma} \right) L_n = \frac{1}{1 - \sigma} \left[\tau + \frac{\beta}{\gamma} \frac{\sigma}{1 - \sigma} \right]$$

and, finally,

$$L_n = \left(\frac{\tau}{1 - \sigma - \tau} \right) \left[1 + \left(\frac{\delta}{\gamma} \right) \frac{\sigma}{1 - \sigma} \right]. \tag{9.64}$$

The condition that L_n be finite (and that a stationary distribution exist) is that

$$\rho = \sigma + \tau < 1.$$

That is, the system traffic intensity ρ must be less than one.

Since the arrival rate for nonpriority customers is β, we have that the mean waiting time for nonpriority customers is given by $W_n = L_n / \beta$.

Some simple numerical studies of (9.45) and (9.64) yield surprising results concerning adding priority to an existing system. Let us consider first a simple $M/M/1$ system with traffic intensity ρ whose mean queue length is given by (9.12) to be $L = \rho/(1 - \rho)$. Let us propose modifying the system in such a way that a fraction $p = \frac{1}{2}$ of the customers have priority. We assume that priority is independent of service time. These assumptions lead to the values $\alpha = \beta = \frac{1}{2}\lambda$ and $\gamma = \delta = \mu$, whence $\sigma = \tau = \rho/2$. Then the mean queue lengths for priority and nonpriority customers are given by

$$L_p = \frac{\sigma}{1 - \sigma} = \frac{\rho/2}{1 - (\rho/2)}$$

and

$$L_n = \left(\frac{\rho/2}{1 - \rho}\right)\left[1 + \frac{\rho/2}{1 - (\rho/2)}\right].$$

The mean queue lengths L, L_p, and L_n were determined for several values of the traffic intensity ρ. The results are

ρ	L	L_p	L_n
.6	1.50	.43	1.07
.8	4.00	.67	3.34
.9	9.00	.82	8.19
.95	19.00	.90	18.05

It is seen that the burden of increased queue length, as the traffic intensity increases, is carried almost exclusively by the nonpriority customers!

Problems 9.4

1. Suppose that incoming calls to an office follow a Poisson process of rate $\lambda = 6$ per hour. If the line is in use at the time of an incoming call, the secretary has a HOLD button that will enable a single additional caller to wait. Suppose that the lengths of conversations are exponentially distributed with a mean length of 5 minutes, that incoming calls while a caller is on hold are lost, and that outgoing calls can be ignored. Apply the results of Section 9.4.1 to determine the fraction of calls that are lost.

2. Consider the two server overflow queue of Section 9.4.4 and suppose the arrival rate is $\lambda = 10$ per hour. The two servers have rates 6 and 4 per hour. Recommend which server should be placed first. That is, choose between

$$\mu_1 = 6 \qquad \text{and} \qquad \mu_1 = 4$$
$$\mu_2 = 4 \qquad\qquad\qquad \mu_2 = 6$$

and justify your answer. Be explicit about your criterion.

3. Consider the preemptive priority queue of Section 9.4.5 and suppose that the arrival rate is $\lambda = 4$ per hour. Two classes of customers can be identified having mean service times of 12 minutes and 8 minutes, and it is proposed to give one of these classes priority over the other. Recommend which class should have priority. Be explicit about your criterion and justify your answer. Assume that the two classes appear in equal proportions and that all service times are exponentially distributed.

4. *Balking* refers to the refusal of an arriving customer to enter the queue. *Reneging* refers to the departure of a customer in the queue before obtaining service. Consider an $M/M/1$ system with reneging such that the probability that a specified single customer in line will depart prior to service in a short time interval $(t, t + \Delta t]$ is $r_n(\Delta t) + o(\Delta t)$ when n is the number of customers in the system. (Note that $r_0 = r_1 = 0$). Assume Poisson arrivals at rate λ and exponential service times with parameter μ, and determine the stationary distribution when it exists.

5. A small grocery store has a single checkout counter with a full-time cashier. Customers arrive at the checkout according to a Poisson process of rate λ per hour. When there is only a single customer at the counter, the cashier works alone at a mean service rate of α per hour. Whenever there is more than one customer at the checkout, however, a "bagger" is added, increasing the service rate to β per hour. Assume that service times are exponentially distributed and determine the stationary distribution of the queue length.

6. A ticket office has two agents answering incoming phone calls. In addition, a third caller can be put on HOLD until one of the agents becomes available. If all three phone lines (both agent lines plus the hold line) are busy, a potential caller gets a busy signal, and is assumed lost. Suppose that the calls and attempted calls occur according to a Poisson process of rate λ, and that the length of a telephone conversation is exponentially distributed with parameter μ. Determine the stationary distribution for the process.

9.5 Open Acyclic Queueing Networks

Queueing networks, comprised of groups of service stations, with the departures of some stations forming the arrivals of others, arise in computer and information processing systems, manufacturing job shops, service industries such as hospitals and airport terminals, and in many other contexts. A remarkable result often enables the steady state behavior of these complex systems to be analyzed component by component.

9.5.1 The Basic Theorem

The result alluded to in the preceding paragraph asserts that the departures from a queue with Poisson arrivals and exponentially distributed service times, in statistical equilibrium, also form a Poisson process. We give the precise statement as Theorem 9.1. The proof is contained in an appendix at the end of this section. See also Problem 5, page 353.

Theorem 9.1 Let $\{X(t),\ t \geq 0\}$ be a birth and death process with constant birth parameters $\lambda_n = \lambda$ for $n = 0, 1, \ldots$, and arbitrary death parameters μ_n for $n = 1, 2, \ldots$. Suppose there exists a stationary distribution $\pi_k \geq 0$ where $\Sigma_k \pi_k = 1$ and that $\Pr\{X(0) = k\} = \pi_k$ for $k = 0, 1, \ldots$. Let $D(t)$ denote the number of deaths in $(0, t]$. Then

$$\Pr\{X(t) = k, D(t) = j\} = \Pr\{X(t) = k\}\Pr\{D(t) = j\}$$

$$= \pi_k \frac{(\lambda t)^j e^{-\lambda t}}{j!} \quad \text{for} \quad k, j \geq 0.$$

Remark The stipulated conditions are satisfied, for example, when $X(t)$ is the number of customers in an $M/M/s$ queueing system that is in steady state wherein $\Pr\{X(0) = j\} = \pi_j$, the stationary distribution of the process. In this case, a stationary distribution exists provided $\lambda < s\mu$, where μ is the individual service rate.

To see the major importance of this theorem, suppose that $X(t)$ represents the number of customers in some queueing system at time t. The theorem asserts that the departures form a Poisson process of rate λ. Furthermore, the number $D(t)$ of departures up to time t is independent of the number $X(t)$ of customers remaining in the system at time t.

We caution the reader that the foregoing analysis applies only if the processes are in statistical equilibrium where the stationary distribution $\pi_k = \Pr\{X(t) = k\}$ applies. In contrast, under the condition that $X(0) = 0$, then neither will the departures form a Poisson process, nor will $D(t)$ be independent of $X(t)$.

9.5.2 Two Queues in Tandem

Let us use Theorem 9.1 to analyze a simple queueing network comprised of two single server queues connected in series as shown in Figure 9.9.

Let $X_k(t)$ be the number of customers in the kth queue at time t. We assume steady state. Beginning with the first server, the stationary distribution (9.11) for a single server queue applies and

$$\Pr\{X_1(t) = n\} = \left(1 - \frac{\lambda}{\mu_1}\right)\left(\frac{\lambda}{\mu_1}\right)^n \quad \text{for} \quad n = 0, 1, \ldots .$$

Theorem 9.1 asserts that the departure process from the first server, denoted by $D_1(t)$, is a Poisson process of rate λ that is statistically independent

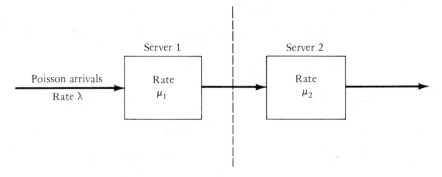

Figure 9.9   Two queues in series in which the departures from the first form the arrivals for the second

of the first queue length $X_1(t)$. These departures form the arrivals to the second server, and therefore the second system has Poisson arrivals and is thus an $M/M/1$ queue as well. Thus, again using (9.11),

$$\Pr\{X_2(t) = m\} = \left(1 - \frac{\lambda}{\mu_2}\right)\left(\frac{\lambda}{\mu_2}\right)^m \qquad \text{for} \quad m = 0, 1, \ldots$$

Furthermore, because the departures $D_1(t)$ from the first server are independent of $X_1(t)$, it must be that $X_2(t)$ is independent of $X_1(t)$. We thus obtain the joint distribution

$$\Pr\{X_1(t) = n \quad \text{and} \quad X_2(t) = m\} = \Pr\{X_1(t) = n\}\Pr\{X_2(t) = m\}$$

$$= \left(1 - \frac{\lambda}{\mu_1}\right)\left(\frac{\lambda}{\mu_1}\right)^n\left(1 - \frac{\lambda}{\mu_2}\right)\left(\frac{\lambda}{\mu_2}\right)^m$$

$$\text{for} \quad n, m = 0, 1, \ldots$$

We again caution the reader that the foregoing analysis applies only when the network is in its limiting distribution. In contrast, if both queues are empty at time $t = 0$, then neither will the departures $D_1(t)$ form a Poisson process nor will $D_1(t)$ and $X_1(t)$ be independent.

9.5.3 Open Acyclic Networks

The preceding analysis of two queues in series applies to more general systems. An *open* queueing network (see Figure 9.10) has customers arriving from and departing to, the outside world. (The repairman model in Section 6.4, is a prototypical *closed* queueing network.) Consider an open network having K service stations, and let $X_k(t)$ be the number of customers in queue k at time t. Suppose

(1) The arrivals from outside the system to distinct servers form independent Poisson processes.

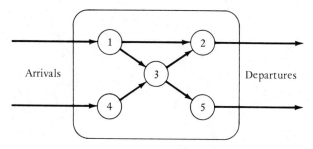

Figure 9.10   An open queueing network

(2) The departures from distinct servers independently travel instantly to other servers, or leave the system, with fixed probabilities.

(3) The service times for the various servers are *memoryless* in the sense that

$$\Pr\{\text{Server } \#k \text{ completes a service in } (t, t + \Delta t] | X_k(t) = n\}$$
$$= \mu_{kn}(\Delta t) + o(\Delta t) \qquad \text{for} \quad n = 1, 2, \ldots, \qquad (9.65)$$

and does not otherwise depend on the past.

(4) The system is in statistical equilibrium (steady state).

(5) The network is *acyclic* in that a customer can visit any particular server at most once. (The case where a customer can visit a server more than once is more subtle, and is treated in the next section.)

Then

(A) $X_1(t), X_2(t), \ldots, X_K(t)$ are independent processes where

$$\Pr\{X_1(t) = n_1, X_2(t) = n_2, \ldots, X_K(t) = n_K\}$$
$$= \Pr\{X_1(t) = n_1\}\Pr\{X_2(t) = n_2\} \cdots \Pr\{X_K(t) = n_K\}. \qquad (9.66)$$

(B) The departure process $D_k(t)$ associated with the kth server is a Poisson process and $D_k(t)$ and $X_k(t)$ are independent.

(C) The arrivals to the kth station form a Poisson process of rate λ_k.

(D) The departure rate at the kth server equals the rate of arrivals to that server.

Let us add some notation so as to be able to express these results more explicitly. Let

$\lambda_{0k} = $ Rate of arrivals to station k from outside the system,

$\lambda_k = $ Rate of total arrivals to station k,

$P_{kj} = $ Probability that a customer leaving station k next visits station j.

Then, the arrivals to station k come from outside the system or from some other station j. The departure rate from j equals the arrival rate to j, the fraction P_{jk} of which go to station k, whence

$$\lambda_k = \lambda_{0k} + \sum_j \lambda_j P_{jk}. \qquad (9.67)$$

Since the network is acyclic, (9.67) may be solved recursively, beginning with stations having only outside arrivals. The simple example that follows will make the procedure clear.

The arrivals to station k form a Poisson process of rate λ_k. Let

$$\psi_k(n) = \pi_{k0} \times \frac{\lambda_k^n}{\mu_{k1}\mu_{k2}\cdots\mu_{kn}} \qquad \text{for} \quad n = 1, 2, \ldots, \qquad (9.68)$$

where

$$\psi_k(0) = \pi_{k0} = \left\{ 1 + \sum_{n=1}^{\infty} \left(\frac{\lambda_k^n}{\mu_{k1}\mu_{k2}\cdots\mu_{kn}} \right) \right\}^{-1}. \qquad (9.69)$$

Referring to (9.39) and (9.40) we see that (9.68) and (9.69) give the stationary distribution for a queue having Poisson arrivals at rate λ_k and memoryless service times at rates μ_{kn} for $n = 1, 2, \ldots$. Accordingly we may now express (9.66) explicitly as

$$\Pr\{X_1(t) = n_1, X_2(t) = n_2, \ldots, X_K = n_k\}$$
$$= \psi_1(n_1)\psi_2(n_2)\cdots\psi_K(n_K). \qquad (9.70)$$

Example   Consider the three station network as shown in Figure 9.11.

The first step in analyzing the example is to determine the arrival rates at the various stations. In equilibrium, the arrival rate at a station must equal its departure rate, as asserted in (D). Accordingly, departures from state 1

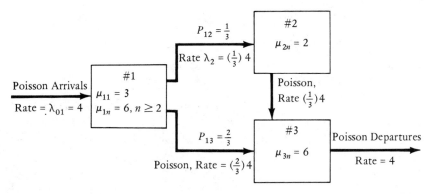

Figure 9.11   A three station open acyclic network. Two servers, each of rate 3, at the first station give rise to the station rates $\mu_{11} = 3$ and $\mu_{1n} = 6$ for $n \geq 2$. Stations 2 and 3 each have a single server of rate 2 and 6, respectively.

occur at rate $\lambda_1 = 4$, and since these departures independently travel to stations 2 and 3 with respective probabilities $P_{12} = \frac{1}{3}$ and $P_{13} = \frac{2}{3}$, we determine the arrival rate $\lambda_2 = (\frac{1}{3})4$. At station 3 the arrivals include both those from station 1 and from station 2. Thus $\lambda_3 = (\frac{2}{3})4 + (\frac{1}{3})4 = 4$.

Having determined the arrival rates at each station, we turn to determining the equilibrium probabilities. Station 1 is an $M/M/2$ system with $\lambda = 4$ and $\mu = 3$. From (9.18), or (9.68) and (9.69), we obtain

$$\Pr\{X_1(t) = 0\} = \pi_0 = \left\{1 + \left(\frac{4}{3}\right) + \frac{(4/3)^2}{2(1/3)}\right\}^{-1} = .2$$

and

$$\Pr\{X_1(t) = n\} = \begin{cases} \left(\frac{4}{3}\right)(.2) & \text{for } n = 1 \\ (.4)\left(\frac{2}{3}\right)^n & \text{for } n \geq 2. \end{cases}$$

Station 2 is an $M/M/1$ system with $\lambda = \frac{4}{3}$ and $\mu = 2$. From (9.11) we obtain

$$\Pr\{X_2(t) = n\} = \left(\frac{1}{3}\right)\left(\frac{2}{3}\right)^n \qquad \text{for } n = 0, 1, \ldots$$

Similarly, station 3 is an $M/M/1$ system with $\lambda = 4$ and $\mu = 6$ so that (9.11) yields

$$\Pr\{X_3(t) = n\} = \left(\frac{1}{3}\right)\left(\frac{2}{3}\right)^n \qquad \text{for } n = 0, 1, \ldots$$

Finally, according to Property A, the queue lengths $X_1(t)$, $X_2(t)$ and $X_3(t)$ are independent so that

$$\Pr\{X_1(t) = n_1, X_2(t) = n_2, X_3(t) = n_3\}$$
$$= \Pr\{X_1(t) = n_1\}\Pr\{X_2(t) = n_2\}\Pr\{X_3(t) = n_3\}.$$

9.5.4 Appendix: Time Reversibility

Let $\{X(t), -\infty < t < +\infty\}$ be an arbitrary countable state Markov chain having a stationary distribution $\pi_j = \Pr\{X(t) = j\}$ for all states j and all times t. Note that the time index set is the whole real line. We view the process as having begun indefinitely far in the past so that it now is evolving in a stationary manner. Let $Y(t) = X(-t)$ be the same process, but with time reversed. The stationary process $\{X(t)\}$ is said to be *time reversible* if $\{X(t)\}$ and $\{Y(t)\}$ have the same probability laws. Clearly $\Pr\{X(0) = j\} = \Pr\{Y(0) = j\} = \pi_j$, and it is not difficult to show that both processes are Markov. Hence, in order to show that they share the same probability laws it suffices to show that they have the same transition probabilities. Let

$$P_{i\,j}(t) = \Pr\{X(t) = j | X(0) = i\},$$

$$Q_{i\,j}(t) = \Pr\{Y(t) = j | Y(0) = i\}.$$

The process $\{X(t)\}$ is reversible if

$$P_{i\,j}(t) = Q_{i\,j}(t) \qquad (9.71)$$

for all states i, j and all times t. We evaluate $Q_{i\,j}(t)$ as follows:

$$
\begin{aligned}
Q_{ij}(t) &= \Pr\{Y(t) = j | Y(0) = i\} \\
&= \Pr\{X(-t) = j | X(0) = i\} \\
&= \Pr\{X(0) = j | X(t) = i\} \qquad \text{(by stationarity)} \\
&= \frac{\Pr\{X(0) = j,\ X(t) = i\}}{\Pr\{X(t) = i\}} \\
&= \frac{\pi_j P_{ji}(t)}{\pi_i}.
\end{aligned}
$$

In conjunction with (9.71) we see that the process $\{X(t)\}$ is reversible if

$$P_{i\,j}(t) = Q_{i\,j}(t) = \frac{\pi_j P_{ji}(t)}{\pi_i}$$

or

$$\pi_i P_{i\,j}(t) = \pi_j P_{ji}(t) \qquad (9.72)$$

for all states i, j and all times t.

As a last step, we determine a criterion for reversibility in terms of the infinitesimal parameters

$$a_{i\,j} = \lim_{t \downarrow 0} \frac{1}{t} \Pr\{X(t) = j | X(0) = i\}, \qquad i \neq j.$$

It is immediate that (9.72) holds when $i = j$. When $i \neq j$, then

$$P_{i\,j}(t) = a_{i\,j} t + o(t) \qquad (9.73)$$

which substituted into (9.72) gives

$$\pi_i[a_{i\,j} t + o(t)] = \pi_j[a_{ji} t + o(t)]$$

and after dividing by t and letting t vanish, we obtain the criterion

$$\pi_i a_{i\,j} = \pi_j a_{ji} \qquad \text{for all} \quad i \neq j. \qquad (9.74)$$

Since the transition probabilities are determined by the infinitesimal parameters, we deduce that the process $\{X(t)\}$ is time reversible whenever (9.74) holds.

All birth and death processes having stationary distributions are time reversible! Because birth and death processes have

$$a_{i,i+1} = \lambda_i,$$

$$a_{i,i-1} = \mu_i,$$

and

$$a_{i,j} = 0 \qquad \text{if } |i - j| > 1,$$

in verifying (9.74) it suffices to check that

$$\pi_i a_{i,i+1} = \pi_{i+1} a_{i+1,i}$$

or

$$\pi_i \lambda_i = \pi_{i+1} \mu_{i+1} \qquad \text{for } i = 0, 1, \dots . \tag{9.75}$$

But [see (6.31) and (6.32)],

$$\pi_i = \pi_0 \left(\frac{\lambda_0 \lambda_1 \cdots \lambda_{i-1}}{\mu_1 \mu_2 \cdots \mu_i} \right) \qquad \text{for } i = 1, 2, \dots ,$$

whence (9.75) becomes

$$\pi_0 \left(\frac{\lambda_0 \lambda_1 \cdots \lambda_{i-1}}{\mu_1 \mu_2 \cdots \mu_i} \right) \lambda_i = \pi_0 \left(\frac{\lambda_0 \lambda_1 \cdots \lambda_i}{\mu_1 \mu_2 \cdots \mu_{i+1}} \right) \mu_{i+1},$$

which is immediately seen to be true.

Proof of Theorem 9.1   Let us consider a birth and death process $\{X(t)\}$ having the constant birth rate $\lambda_k = \lambda$ for $k = 0, 1, \dots$, and arbitrary death parameters $\mu_k > 0$ for $k = 1, 2, \dots$. This process corresponds to a memoryless server queue having Poisson arrivals. A typical evolution is illustrated in Figure 9.12. The arrival process for $\{X(t)\}$ is a Poisson process of rate λ. The reversed time process $Y(t) = X(-t)$ has the same probabilistic laws as does $\{X(t)\}$, so the arrival process for $\{Y(t)\}$ also must be a Poisson process of rate λ. But the arrival process for $\{Y(t)\}$ is the departure process for $\{X(t)\}$ (see Figure 9.12). Thus it must be that these departure instants also form a Poisson process of rate λ. In particular, if $D(t)$ counts the departures in the $X(\cdot)$ process over the duration $(0, t]$, then

$$\Pr\{D(t) = j\} = \frac{(\lambda t)^j e^{-\lambda t}}{j!} \qquad \text{for } j = 0, 1, \dots \tag{9.76}$$

Moreover, looking at the reversed process $Y(-t) = X(t)$, the "future" arrivals for $Y(-t)$ in the Y duration $[-t, 0)$ are independent of $Y(-t) = X(t)$. (See Figure 9.12.) These future arrivals for $Y(-t)$ are the departures for $X(\cdot)$ in the interval $(0, t]$. Therefore these departures and $X(t) = Y(-t)$ must be independent. Since $\Pr\{X(t) = k\} = \pi_k$, by the assumption of stationarity, the independence of $D(t)$ and $X(t)$ and (9.76) give

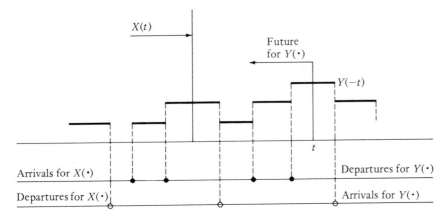

Figure 9.12 A typical evolution of a queueing process. The instants of arrivals and departures have been isolated on two time axes below the graph.

$$\Pr\{X(t) = k,\ D(t) = j\} = \Pr\{X(t) = k\}\Pr\{D(t) = j\}$$

$$= \frac{\pi_k e^{-\lambda t}(\lambda t)^j}{j!}$$

and the proof of Theorem 9.1 is complete. $\square$

Problems 9.5

1. Consider the three server network pictured here:

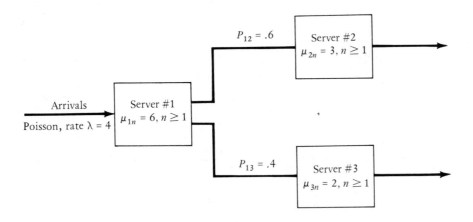

In the long run, what fraction of time is server #2 idle while, simultaneously, server #3 is busy? Assume that all service times are exponentially distributed.

2. Suppose three service stations are arranged in tandem so that the departures from one form the arrivals for the next. The arrivals to the first station are a Poisson process of rate $\lambda = 10$ per hour. Each station has a single server and the three service rates are $\mu_1 = 12$ per hour, $\mu_2 = 20$ per hour and $\mu_3 = 15$ per hour. In-process storage is being planned for station 3. What capacity C_3 must be provided if, in the long run, the probability of exceeding C_3 is to be less than or equal to 1 percent? That is, what is the smallest number $C_3 = c$ for which $\lim_{t\to\infty} \Pr\{X_3(t) > c\} \le .01$?

9.6 General Open Networks

The preceding section covered certain memoryless queueing networks in which a customer could visit any particular server at most once. With this assumption, the departures from any service station formed a Poisson process that was independent of the number of customers at that station in steady state. As a consequence the numbers $X_1(t)$, $X_2(t)$, . . . , $X_K(t)$ of customers at the K stations were independent random variables and the product form solution expressed in (9.66) prevailed.

The situation where a customer can visit a server more than once is more subtle. On the one hand, many flows in the network are no longer Poisson. On the other hand, rather surprisingly, the product form solution of (9.66) remains valid.

Example To begin our explanation, let us first reexamine the simple feedback model of Section 9.4.3. The flow is depicted in Figure 9.13. The arrival process is Poisson, but the input to the server is not. (The distinction be-

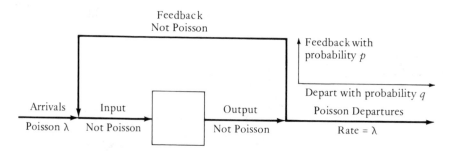

Figure 9.13   A single server with feedback

tween the arrival and input processes is made on pp. 344 and 362.) The output process, as shown in Figure 9.13, is not Poisson, nor is it independent of the number of customers in the system. Recall that each customer in the output is fed back with probability p and departs with probability $q = 1 - p$. In view of this non-Poisson behavior, it is remarkable that the distribution of the number of customers in the system is the same as that in a Poisson $M/M/1$ system whose input rate is λ/q and whose service rate is μ, as verified in (9.41).

Example Let us verify the product form solution in the slightly more complex two server network depicted in Figure 9.14.

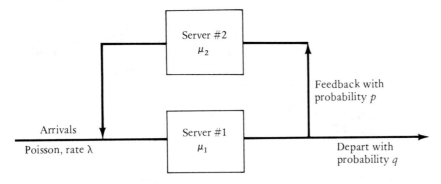

Figure 9.14   A two server feedback system. For example, server #2 in this system might be an inspector returning a fraction p of the output for rework.

If we let $X_i(t)$ denote the number of customers at station i at time t, for $i = 1, 2$, then $\mathbf{X}(t) = [X_1(t), X_2(t)]$ is a Markov chain whose transition rates are given in the following table:

From State	To State	Transition Rate	Description
(m, n)	$(m + 1, n)$	λ	Arrival of new customer
(m, n) $n \geq 1$	$(m + 1, n - 1)$	μ_2	Input of feedback customer
(m, n) $m \geq 1$	$(m - 1, n)$	$q\mu_1$	Departure of customer
(m, n) $m \geq 1$	$(m - 1, n + 1)$	$p\mu_1$	Feedback to server #2

Let $\pi_{m,n} = \lim_{t\to\infty} \Pr\{X_1(t) = m,\ X_2(t) = n\}$ be the stationary distribution of the process. Reasoning analogous to that of (6.68) and (6.69) of Chapter 6 (where the theory was developed for finite state Markov chains) leads to the following equations for the stationary distribution:

$$\lambda\pi_{0,0} = q\mu_1\pi_{1,0} \tag{9.77}$$

$$(\lambda + \mu_2)\pi_{0,n} = p\mu_1\pi_{1,n-1} + q\mu_1\pi_{1,n}, \qquad\qquad n \geq 1 \tag{9.78}$$

$$(\lambda + \mu_1)\pi_{m,0} = \lambda\pi_{m-1,0} + q\mu_1\pi_{m+1,0} + \mu_2\pi_{m-1,1},$$
$$m \geq 1 \tag{9.79}$$

$$(\lambda + \mu_1 + \mu_2)\pi_{m,n} = \lambda\pi_{m-1,n} + p\mu_1\pi_{m+1,n-1} + q\mu_1\pi_{m+1,n}$$
$$+ \mu_2\pi_{m-1,n+1}, \qquad m, n \geq 1. \tag{9.80}$$

The mass balance interpretation as explained following (6.69) on page 256 in Chapter 6 may help motivate (9.77) through (9.80). For example, the left side in (9.77) measures the total rate of flow out of state $(0, 0)$ and is jointly proportional to $\pi_{0,0}$, the long run fraction of time the process is in state $(0, 0)$, and λ, the (conditional) transition rate out of $(0, 0)$. Similarly, the right side of (9.77) measures the total rate of flow into state $(0, 0)$.

Using the product form solution in the acyclic case, we will "guess" a solution, and then verify that our guess indeed satisfies (9.77) through (9.80). First we need to determine the input rate, call it λ_1, to server #1. In equilibrium, the output rate must equal the input rate, and of this output, the fraction p is returned to join the new arrivals after visiting server #2. We have

$$\text{Input rate} = \text{New Arrivals} + \text{Feedback}$$

which translates into

$$\lambda_1 = \lambda + p\lambda_1,$$

or

$$\lambda_1 = \frac{\lambda}{1 - p} = \frac{\lambda}{q}. \tag{9.81}$$

The input rate to server #2 is

$$\lambda_2 = p\lambda_1 = \frac{p\lambda}{q}. \tag{9.82}$$

The solution that we guess is to treat server #1 and server #2 as independent $M/M/1$ systems having input rates λ_1 and λ_2, respectively (even though we know from our earlier discussion that the input to server #2, while of rate λ_2, is not Poisson). That is, we attempt a solution of the form

$$\pi_{m,n} = \left(1 - \frac{\lambda_1}{\mu_1}\right)\left(\frac{\lambda_1}{\mu_1}\right)^m\left(1 - \frac{\lambda_2}{\mu_2}\right)\left(\frac{\lambda_2}{\mu_2}\right)^n$$

$$= \left(1 - \frac{\lambda}{q\mu_1}\right)\left(\frac{\lambda}{q\mu_1}\right)^m\left(1 - \frac{p\lambda}{q\mu_2}\right)\left(\frac{p\lambda}{q\mu_2}\right)^n \quad \text{for} \quad m, n \geq 1. \quad (9.83)$$

It is immediate that

$$\sum_{m=0}^{\infty}\sum_{n=0}^{\infty} \pi_{m,n} = 1,$$

provided $\lambda_1 = (\lambda/q) < \mu_1$ and $\lambda_2 = p\lambda/q < \mu_2$.

We turn to verifying (9.77) through (9.80). Let $\theta_{m,n} = (\lambda/q\mu_1)^m \times (p\lambda/q\mu_2)^n$. It suffices to verify that $\theta_{m,n}$ satisfies (9.77) through (9.80) since $\pi_{m,n}$ and $\theta_{m,n}$ differ only by the constant multiple $\pi_{0,0} = (1 - \lambda_1/\mu_1) \times (1 - \lambda_2/\mu_2)$. Thus we proceed to substitute $\theta_{m,n}$ into (9.77) through (9.80) and verify that equality is obtained.

We verify (9.77):

$$\lambda = q\mu_1\left(\frac{\lambda}{q\mu_1}\right) = \lambda.$$

We verify (9.78):

$$(\lambda + \mu_2)\left(\frac{p\lambda}{q\mu_2}\right)^n = p\mu_1\left(\frac{\lambda}{q\mu_1}\right)\left(\frac{p\lambda}{q\mu_2}\right)^{n-1} + q\mu_1\left(\frac{\lambda}{q\mu_1}\right)\left(\frac{p\lambda}{q\mu_2}\right)^n$$

or, after dividing by $(p\lambda/q\mu_2)^n$ and simplifying,

$$\lambda + \mu_2 = \left(\frac{p\lambda}{q}\right)\left(\frac{q\mu_2}{p\lambda}\right) + \lambda = \lambda + \mu_2.$$

We verify (9.79):

$$(\lambda + \mu_1)\left(\frac{\lambda}{q\mu_1}\right)^m = \lambda\left(\frac{\lambda}{q\mu_1}\right)^{m-1} + q\mu_1\left(\frac{\lambda}{q\mu_1}\right)^{m+1} + \mu_2\left(\frac{\lambda}{q\mu_1}\right)^{m-1}\left(\frac{p\lambda}{q\mu_2}\right)$$

which, after dividing by $(\lambda/q\mu_1)^m$, becomes

$$(\lambda + \mu_1) = \lambda\left(\frac{q\mu_1}{\lambda}\right) + q\mu_1\left(\frac{\lambda}{q\mu_1}\right) + \mu_2\left(\frac{q\mu_1}{\lambda}\right)\left(\frac{p\lambda}{q\mu_2}\right),$$

or

$$\lambda + \mu_1 = q\mu_1 + \lambda + p\mu_1 = \lambda + \mu_1.$$

The final verification, that $\theta_{m,n}$ satisfies (9.80), is left to the reader as Problem 1 at the end of this section.

The General Open Network

Consider an open queueing network having K service stations, and let $X_k(t)$ denote the number of customers at station k at time t. We assume that

(1) The arrivals from outside the network to distinct servers form independent Poisson processes, where the outside arrivals to station k occur at rate λ_{0k}.

(2) The departures from distinct servers independently travel instantly to other servers, or leave the system, with fixed probabilities, where the probability that a departure from station j travels to station k is P_{jk}.

(3) The service times are *memoryless* or *Markov* in the sense that

$$\Pr\{\text{Server } \#k \text{ completes a service in } (t, t + \Delta t] | X_k(t) = n\}$$

$$= \mu_{kn}(\Delta t) + o(\Delta t) \qquad \text{for} \quad n = 1, 2, \ldots, \tag{9.84}$$

and does not otherwise depend on the past.

(4) The system is in statistical equilibrium (stationary).

(5) The system is completely open in that all customers in the system eventually leave.

Let λ_k be the rate of input at station k. The input at station k is comprised of customers entering from outside the system, at rate λ_{0k}, plus customers traveling from (possibly) other stations. The input to station k from station j occurs at rate $\lambda_j P_{jk}$, whence, as in (9.67),

$$\lambda_k = \lambda_{0k} + \sum_{j=1}^{K} \lambda_j P_{jk} \qquad \text{for} \quad k = 1, \ldots, K. \tag{9.85}$$

The condition (5), that all entering customers eventually leave, ensures that (9.85) has a unique solution.

With $\lambda_1, \ldots, \lambda_K$ given by (9.85), the main result is the product form solution:

$$\Pr\{X_1(t) = n_1, X_2(t) = n_2, \ldots, X_k(t) = n_K\}$$

$$= \psi_1(n_1)\psi_2(n_2) \cdots \psi_K(n_K), \tag{9.86}$$

where

$$\psi_k(n) = \frac{\pi_{k0}\lambda_k^n}{\mu_{k1}\mu_{k2} \cdots \mu_{kn}} \qquad \text{for} \quad n = 1, 2, \ldots, \tag{9.87}$$

and

$$\psi_k(0) = \pi_{k0} = \left\{ 1 + \sum_{n=1}^{\infty} \frac{\lambda_k^n}{\mu_{k1}\mu_{k2} \cdots \mu_{kn}} \right\}^{-1} \tag{9.88}$$

Example   The example of Figure 9.13 (see also Section 9.4.3) corresponds to $K = 1$ (a single service station) for which $P_{11} = p < 1$. The external arrivals are at rate $\lambda_{01} = \lambda$ and (9.85) becomes

$$\lambda_1 = \lambda_{01} + \lambda_1 P_{11} \quad \text{or} \quad \lambda_1 = \lambda + \lambda_1 p$$

which solves to give $\lambda_1 = \lambda/(1 - p) = \lambda/q$. Since the example concerns a single server, then $\mu_{1n} = \mu$ for all n and (9.87) becomes

$$\psi_1(n) = \pi_{10}\left(\frac{\lambda_1}{\mu}\right)^n = \pi_{10}\left(\frac{\lambda}{q\mu}\right)^n$$

where

$$\pi_{10} = \left(1 - \frac{\lambda}{q\mu}\right),$$

in agreement with (9.41).

Example Consider next the two server example depicted in Figure 9.14. The data given there furnish the following information:

$$\lambda_{01} = \lambda \qquad \lambda_{02} = 0$$
$$P_{11} = 0 \qquad P_{12} = p$$
$$P_{21} = 1 \qquad P_{22} = 0$$

which substituted into (9.85) gives

$$\lambda_1 = \lambda + \lambda_2(1)$$
$$\lambda_2 = 0 + \lambda_1(p)$$

which readily solves yielding

$$\lambda_1 = \frac{\lambda}{q} \quad \text{and} \quad \lambda_2 = \frac{p\lambda}{q}$$

in agreement with (9.81) and (9.82). It is readily seen that the product solution of (9.86) through (9.88) is identical with (9.83), which was directly verified as the solution in this example.

Problems 9.6

1. In the case $m \geq 1$, $n \geq 1$, verify that $\theta_{m,n}$ as given following (9.83) satisfies the equation for the stationary distribution, (9.80).

2. Consider the three server network pictured here:

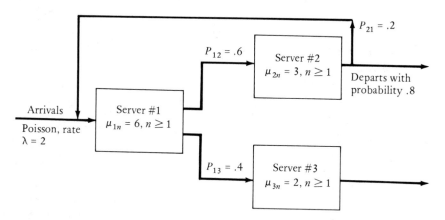

In the long run, what fraction of the time is server #2 idle while, simultaneously, server #3 is busy? Assume the system satisfies Assumptions 1 through 5 of a general open network.

Further Readings

Elementary Textbooks

Breiman, L. *Probability and Stochastic Processes with a View Toward Applications.* Boston: Houghton Mifflin, 1969.

Cinlar, E. *Introduction to Stochastic Processes.* Englewood Cliffs, N.J.: Prentice-Hall, 1975.

Cox, D. R., and H. D. Miller. *The Theory of Stochastic Processes.* New York: John Wiley & Sons, 1965.

Hoel, R. G., S. C. Port, and C. J. Stone. *Introduction to Stochastic Processes.* Boston: Houghton Mifflin, 1972.

Kemeny, J. G., and J. L. Snell. *Finite Markov Chains.* New York: Van Nostrand Reinhold, 1960.

Ross, S. M. *Introduction to Probability Models.* New York: Academic Press, 1972.

Intermediate Textbooks

Bhat, U. N. *Elements of Applied Stochastic Processes.* New York: John Wiley & Sons, 1972.

Breiman, L. *Probability.* Reading, Mass.: Addison Wesley, 1968.

Dynkin, E. B., and A. A. Yushkevich. *Markov Processes: Theorems and Problems.* New York: Plenum, 1969.

Feller, W. *An Introduction to Probability Theory and Its Applications.* 2 vols. New York: John Wiley & Sons, 1966 (vol. 2), 1968 (vol. 1, 3rd ed.).

Karlin, S. and H. M. Taylor. *A First Course in Stochastic Processes.* New York: Academic Press, 1975.

Karlin, S. and H. M. Taylor. *A Second Course in Stochastic Processes.* New York: Academic Press, 1981.

Kemeny, J. G., J. L. Snell, and A. W. Knapp. *Denumerable Markov Chains.* New York: Van Nostrand Reinhold, 1966.

Ross, S. M. *Stochastic Processes.* New York: John Wiley & Sons, 1983.

Renewal Theory

Kingman, J. F. C. *Regenerative Phenomena*. New York: John Wiley & Sons, 1972.

Queueing Processes

Kleinrock, L. *Queueing Systems*. Vol. 1, *Theory*. Vol. 2, *Computer Applications*. New York: John Wiley & Sons, Interscience, 1976.

Branching Processes

Athreya, K. B., and P. Ney. *Branching Processes*. New York: Springer-Verlag, 1970.

Harris, T. *The Theory of Branching Processes*. New York: Springer-Verlag, 1963.

Stochastic Models

Bartholomew, D. J. *Stochastic Models for Social Processes*. New York: John Wiley & Sons, 1967.

Bartlett, M. S. *Stochastic Population Models in Ecology and Epidemiology*. New York: John Wiley & Sons, 1960.

Goel, N. S., and N. Richter-Dyn. *Stochastic Models in Biology*. New York: Academic Press, 1974.

Point Processes

Lewis, P. A. *Stochastic Point Processes: Statistical Analysis, Theory, and Applications*. New York: John Wiley & Sons, Interscience, 1972.

Solutions to Selected Problems

Chapter 1

Problems 1.2

7. (a) $F_X(x) = x^R$ for $0 \le x \le 1$.
   (b) $E[X] = R/(R + 1)$.
   (c) $\text{Var}[X] = R/(R + 2)(R + 1)^2$.
11. $E[\mathbf{1}(A_k)] = \frac{1}{13}$ whence $E[N] = 1$.
15. (a) $\Pr\{A \text{ wins}\} = p/[p + q - pq]$.
    (b) $E[N|A \text{ wins}] = 1/p$.
22. $\text{Cov}[X, Y] = -\sigma^2$.
23. $f_{U,V}(u, v) = 2$ for $0 \le u < v \le 1$.

Problems 1.3

4. $\Pr\{X \text{ is odd}\} = (1 - e^{-2\lambda})/2$.
13. $\Pr\{X > 6\} = 1 - \Pr\{X \le 6\} = 0.1107$.

Problems 1.5

1. $\Pr\{N = n\} = [1 - F(\xi)]F(\xi)^{n-1}$ for $n = 1, 2, \ldots$
2. $F_Z(z) = 1 - e^{-n\lambda z}$ for $z \ge 0$.
7. $F_V(v) = 1 - \exp\{-(\lambda_1 + \ldots + \lambda_n)v\}$ for $v \ge 0$.

Chapter 2

Problems 2.1

1. $E[X|X \text{ is odd}] = \lambda(1 + e^{-2\lambda})/(1 - e^{-2\lambda})$.
2. (a) $P_{U,Z}(u, z) = \rho^2(1 - \rho)^z$ for $0 \le u \le z$.
   (b) $p_{U|Z}(u|n) = 1/(n + 1)$ for $0 \le u \le n$.

Problems 2.3

3. (a) $E[Z] = 0$, $\text{Var}[Z] = E[N] = (1 - \alpha)/\alpha$.
   (b) $E[Z^3] = 0$, $E[Z^4] = E[3N^2 - 2N] = (1 - \alpha)(6 - \alpha)/\alpha^2$.
5. $E[Z] = \mu(1 + \nu)$, $\text{Var}[Z] = \sigma^2(1 + \nu) + \mu^2\tau^2$.

Problems 2.4

1. $f_Z(z) = 1/(1 + z)^2$ for $z > 0$.
5. (a) $\Pr\{X = k\} = \theta/(1 + \theta)^{k+1}$ for $k = 0, 1, \ldots$
   (b) $f(\lambda|k) = (1 + \theta)^{k+1}\lambda^k e^{-(1+\theta)\lambda}/k!$.

Chapter 3

Problems 3.1

1. $(.3)(.2)(0) = 0$.
2. (a) $(1 - \alpha)^2$; (b) $(1 - \alpha)^2 + \alpha^2$.

Problems 3.2

3. $[1 + (1 - 2\alpha)^5]/2$.

Problems 3.3

2. $P_{k,k+1} = (1 - k/N)q$; $P_{k,k-1} = (k/N)p$.

Problems 3.4

1. $v_0 = \frac{20}{3}$.
4. $\frac{1}{3}$.

Problems 3.5

2. (a) .8044; (b) .3578.
5. $p_k = a_{k+1}/(a_{k+1} + a_{k+2} + \ldots)$.

Problems 3.6

2. $\Pr\{X_T = 0 | X_0 = 1\} = .65$.
4. $v_0 = (1 + \beta)/\beta^2$.

Chapter 4

Problems 4.1

1. $\pi_0 = \frac{5}{14}, \pi_1 = \frac{6}{14}, \pi_2 = \frac{3}{14}$.
2. $\pi_0 = .32, \pi_1 = .09, \pi_2 = .31, \pi_3 = .29$.
5. $\pi_k = 1/(N + 1)$ for $k = 0, \ldots, N$.
6. (a) π_j; (b) $\pi_k P_{kj}$.

Problems 4.2

1. (b) $\pi_{(s,s)} + \pi_{(s,c)} = .4$.

Problems 4.3

1. $\{0\}, \{1\}, \{2, 3, 4, 5\}$.

Problems 4.5

2. (a) .275, (b) 0, (c) .0606, (d) .22, (e) $\frac{3}{11}$, (f) Does not exist, (g) .33, (h) .148.

Chapter 5

Problems 5.1

2. $f(t) = \lambda^k t^{k-1} e^{-\lambda t}/(k - 1)!$ for $t \geq 0$.
4. $E[X(t)X(t + s)] = \lambda t + \lambda^2 t(t + s)$.
5. $[1 - e^{-2\lambda t}]/2$.
6. (a) $E[X(T)|T = t] = \lambda t; E[X(T)^2|T = t] = \lambda t + \lambda^2 t^2$.
   (b) $E[X(T)] = 3\lambda/2; \text{Var}[X(T)] = \lambda^2/12 + 3\lambda/2$.
10. (a) K; (b) $c\lambda T^2/2$; (c) $(K/T) + c\lambda T/2$; (d) $T^* = (2K/c\lambda)^{1/2}$.

Problems 5.2

2. $\exp\{-3 \times 240/600\}$.

3. $p(k) = \binom{N}{k}\left(\dfrac{a}{A}\right)^k\left(1 - \dfrac{a}{A}\right)^{N-k}$

Problems 5.3

1. $f(w_r|n) = \dfrac{n!}{(r-1)!(n-r)!}\left(\dfrac{w}{t}\right)^{r-1}\left(\dfrac{1}{t}\right)\left(1 - \dfrac{w}{t}\right)^{n-r}$

2. (a) $p = \lambda_1/(\lambda_1 + \lambda_2)$; (b) $p^2 + 2p^2(1-p)$.

3. $\Pr\{T \le t\} = (1 - e^{-\lambda t})^n$.

4. $\Pr\{T \le t\} = 1 - \sum_{j=1}^{k} c_j \exp\{-\lambda_j p_j t\}$.

Problems 5.4

1. $E[Z(t)] = (\lambda/\alpha)E[\xi_1](1 - e^{-\alpha t})$.

Problems 5.5

5. $F_D(x) = 1 - \exp\{-v\pi x^2\}$; $E[D] = 1/(2\sqrt{v})$.

Chapter 6

Problems 6.1

2. $P_0(t) = [\alpha + \beta e^{-(\alpha+\beta)t}]/(\alpha + \beta)$.

3. $E[N(t)] = \dfrac{2\alpha\beta}{\alpha + \beta}t + \dfrac{\alpha - \beta}{(\alpha + \beta)^2}[e^{-(\alpha+\beta)t} - 1]$.

Problems 6.4

3. $\pi_n = (1 - \lambda/\mu)(\lambda/\mu)^n$ for $n = 0, 1, \ldots$ if $\lambda < \mu$.

4. $\pi_k = (1 - \theta)^2(k + 1)\theta^k$ for $k = 0, 1, \ldots$.

5. $\pi_k = e^{-(\lambda/\mu)}(\lambda/\mu)^k/k!$ for $k = 0, 1, \ldots$.

Problems 6.6

1. With $P_{ij}(t)$ given by (6.11) through (6.14), then

$$\mathbf{P}^{\#}(t) = \left\| \begin{array}{ccc} P_{00}(t)^2 & 2P_{00}(t)P_{01}(t) & P_{01}(t)^2 \\ P_{00}(t)P_{10}(t) & P_{00}(t)P_{11}(t) + P_{01}(t)P_{10}(t) & P_{01}(t)P_{11}(t) \\ P_{10}(t)^2 & 2P_{10}(t)P_{11}(t) & P_{11}(t)^2 \end{array} \right\|$$

Chapter 7

Problems 7.1

1. (a) True; (b) False; (c) False
3. $\Pr\{\gamma_t > y | \delta_t = x\} = \dfrac{1 - F(x + y)}{1 - F(x)}$ for $y \geq 0$; $E[\gamma_t | \delta_t = x] = \int\limits_0^\infty [1 - F(x + y)]dy/[1 - F(x)]$.

Problems 7.2

3. $K^\star = 4$.
4. $M(n) = n\beta$ for $n = 0, 1, \ldots$

Problems 7.3

2. $p(t) = \begin{cases} e^{-\lambda t} & \text{for } t \leq \tau, \\ e^{-\lambda \tau} & \text{for } t > \tau. \end{cases}$
3. $E[L] = \frac{2}{3}$.

Problems 7.4

3. $\lim_{t \to \infty} \Pr\{\gamma_t \leq x\} = 2x(1 - \frac{1}{2}x)$.

Problems 7.5

1. $2/(2 + 7)$.
2. $E[X - x | X > x] = \int_0^\infty [1 - F(x + y)]dy/[1 - F(x)]$.
3. $W(t) = \sum_{k=1}^{N(t)} Y_k$; $\lim_{t \to \infty} E[W(t)]/t = \nu/\mu$ where $\nu = \int y dG(y)$ and $\mu = \int x dF(x)$.

Chapter 8

Problems 8.2

1. $u_\infty = .82$.
3. $\phi(s) = p + qs^N$.
6. (a) $(\frac{1}{4})^{2^n - 1}$; (b) $\frac{1}{3}$.

Problems 8.4

1. (a) u_∞ is the smallest solution to $u = f[g(u)]$;
   (b) $[f'(1)g'(1)]^{n/2}$ when $n = 0, 2, 4, \ldots$
   $f'(1)^{(n+1)/2}g'(1)^{(n-1)/2}$ when $n = 1, 3, \ldots$;
   (c) Yes, both.

Chapter 9

Problems 9.2

2. $\pi_0 = 1/\left\{1 + \dfrac{\lambda}{\mu} + \dfrac{(\lambda/\mu)^2}{2(1 - \lambda/2\mu)}\right\}.$

   $L = \dfrac{\lambda}{\mu} + \dfrac{1}{2}\left(\dfrac{\lambda}{\mu}\right)^2 \dfrac{(\lambda/2\mu)^2}{[1 - (\lambda/2\mu)]^2}\, \pi_0.$

4. (a) $\lambda_0 = \lambda_1 = \lambda_2 = \lambda$, $\mu_1 = \mu_2 = \mu_3 = \mu$; $\lambda_3 = \mu_0 = 0$;

   (b) $\pi_0 = \left\{1 + \dfrac{\lambda}{\mu} + \left(\dfrac{\lambda}{\mu}\right)^2 + \left(\dfrac{\lambda}{\mu}\right)^3\right\}^{-1}$;

   (c) $\pi_3 = \left(\dfrac{\lambda}{\mu}\right)^2 \pi_0.$

Problems 9.4

2. Choose $\mu_1 = 6$, $\mu_2 = 4$ to minimize W.
3. Give priority to the class having mean service time of 8 minutes.

Index

A 4
B 5
C 6
D 7
E 8
F 9
G 0
H 1
I 2
J 3